NATO ASI Series

Advanced Science Institutes Series

A series presenting the results of activities sponsored by the NATO Science Committee, which aims at the dissemination of advanced scientific and technological knowledge, with a view to strengthening links between scientific communities.

The Series is published by an international board of publishers in conjunction with the NATO Scientific Affairs Division

A	Life Sciences	Plenum Publishing Corporation
B	Physics	London and New York
C	Mathematical and Physical Sciences	Kluwer Academic Publishers
D	Behavioural and Social Sciences	Dordrecht, Boston and London
E	Applied Sciences	
F	Computer and Systems Sciences	Springer-Verlag
G	Ecological Sciences	Berlin Heidelberg New York
H	Cell Biology	London Paris Tokyo Hong Kong
I	Global Environmental Change	Barcelona Budapest

PARTNERSHIP SUB-SERIES

1. Disarmament Technologies	Kluwer Academic Publishers
2. Environment	Springer-Verlag
3. High Technology	Kluwer Academic Publishers
4. Science and Technology Policy	Kluwer Academic Publishers
5. Computer Networking	Kluwer Academic Publishers

The Partnership Sub-Series incorporates activities undertaken in collaboration with NATO's Cooperation Partners, the countries of the CIS and Central and Eastern Europe, in Priority Areas of concern to those countries.

NATO-PCO DATABASE

The electronic index to the NATO ASI Series provides full bibliographical references (with keywords and/or abstracts) to about 50 000 contributions from international scientists published in all sections of the NATO ASI Series. Access to the NATO-PCO DATABASE compiled by the NATO Publication Coordination Office is possible in two ways:

- via online FILE 128 (NATO-PCO DATABASE) hosted by ESRIN,
  Via Galileo Galilei, I-00044 Frascati, Italy.

- via CD-ROM "NATO Science & Technology Disk" with user-friendly retrieval software in English, French and German (© WTV GmbH and DATAWARE Technologies Inc. 1992).

The CD-ROM can be ordered through any member of the Board of Publishers or through NATO-PCO, Overijse, Belgium.

Series F: Computer and Systems Sciences, Vol. 144

Springer
Berlin
Heidelberg
New York
Barcelona
Budapest
Hong Kong
London
Milan
Paris
Tokyo

The Biology and Technology of Intelligent Autonomous Agents

Edited by

Luc Steels

Artificial Intelligence Laboratory, Department of Computer Science
University of Brussels (Vrije Universiteit Brussel),
Pleinlaan 2, B-1050 Brussels, Belgium

Springer
Published in cooperation with NATO Scientific Affairs Division

Proceedings of the NATO Advanced Study Institute on The Biology and
Technology of Intelligent Autonomous Agents, held in Castel Ivano, Trento, Italy,
March 1–12, 1993

Library of Congress Cataloging-in-Publication Data

The biology and technology of intelligent autonomous agents / edited by Luc Steels.
p. cm. – (NATO ASI series. Series F, Computer and system sciences; no. 144)
"Proceedings of the NATO Advanced Study Institute on the Biology and Technology of Intelligent
Autonomous Agents, held in Castel Invano, Trento, Italy, March 1–12, 1993" – T. p. verso.
"Published in cooperation with NATO Scientific Affairs Division." Includes bibliographical references.
ISBN 3-540-59052-8 (Springer-Verlag Berlin Heidelberg New York: alk. paper)
1. Intelligent control systems. 2. Artificial intelligence. I. Steels, Luc. II. North Atlantic Treaty Or-
ganization. Scientific and Environmental Affairs Division. III. Advanced Study Institute on the
Biology and Technology of Intelligent Autonomous Agents (1993: Trento, Italy) IV. Series: NATO ASI
series. Series F, Computer and system sciences; no. 144.b
TJ217.5B56   1995   006.3–dc20   95-20211   CIP

CR Subject Classification (1991): I.2, J.2–3

ISBN 3-540-59052-8 Springer-Verlag Berlin Heidelberg New York

© Springer-Verlag Berlin Heidelberg 1995
Printed in Germany

Typesetting: Camera-ready by editor
Printed on acid-free paper
SPIN: 10486096   45/3142 – 5 4 3 2 1 0

Preface

The NATO sponsored Advanced Study Institute 'The Biology and Technology of Intelligent Autonomous Agents' was an extraordinary event. For two weeks it brought together the leading proponents of the new behavior-oriented approach to Artificial Intelligence in Castel Ivano near Trento. The goal of the meeting was to establish a solid scientific and technological foundation for the field of intelligent autonomous agents with a bias towards the new methodologies and techniques that have recently been developed in Artificial Intelligence under the strong influence of biology. Major themes of the conference were: bottom-up AI research, artificial life, neural networks and techniques of emergent functionality.

The meeting was such an extraordinary event because it not only featured very high quality lectures on autonomous agents and the various fields feeding it, but also robot laboratories which were set up by the MIT AI laboratory (with a lab led by Rodney Brooks) and the VUB AI laboratory (with labs led by Tim Smithers and Luc Steels). This way the participants could also gain practical experience and discuss in concreto what the difficulties and achievements were of different approaches. In fact, the meeting has been such a success that a follow up meeting is planned for September 1995 in Monte Verita (Switzerland). This meeting is organised by Rolf Pfeifer (University of Zurich).

This book is a document of the event. The contributions have been brought up to date based on lectures, research contributions, and experiments conducted at the institute. There are three parts:

Part I. Lectures

This consists of papers based on lectures given by some of the lecturers at the institute. They are organised around the different themes of intelligent autonomous agents research: autonomy, intelligence, cognition, biology, learning, cognitive architectures. Some of the papers give an overview of the subfield before focusing on the line of thought pursued by the author. Other papers go in the direction of original research contributions.

Part II. Research Papers

This section consists of papers by participants based on presentations made at the institute or experiments conducted at the institute. These papers give a good idea of how the behavior-oriented approach is showing up in the current practice of research, and the nature of the results that are being achieved.

Part III. Research Notes

This section contains shorter contributions: proposals for research, small experiments, ideas for new approaches. These contributions were discussed at the Trento gathering or generated as a result of the meeting.

The organisation of such a complex event as a NATO Advanced Study Institute would not be possible without the help of many people. First of all thanks are due to Dr. Staudinger the owner of the Castel Ivano and his family and staff, particularly Paula Nicoletti. They received us with extraordinary hospitality at the Castel Ivano near Trento. It was an honor to stay at this castle which is filled with history. We also thank the Bosco family for the hospitality at their superb guesthouse Hotel Romando in Levico Terme and Oliviero Stock from IRST (Trento) who helped make the connections for the local organisation.

An efficient administrative staff was set up at the VUB AI laboratory with Jenneke Christiaens, Karina Bergen, and Siegried d'Haeseleer. Jacqueline Van Lierde acted as special secretary in Trento itself. Without their incredible support the institute would not have been possible. Ann Sjostrom and Ugo Piazzalunga played special roles as liaison between Italy and Belgium. We also thank Kristina Bold who gave a superb piano recital on one of the social occasions and Walter Van de Velde who organised the ski adventures. Van de Velde, who won with his team the first prize at the Robot talent show, also played many other important organisational roles.

For the technical support a number of people have contributed in extraordinary ways. From the VUB AI laboratory, Peter Stuer, Danny Vereertbrugghe, and Filip Vertommen took care of many of the practical arrangements at the technical level. From the MIT AI laboratory Rodney Brooks and Maja Matarić contributed heavily in the set up of the robot laboratory.

The present proceedings would never have come into existence without the help of Brigitte Hönig who managed to convince the authors to make the final push needed to bring the volume together.

Finally we thank Dr. L. Veiga da Cunha, director of the NATO ASI programme for the support throughout the whole administrative process associated with the NATO funding. This funding is hereby gratefully acknowledged.

Brussels   Luc Steels
March 1995   ASI Director

Table of Contents

Part I. Lectures

The Biology of Behavior - Criteria for Success in Animals and Robots 1
David McFarland

Autonomy - On Quantitative Performance Measures of Robot Behaviour 21
Tim Smithers

Robot Adaptivity 53
Carme Torras

Intelligence - Dynamics and Representations 72
Luc Steels

An Introduction to Reinforcement Learning 90
Leslie Pack Kaelbling, Michael L. Littman, Andrew W. Moore

Cognition - Perspectives from Autonomous Agents 128
Rolf Pfeifer

Lifelong Robot Learning 165
Sebastian Thrun, Tom M. Mitchell

Cognitive Architectures - From Knowledge Level to Structural Coupling 197
Walter Van de Velde

Part II. Research Papers

Circle in the Round: State Space Attractors for Evolved Sighted Robots 222
Philip Husbands, Inman Harvey, Dave Cliff

Animal and Robot Navigation 258
Ulrich Nehmzow

From Local Interactions to Collective Intelligence 275
Maja J. Matarić

The Mobile Robot of MAIA:
Actions and Interactions in a Real Life Scenario 296
Giulio Antoniol, Bruno Caprile, Alessandro Cimatti, Roberto Fiutem

Of Elephants and Men 312
Johan M. Lammens, Henry H. Hexmoor, Stuart C. Shapiro

How Do You Choose Your Agents?
How Do You Distribute Your Processes? 345
Miles Pebody

The Reactive Accompanist:
Adaptation and Behavior Decomposition in a Music System 365
Joanna Bryson

Behavior-Based Architecture with Distributed Selection 377
Luís Correia, A. Steiger-Garção

Reporting Experiments on Integration of Learning Algorithms and
Reactive Behaviour-Oriented Control Systems on a Real Mobile Robot 390
Filip Vertommen

Distributed Reinforcement Learning 415
Gerhard Weiß

Part III. Research Notes

A New Three-Degree-of-Freedom Spatially Mobile Robot Topology 429
Martin Nilsson

How Swarms Build Cognitive Maps 439
Dante R. Chialvo, Mark M. Millonas

Multiple Neural Experts for Improved Decision Making 451
Ethem Alpaydin

Understanding Complex Systems: What can the Speaking Lion Tell us? 459
Erich Prem

Multi-Modal Active Sensing for a Simple Mobile Agent 475
Kristian T. Simsarian

Evaluating an Active Camera Controlled by a Subsumption Architecture 484
Claudio S. Pinhanez

AMOS: Basic Autonomy via Integrating Symbolic and
Subsymbolic Mechanisms 501
Christian Schlegel, Manfred Knick

List of Contributors 513

The Biology of Behavior - Criteria for Success in Animals and Robots

David McFarland

Department of Zoology and Balliol College, Oxford, UK

The aim of this paper is to discuss "doing the right thing". The question is what is the right thing? In animal behaviour studies there is a consensus on this question, and I will start by outlining this. In animal behaviour studies there is no consensus as to how optimal behaviour is achieved, and I will outline some of the problems. In robotics there is no consensus about doing the right thing and I will ask what can be learned from animal behaviour.

1 The Criteria for Biological Success

Animals are well designed. Let us take a simple example from the work of Tinbergen and his coworkers (Hoogland, et al., 1957). They investigated the function of the spines of sticklebacks. Two species of stickleback occur in European fresh water. They are the three-spined stickleback *(Gasterosteus aculeatus)* and the ten-spined stickleback *(Pygosteus pungitius).* They are preyed upon by pike *(Esox lucius).*

The spines of sticklebacks are thought to give some protection against predators, and to test this possibility Tinbergen and his co-workers carried out experiments in which they compared sticklebacks with fish of similar size, which have no spines, such as minnows *(Phoxinus phoxinus),* as shown in Figure 1.

Three-spined Stickleback

Ten-spined Stickleback

Minnow

Figure 1. Fish used in experiments on prey capture by pike (After Hoogland et al., 1957)

The experiments were conducted in large aquariums and involved observations of the predatory behaviour of pike and the reactions of the prey fish. The researchers observed that pike usually attempt to take their prey head first but that they are not always successful. They can swallow minnows tail first without difficulty, but they cannot cope with a stickleback in this way and usually spit it out and try to catch it again immediately. When attacked, sticklebacks raise their spines and attempt to keep their head away from the predator. Once caught, some sticklebacks manage to escape, especially if the predator attempted to manouver its prey in order to swallow it head first (see Figure 2).

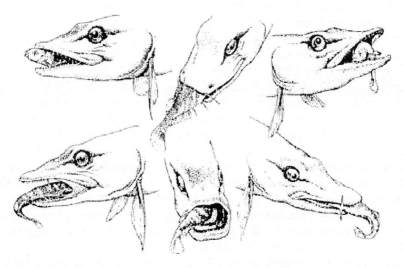

Figure 2. Drawings made from a film showing the variety of positions of a stickle-back in the mouth of a pike (After Hoogland et al, 1957)

When presented with mixed schools of minnows and sticklebacks, pike ate the minnows more quickly than the sticklebacks, partly the result of the predators' preferring to attack prey without spines and partly a result of sticklebacks being caught and then rejected or being caught and then escaping.

In some tests, sticklebacks were offered to perch and pike after their spines had been cut off. Such fish survived much less well than intact sticklebacks. This supports the main conclusion gained from these experiments, which is that the spines of sticklebacks do indeed provide some protection against predators.

Three-spined sticklebacks survived much longer than ten-spined sticklebacks, presumably because their spines are much larger.

How, then, are the ten-spined sticklebacks, or any fish without protective spines, able to survive in nature? Comparison of the habitats and behaviour of three-spined and ten-spined sticklebacks gives some clues. The male three-spined stickleback builds its nests on a substratum relatively free of weeds, and it has bright red nuptial colouration. The male ten-spined stickleback builds its nest among thick weeds and has black colouration during the breeding season. Thus, the ten-spined stickleback apparently compensates for its relative lack of protection by its cryptic habitat, colouration and behaviour.

In other words, the design feature of large spines enables the male three spined stickleback to gain a reproductive advantage, because he can show off his bright red breast which is attractive to females, without incurring a high predation penalty. Although the red breast will also attract predators, it probably also serves as warning coloration.

In nature design is done by the process of natural selection, as first explained by Charles Darwin (1859). The elements of Darwin's theory of natural selection are as follows: (1) There is considerable variation among individuals belonging to any population of animals of the same species. (2) Much of this variation is genetically inherited. (3) There are many more individuals born in each generation than can survive to maturity. From these facts it follows that different individuals will not have the same likelihood of survival. Those whose characteristics are best suited to the environment will be more likely to survive, and pass on their beneficial characteristics to the next generation. Thus certain characteristics will tend to be perpetuated within a population. In other words, certain features of animals are selected.

It is important to distinguish between the logic of the theory of natural selection and the evidence that the process of natural selection is actually effective in the natural environment. Logically, provided it is true that animals reproduce in larger numbers than can survive, and that there are inherited variations among the offspring, then it is inevitable that the variants that survive will tend to be those best fitted to, or selected by, the environment.

Darwin spent a considerable amout of time obtaining evidence to support his theory, and since his time the evidence has become overwhelming. I do not propose to discuss it here.

Natural selection acts upon the *phenotype* (bodily features and behaviour) of an individual and its effectiveness in changing the nature of a population depends upon the degree to which the phenotypic characteristics are controlled *genetically*. Thus, the effectiveness of natural selection depends upon the genetic influence that an individual can exert upon the population as a whole. Obviously, an individual that has no offspring during its lifetime will exert no genetic influence upon the population, however great its ability to survive in the natural environment.

So we have two seperate concepts to consider - *survival value* and *reproductive success*. These are combined in the concept of *fitness*. Fitness is a measure of the ability of genetic material to perpetuate itself in the course of evolution. It depends not only upon the individual's ability to survive but also upon its rate of reproduction and the viability of its offspring.

So far, we can see that the extent to which a genetic trait is passed from one generation to the next, in a wild population, is determined by the breeding success of individuals of the parent generation and the value of the trait in enabling the animals to overcome natural hazards such as food shortage, predators and sexual rivals. Such environmental pressures can be looked upon as selecting those traits that fit the animal to the environment. The capacity to produce offspring is called *Darwinian Fitness.*

In any population of animals there is variation among individuals, and consequently, some have greater reproductive success than others. Individuals that produce a higher number of viable offspring are said to have greater Darwinian fitness.

An individual's fitness depends upon its ability to survive to reproductive age, its success in mating, the fecundity of the mated pair and the probability of survival to reproductive age of the resulting offspring.

Reproduction incurs costs. It may expose an animal to risks, thus reducing the chances of subsequent survival and reproduction. In addition, the more time and energy a parent expends upon a particular offspring, the fitter that offspring will be. There is often an inverse relationship between the total number of offspring produced and their average fitness. To maximize fitness the best reproductive strategy must be a compromise between having a large number of progeny and attaining a high individual fitness for each offspring.

Although an animal's individual, or Darwinian, fitness is a measure of its ability to leave viable offspring, the effectiveness of natural selection depends upon the mixture of genotypes in the population. Thus, the *relative* fitness of a genotype depends upon the other genotypes present in the population, as well as upon other environmental conditions.

The concept of fitness can be applied to *individual genes* by considering the survival of particular genes in the gene pool from one generation to another. A gene that can enhance the reproductive success of the animal carrying it thereby will increase its representation in the gene pool. It could do this by influencing the animal's morphology or physiology, making it more likely to survive climatic and other hazards or by influencing its behaviour; making the animal more successful in courtship or raising young. A gene that influences parental behaviour will probably be represented in the offspring so that by facilitating parental care, the gene itself is likely to appear in other individuals. Indeed, a situation could arise in which the gene could have a deleterious effect upon the animal carrying it but increase its probability of survival in the offspring. An obvious example is a gene that leads the parent to endanger its own life in attempts to preserve the lives of its progeny. This is an example of apparant *altruism.*

William Hamilton (1964) was the first to enunciat the general principle that natural selection tends to maximize not individual fitness but *inclusive fitness*; that is, an animal's fitness depends upon not only its own reproductive success but also that of its kin. The inclusive fitness of an individual depends upon the survival of its descendants and of its collateral relatives. Thus even if an animal has no offspring its inclusive fitness may not be zero, because its genes will be passed on by neices, nephews and cousins.

Inclusive fitness is the proper measure of *biological success.* All that matters is the representation of genes in the gene pool. It does not matter how the success is achieved. It can be achieved by honest toil, by sexual attractiveness, by exploiting other species, or whatever.

For example, one of the most successful north american species this century is the turkey. The turkey population has grown enormously as a result of genetic changes that have enabled turkeys to exploit humans. Turkeys are now white rather than black. They mature much more quickly than they used, and they are shorter and heavier in the body. The genetic changes responsible for these adaptations have been crucial in the success of the turkey. If the white coloration had not been genetically possible, the turkey would not look so attractive when plucked. If shortening of the body had not been genetically possible then a turkey of a decent weight would not fit in the normal household oven.

You may think that this example is not valid because it involves humans as a selective agent. But such relationships between species are commonplace in the animal kingdom. For example, many types of ants farm aphids and milk them for their honeydew, as shown in Figure 3.

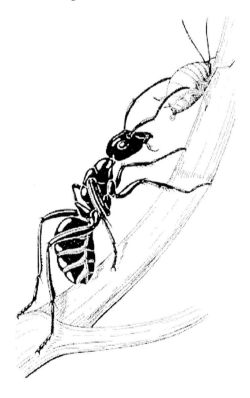

Figure 3. Wood ant worker "milking" a greenfly (After Dumpert, 1981)

Like turkeys the aphids have undergone genetic adaptations to this role. Like turkeys, the aphids benefit from protection from predators. Some ants build structures to protect their aphids from the weather. They may move them to new pastures when necessary, and like turkeys, the aphid eggs may be specially cared for. Some ants collect aphid eggs are bring them into their nest for the winter (Dumpert, 1981).

From the biological point of view, there is no difference in principle between these two cases. Both are the result of evolution by natural selection.

Implicit in our discussion of inclusive fitness are two alternate ways of describing natural selection. In population genetics, the unit of selection is the gene, and replication of the gene is the quantity maximized by natural selection. While this approach has an appealing logic that can sometimes be used to illuminate aspects of animal behaviour (e.g., Dawkins, 1976), it is not really convenient for those interested in the behaviour of individual animals. An equivalent approach is to regard the individual animal as the unit of selection and inclusive fitness as the quantity maximized by natural selection.

2 Survival and Reproduction in Animals

In recent years attempts have been made to relate the temporal organisation of behaviour to the theory of natural selection. In considering the medium-term temporal organisation of behaviour the most obvious design criteria are minimising risk, maximising opportunities, and neglecting no important behaviour. Some of these points can be illustrated by studies of herring gulls conducted by Tinbergen (1953), McFarland (1977) and their coworkers.

Herring gulls *(Larus argentatus)* are widespread in the northern hemisphere. In the UK they nest in large colonies within which each mated pair holds a small territory. The female lays three eggs in a crude nest on the ground, and the male and female take it in turns to incubate and guard the eggs for the one month incubation period.

The problems facing the birds are formidable. If incubation is disrupted the eggs may not develop properly, or may be stolen. As well as incubating and guarding the eggs the birds must obtain food for themselves in a complex and competitive world.

These aspects of the herring gull life can be investigated experimentally. For example, the body weight of an incubating bird can be measured by means of a specially designed balance placed in a hole in the ground under the nest (McFarland, 1985).

The amount of fat a bird is carrying can be calculated from its weight in relation to skeletal size. The measurements of the skeleton can be made when the bird initially is caught and marked. The amount of food that a herring gull obtains from foraging can be estimated by the change in weight measured on the nest balance before and after foraging trips. The quality of the food can be estimated from analysis of the fecal remains gathered from around the nest and from observations made on the bird while foraging. Herring gulls may fly a number of miles to obtain food, and to obtain observations on the foraging of particular birds, they are fitted with radio transmitters. The foraging bird then can be tracked down by means of a directional radio receiver. Thus, by using a variety of techniques it is sometimes possible to arrive at fairly accurate estimates of the costs and benefits incurred by animals living a normal life in a natural environment.

Now consider the situation of a herring gull incubating its eggs. Normally, both parents incubate in turn and a nest left unattended is soon subject to predation by neighbours. A sitting bird will not quit the nest until relieved by its partner, unless it is flushed from the nest by a predator, such as a fox or human. Even then it remains in the vicinity, and often attacks the intruder. Usually the partner leaves the nest to forage for food and returns to the territory within a few hours. Sometimes, however, the return is delayed as a result of some mishap, injury, or capture by a scientist. The sitting bird cannot feed, and its increasing hunger should eventually cause it to desert the nest. Herring gulls breed in many successive seasons and it is not in the genetic interests of an individual to endanger its life for a single clutch of three eggs, which are likely to give rise to only 0.8 fledglings.

The basic dilemma for the incubating bird lies between the possibility that its partner may return at any time, and the increasing necessity to obtain food. The situation is complicated by the fact that the gull must assess the return on time invested in foraging, which vary considerably with the weather, the state of the tide, and the time of day. Moreover, individual gulls often specialise in particular types of foraging, so that the sitting bird should take account of its own foraging skills in assessing the likely returns from the time spent foraging in different places. It may seem that the best course of action requires quite a bit of computation on the part of our herring gull. It is not likely that all this is done cognitively. Rather, it is more likely that the bird relies on rules of thumb, based upon its normal daily routine and habitual behaviour.

Sibly and McCleery (1985) conducted a sophisticated functional analysis of the herring gull activity selection. They describe a procedure for testing whether a particular set of decision rules maximise fitness and they use it to identify optimal decision rules governing feeding, incubation, and the presence on territory, of herring gulls during incubation. The characteristics of the optimal decision rules depend upon the risks attendant upon each aspect of behaviour. Therefore in calculating the rules it is necessary to measure the risks an animal takes in carrying out its various activities. Using such estimates they identify the characteristics of the optimal decision rules and use them as a basis for evaluating the fitness of the observed behaviour. The method is a development of that first proposed by Sibly and McFarland (1976).

The main risks of reproductive failure during the incubation period are of egg death due to exposure or predation, and of parent death as a result of accident, disease, predation or starvation. These risks can be quantitatively evaluated by experiment (Sibly and McCleery, 1985). The opportunities relate mainly to feeding. Figure 4 shows the availability of the major foods taken by herring gulls at Walney. The various colony members specialise (to the extent of having preferred sources of food) in feeding on different rhythmically available resources. These include refuse and earthworms, available on a circadian basis, and mussels and starfish available on a tidal cycle. Each gull has to adjust its temporal pattern of incubation to permit both itself and its partner to take advantage of the feeding opportunity profiles that each is specialised in exploiting, without leaving the nest unattended. An opportunity profile is a function of time that represents the returns that the animal would obtain if it foraged at that time. An example is illustrated in Figure 5. Successful incubation by a breeding pair requires complex cooperation and behavioural synchronisation with each other and prevailing circadian, tidal, and social rhythms.

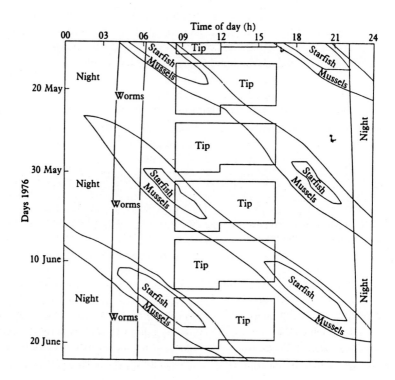

Figure 4. The availability of the major foods of herring gulls at Walney in 1977. 'Worms' indicates the availability of invertebrates (mainly earthworms) on pasture fields on the mainland within an hour of sunrise; 'Tip' indicates the availability of domestic waste at refuse tips 08.30-16.30 Monday to Friday, 08.30-12.00 on Saturdays; 'Mussels' indicates the availability of first year mussels and other invertebrates on mussel skears when the tide was below 1.9 m below O.D.; 'Starfish' indicates the availability of starfish and other invertebrates on mussel skears when the tide was below 3.1 m below O.D.; 'Night' was from one hour after sunset to one hour before sunrise, when food was assumed to be unavailable (After Sibly and McCleery, 1985)

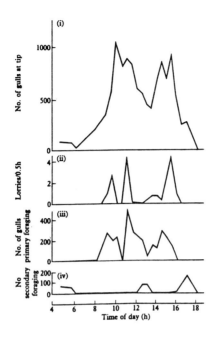

Figure 5. Gull activity at Walney refuse tip on weekdays, April-June 1976. About 90% were adult herring gulls, 10% were adult lesser black-backed gulls. (i) Average number of gulls at the tip. (ii) Number of trucks carrying domestic waste that unloaded within half an hour. (iii) Average number of gulls engaged in primary foraging. (iv) Average number of gulls engaged in secondary foraging. The opportunity profile is indicated by (ii). The animal response to the opportunity is indicated by (iii). (After Sibly and McCleery (1983)

Sibly and McCleery (1985) found that the feeding preferences of individual gulls were very important determinants of fitness. They constructed a fitness landscape for each breeding pair (an example is shown in Figure 6), based on two important hunger variables, which they called normal feeding theshold and desperation feeding theshold. The former is the degree of hunger at which a bird would leave the territory to forage if one of its preferred foods became available, provided that its partner was on territory to take over incubation. The latter is the degree of hunger at which the bird would leave its nest to feed when a suitable opportunity arose, even if its mate was not available for incubation duty. On the basis of the fitness calculations Sibly and McCleery (1985) identified various characterisitics of the optimal behaviour strategies. In particular, they predicted (1) that energy reserves should be maintained between 500 and 1200 kcal, (2) the members of a mated pair should have complementary food preferences, with at least one being a preference for feeding at the local refuse tip; (3) the parent should desert the offspring if the energy reserves fall below 200 kcal. They were able to test these predictions by observation and experiment.

This example shows how functional theories about the temporal organisation of behaviour can be made rigorous, and can be emprically tested. It highlights the importance of measuring the risks and opportunities associated with behaviour, but it does not, strictly, provide a theory of behavioural choice mechanisms.

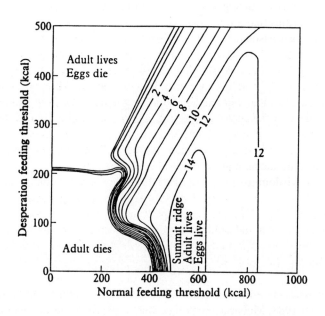

Figure 6. Fitness map with desperation feeding threshold and normal feeding threshold as axes. The lines represent contours of equal fitness. Maximum fitness is attained along the summit ridge, where the adult lives and eggs hatch. Fitness declines to the right of the ridge, because of the cost of carrying extra weight. Minimum fitness occurs when the adult dies. On the plateau at top left the adult lives but the eggs die (After Sibly and McCleery, 1983)

3 Activity Selection and Decision-Making in Animals

Animals are often faced with alternative courses of action among which a trade-off exists in terms of costs or benefits. The best course for an animal to take is the one that provides the least cost or greatest benefit. From evolutionary considerations we expect well-designed animals to behave in a way which will maximize inclusive fitness. Thus, the concepts of cost and benefit are related to the concept of fitness. Animals incur costs and benefits in a great variety of ways, which correspond to the selective pressures of the natural environment. Natural selection acts upon the animal as a whole, but in analyzing the costs and benefits of behavior, we have to separate out the costs (decrements in fitness) and benefits (increments in fitness) that can be ascribed to each aspect of the animal's internal state and behavior. The total specification of the animal in these terms is called the *cost function*. This can be defined as the specification of the instantaneous level of risk incurred by (and reproductive benefit available to) an animal in a particular internal state, engaged in a particular activity in a particular environment (McFarland, 1977).

In general, the cost function is made up of two main parts: (1) the costs associated with the activity occurring at a particular time and (2) the costs associated with all the other activities that are not occurring at that time. For example, an incubating gull incurs costs associated with incubation that include the physiological cost involved in maintaining the eggs at a certain temperature and behavior costs including costs arising from the risk of being caught by a predator while sitting. At the same time, some costs arise because incubation is incompatible with other activities. These costs are usually connected with the animal's state. For example, while incubating, the animal's state of hunger inevitably rises. Being in a certain state of hunger involves a risk of starvation because it may happen that food is unavailable when the animal decides to start foraging. As we shall see, such considerations are related intimately to the animal's ecological circumstances.

Conventional wisdom assumes that animals do that activity which they are most motivated to do at the time. Thus if an animal is more hungry than aggressive it will eat rather than fight. If a predator appears fear will quickly dominate all other motivations, and the animal will flee or hide. This simple model of *motivational competition* raises an important and fundamental question: *How are animals so organised that they are most motivated to do what they ought to do at a particular time?*

Much effort has gone into calculating what animals ought to do (e.g. Houston and McNamara, 1988; McFarland and Houston, 1981; Stephens and Krebs, 1986).

The survival and reproductive success of individual animals depends largely upon their use of resources, such as food, territory, mates, etc. At any particular time an animal may have alternative means of possible resource utilising activity. Every activity will have associated costs and benefits, in terms of the ultimate reproductive success, or fitness, of the animal.

In studying activity selection in animals we are not primarily concerned with what an animal ought to do, but rather how it does achieve the optimal outcome (if it does). This is as an exercise in theory and experiment on motivational mechanisms. Of course, if we know what a mechanism is supposed to do, it is much easier to discover how it works. To many ethologists, this is the main attraction of functional studies.

There are some general points that can be made about activity selection.

1. Animals are not always motivated to perform all activities in their repertoire. Sometimes the necessary internal conditions are absent. For example, outside the breeding season many animals lack the hormonal balance necessary for the motivation of sexual behaviour. Sometimes necessary external stimuli are

absent. Thus an animal cannot eat if there is no food, and may have no incentive to search for food if there are absolutely no external stimuli indicating food availability.

Often, however, the necessary motivational conditions for many incompatible behaviour patterns will exist simultaneously. These conditions will generally include some aspects of the animal's internal state, and some aspects of its assessment of external stimuli. The consensus amongst ethologists is that both internal and external factors contribute to the *tendency* to perform a particular activity.

2. Somehow, the animal has to make a 'decision' among the different possible courses of action. Moreover, such decisions must be made in terms of a common currency, so that the animal has some means of comparing the 'merits' of sleeping, courting, feeding, etc. In other words, there must be some trade-off process built in to the mechanisms controlling behaviour.

3. The main assumptions behind these conclusions are (a) that the strongest tendency determins the activity that the animal will perform, (b) that all the candidates have equal status in the final common path, and (c) that the choices among candidate tendencies are transitive (see McFarland and Sibly, 1975).

Operationally, *motivational competition* can be defined as follows: "A changeover due to competition can in practice be recognized when a change in the level of causal factors for a second-in-priority activity results in an alteration in the temporal position of the occurrence of that activity" (McFarland, 1969a). Thus, it is envisaged that the systems controlling activities A and B compete with each other. The system with the higher tendency is the one that wins the competition, and it is that activity that is observed. This may all seem very obvious, but it has interesting implications in a situation in which manipulation of second-priority causal factors does not alter the time of occurrence of the relevant behavior.

The occurrence of a second-priority activity is inhibited by the occurrence of a top-priority activity. If this inhibition were removed, then the second-priority activity would be disinhibited. This idea leads to an operational definition of *disinhibition*.

"The time of occurrence of a disinhibited activity is independent of the level of causal factors relevant to that activity".

The important point is that the B tendency is not instrumental in the sudden change in the A tendency, and therefore the strength of B tendency does not determine the time of occurrence of activity B. Of course, if the B tendency were raised sufficiently, it could become greater than the A tendency and an appropriate change in behavior would be observed. Activity B then would not be disinhibited.

Having outlined a simple model of motivational competition, we now have to consider whether it will work. There is plenty of evidence that motivational competition does occur, but, at the same time there are problems with the idea of competition as the sole principle of activity selection in animals.

Let us end this section with a simple example. During feeding behavior, doves *(Streptopelia)* typically pause for a few seconds and then resume feeding. If water is made available, the birds may drink during the pause but the time of occurrence of the pause is not affected by the presence or absence of water. Similarly, if paper clips are fixed to the primary wing feathers, the birds will try to remove them during the pause, but again the timing of the pause is not affected by the presence or absence of the paper clips (McFarland, 1970). The water dish and the paper clips are examples of stimuli that alter the strength of the second-priority activities of drinking and preening respectively. However, manipulation of these stimuli does not affect the time at which these activities occurred, as is illustrated in Figure 7. According to the operational definition of disinhibition (above) these activities have been disinhibited. Such disinhibition has been shown to occur in a variety of situations (e.g. Brown and McFarland, 1979; Cohen and McFarland; 1979, McFarland and L'Angellier, 1966; McFarland, 1974).

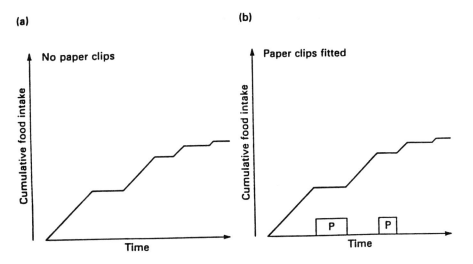

Figure 7. The feeding pattern of hungry doves is not affected by putting paper clips on to the wing feathers, although the birds try to remove the paper clips by preening during the pauses in feeding

4 The Criteria for Success in Robots

Success in animals is measured in terms of fitness. What about robots? From the behavioural point of view, what are the fundamental differences between animals and robots? There are obvious though behaviourally trivial differences, such as that animals are made from different hardware, are more sophisticated, have better sensors, etc. As scientists, we are not inclined to say that robots are machines whereas animals are not, or that the behaviour of robots is predetermined whereas that of animals is not. As scientists we believe that a deterministic philosophy applies to both animals and robots, and this attitude enables us to entertain the notion that robots may, one day, be analogues of animals.

In some respects robots already are analogues of animals. Like other man-made machines they evolve through a system trial and error, research and development, and natural selection in the marketplace. The analogy between the life history of animals and robots can be developed as a proper mathematical analogy (McFarland, 1991).

I am not suggesting that robots should be modelled directly on animals or humans, but rather that they should be designed in terms of what is best for their own ecological circumstances. The robots will be bought and sold and will therefore have to compete in the marketplace against other robots and against humans willing to carry out the same tasks. This ecological competition will lead to the evolution of certain attributes (McFarland, 1991), among which will be robustness, speed of reaction, self-sufficiency and autonomy. Each type of robot will be tailored to its niche.

In animal ecology, niche occupancy usually implies ecological competition. When animals of different species use the same resources or have certain habitat preferences or tolerance ranges in common, niche overlap occurs. This leads to competition between species, especially when resources are in short supply. There is a competitive exclusion principle which states that two species with identical niches cannot live together in the same place at the same time when resources are limited. The corollary is that, if two species coexist, there must be ecological differences between them.

The most widely accepted definition of the *niche* is that of Hutchinson (1957), for whom the niche is "a multidimensional hypervolume, defined by the sum of all the interactions of an organism and its (abiotic and biotic) environment." Along a single dimension, the niche is defined by the set of resources used by the organism, and is constrained by the organism's physiological limitations. The entire set of resources that the organism is physiologically capable of utilising is known as its *fundamental niche*. This is primarily determined by abiotic factors, but in practice, animals have also to contend with biotic factors, such as competition, predation, and food availability. These usually combine to restrict the animal to a smaller *realised niche*.

Suppose a robot is to be employed as a bomb-disposer, then it must be able to compete in the market place with human bomb disposers. Of course, the two species of bomb-disposer will not be alike in every (bomb-disposing) respect. Employers would value one for certain qualities, and the other for other qualities. In other words there will be only partial niche overlap. The basic capabilities of the bomb-disposing robot, including its refuelling, bomb-disposing and danger avoiding capabilities, will determine its fundamental niche. Its realised niche will depend upon the robots ability to compete in the marketplace against other forms of bomb disposal. Thus employers will apply roughly the same cost-efficiency criteria to each species, and these criteria will supply the selective pressures characteristic of the bomb-disposer niche. These selective pressures provide the main ingredients of a mathematical *cost function* that is characteristic of the environment in which the robot is to operate.

The analogy between animal and product design can be formulated as an exact mathematical analogy (see Figure 8). Briefly, the success of a biological design is measured by the success of the genes that produce it, and this depends upon the ability of those genes to increase their representation in the population in the face of competition from rival genes. How does this relate to product design? Suppose that a variety of products is under consideration. They vary in the period required for product development, in the chance of failure in the market place, and in the expected returns from sales if the product is successful. The development period refers to the period before any return is achieved on investment. For animals this is the period between birth and reproduction, and for products it is the period prior to time that financial return accrues to the investor. The success of a design is evaluated by the net rate of increase of the genes coding for it (i.e. the return on investment) in the animal case, or, in the case of a product launched into the marketplace, of the money invested in it.

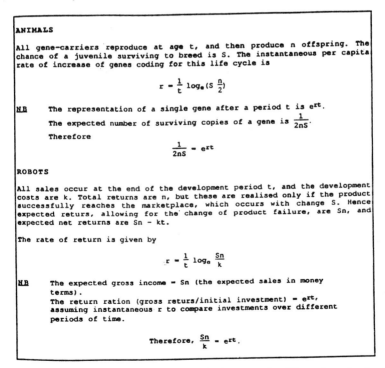

Figure 8. Outline of the mathematical analogies between the life cycles of animals and robots (From McFarland, 1991)

The analogies between the design cycles of animals and robots are summarised in Figure 9. Development period t refers to the period before any return is achieved on investment. For animals this is the period between birth and reproduction, and for products it is the period before any financial returns accrue to the investor. S measures the chance of a product (animate or inanimate) surviving to enter the gene pool or marketplace. n refers to the return on investment. For animals this is the number of offspring produced when the individual reproduces (i.e. when the gene is copied into half the offspring on average, in accordance with Mendel's laws). In the case of man-made products, n refers to the income from sales. The success of the design is evaluated by the net rate of increase of genes coding for it, in the animal case, or the money invested in it, in the case of man-made products.

ANALOGIES BETWEEN ANIMAL AND ROBOT LIFE CYCLES		
return on investments	n number of offspring	gross sales income assuming no failures
reproductive probability	S chance of juvenile surviving to breed	chance of product reaching market
development period	t age at breeding	development cost = k
design success (rate of return)	net rate of increase of genes coding for design ("fitness")	net rate of increase of money invested in design (instantaneous interest rate)

Figure 9. Table of analogies between animal and robot life cycles (From McFarland, 1991)

The cases of longer adult life of sales period, can be analysed by an extension of the above method. In the simplest such case the adult achieves a constant breeding performance, producing n offspring every time it breeds, which occurs at intervals t apart. Similarly a product might achieve sales at a constant rate over a given period. If the animal is subject to mortality at a constant rate, the expected breeding performance is likely to follow a negative exponential curve. Similarly expected product sales might decline with time, taking into account the chances of failure to sell as a result of market competition or loss of market appeal.

This approach may seem premature, but already we can see the beginnings of the process. If we look towards the future we can see that the next development is likely to be self-sufficiency in robots, because a self-sufficient robot will generally have an advantage over one that depends upon human intervention for refuelling etc. The cost of the human labour involved in servicing the dependent robot will soon outweigh the difference in the capital costs of the two types of robot. So we can expect useful robots to evolve towards self-sufficiency.

Self-sufficiency and *autonomy* are not the same thing (McFarland, 1995). To be self-sufficient, a robot must not incur any debts of vital resources that it cannot repay through its own behaviour. Such debts may include energy debts (i.e. the robot must not get itself into a position where it cannot regain energy balance by foraging, recharging, or whatever), and task debts which relate to those features of robot behaviour that are designed to please the customer (McFarland and Boesser, 1993). For example, suppose we design a robot to fly in after a nuclear catastrophe and measure the radiation at various locations. It is no good the robot coming back and reporting that it was so busy refuelling that it had no time to gather any data.

Self-sufficiency is largely a matter of *behavioural stability*, a concept that can be mathematically defined (McFarland, 1995; McFarland and Boesser, 1993). To maintain stability, the *resilience* of the various activities must be suitably tuned.

In animal behaviour studies, *resilience* is a measure of the cost to the animal of abstaining from each activity in its natural repertoire (McFarland and Houston, 1981). If an animal did no feeding, for example, the cost would be high, but if it abstained from groomimg the cost would probably be relatively low. An animal that had equal motivation to feed and groom would incur less cost (in terms of fitness decrement) if it devoted its available time to feeding. If it used up much of its time feeding, there would be less time for grooming. In terms of the time budget, feeding would have high resilience while grooming would have low resilience to being pushed out by other activities. Thus behavioural resilience is a measure of the extent to which an activity can be squashed in terms of time by other activities. It reflects the importance of an activity in a long-term sense. When given mathematical expression, behavioural resilience can be shown to be important in the optimal allocation of behaviour. It is also closely related to the economic demand functions that are a common feature of animal behaviour (McFarland and Houston, 1981).

The self-sufficient robot must also have the appropriate *motivation* for debt recovery. That is, it must have the motivation to regain lost energy, to avoid temperature extremes, to avoid predators and traps, and to carry out its tasks. For example, our flying robot must want to find radiation high spots. It must want to map them. It must want to remain viable, etc. These various aspects of motivation have to be satisfied simultaneously, but the robot can do only one thing at a time, so it needs some kind of activity selection mechanism (see below).

The self-sufficient robot must have some degree of *autonomy* because it must have the freedom to do that behaviour that is in its own vital interests. Autonomy implies freedom from control. If A controls B then A can make B behave in the way that A wants. To do this A must have sufficient knowledge of B.

If B is autonomous then B must be self-controlling. That is, B must want B to behave in a certain way and B must have sufficient knowledge of B to be able to do this. In other words an autonomous agent must have some degree of motivation and cognition, organised in such a way that an outside agent cannot obtain sufficient knowledge to control the autonomous agent.

Once an autonomous agent comes under the control of another agent it autonomy is lost. Our autonomous flying robot must be able to make its own decisions about what to do next, and must be able to resist interference from outside agents such as humans and other robots.

5 Animal Robotics

The discipline that treats robots and animals as similar may be called *animal robotics*. The viewpoint of animal robotics is different from that of conventional robotics. Most roboticists aspire to make robots that will carry out particular tasks that are tedious or difficult for humans to carry out. In other words, they try to make robots that will do what humans want them to do. The point of animal robotics is that instead of attempting to control the behaviour of the robot, and make it do what we want it to do, we design the robot to want to do what we want it to do.

In animal robotics, the relationship between the human and the robot is similar to that between a human and an animal. The human may aspire to influence the behaviour of the animal, but cannot hope to have complete control over the animal's behaviour. This view sees robots as similar to animals in being autonomous and self-sufficient.

Of course, we humans would hope to be able to influence the behaviour of such robots, much as we hope to influence the behaviour of a sheep dog or guide dog. Nevertheless, the dog and the robot remain autonomous, capable of overriding our aspirations at any moment.

The autonomous robot will have simultaneous motivational pressures, but will be able to carry out only one activity at a time. We are now back with the problem of *activity selection*. The most important feature of the activity selection system of a self-sufficient autonomous robot is that it should be sensitive to opportunity profiles.

An *opportunity profile* is a function of time which indicates what benefit the agent will derive if it performs a particular action at a particular time. An example is shown in Figure 5. The magnitude of an opportunity is determined partly by environmental circumstances and partly by motivational state. For example, for an agent that is hungry for fuel there will be a high opportunity value when stimuli indicating high fuel availability are present in the environment. For an agent that is less hungry in the same environment the opportunity will be less, because foraging for fuel will be less beneficial. Similarly, at a given hunger an agent that senses high fuel availability will have a higher opportunity that one that senses low availability, because foraging when high availability is signalled is likely to be more beneficial.

The units of benefit will vary from agent to agent. If the agent is an animal, then the benefit will be related to fitness. We must be careful here to distinguish between the real benefit as determined by environmental circumstances, and the notional benefit as perceived by the agent. In animal behaviour studies the former is sometimes called cost (the negative of the benefit) and the latter is sometimes called utility (McFarland and Houston, 1981). In robotics, the former is a measure of marketability, which I will call benefit, and the latter may be called value (McFarland and Boesser, 1993). Thus the *value* of an opportunity to a robot is its evaluation of the benefit in terms of future change of state that the robot would gain if it engaged in the appropriate behaviour. There are various ways (both implicit and explicit) in which the robot may measure the value (McFarland, 1992b), but the details need not concern us here.

Obviously, a robot that is able to exploit opportunities will have higher marketability that one that allows opportunities to pass it by. The former will appear to be more efficient that the latter. Pushed to its logical conclusion, this line of thinking results in the idea of *optimal activity selection* based upon trade-off amongst the alternative courses of action at any one time. Thus we arrive at a principle similar to the *trade-off principle* of animal behaviour (Stephens and Krebs, 1986; McFarland, 1989).

The trade-off principle implies that optimal activity selection can be achieved only if the activity selection mechanism involves a 'level playing field' in which all players (i.e. all candidates for behavioural expression) have equal status. This means that all variables in the trade-off have equal status, there being no bias in favour of any particular activity (including the ongoing activity) (McFarland and Sibly, 1975).

To summarise, for a robot to be well adapted to its ecological niche, it must be capable of optimal allocation of its time and energy resources. To do this it must somehow trade-off amongst the various opportunities open to it at any given time. This a design criterion based upon an answer to the question "what is the right thing to do?". To meet this design criterion the robot's activity selection mechanism must involve a level-playing field on which all players have equal status.

References

Brown, R.E. and McFarland, D.J. (1979) Interaction of hunger and sexual motivation in the male rat: a time-sharing approach. Animal behaviour, 27, 887-896

Cohen, S. and McFarland, D.J. (1979) Time-sharing as a mechanism for the control of behaviour sequences during the courtship of the three-spined stickleback (Gasterosteus Aculeatus). Animal behaviour, 27, 270-283

Dawkins R. (1976) The selfish gene. Oxford University Press: Oxford

Dumpert, K. (1981) The social biology of ants. Pitman Books, London

Hamilton, W.D. (1964) The genetical theory of social behaviour (I and II) Journal of Theoretical Biology 7:1-16; 17-32

Hoogland, R., Morris, D. and Tinbergen, N. (1957) The spines of sticklebacks (Gasterosteus and Pygosteus) as means of defence against predators (Perca and Essox). Behaviour, 10, 205-236

Houston, A. and McNamara, J.M. (1988) A framework for the functional analysis of behavior. Behavioral and Brain Sciences, 11, 117-154

Hutchinson, G.E. (1957) Concluding remarks. Cold Spring Harbor Symposium on Quantitative Biology 22: 415-427

McFarland, D.J. (1969a) Mechanisms of behavioural disinhibition. Animal Behaviour, 17: 238-242

McFarland, D.J. (1970b) Adjunctive behaviour in feeding and drinking situations. Revue comportement animal 4: 64-73

McFarland, D.J. (1974) Time-sharing as a behavioral phenomenon. In Advances in The Study of Behavior. (eds.) D.S. Lehrman, J.S. Rosenblatt, R.A. Hinde and E. Shaw. Academic Press, N.Y.

McFarland, D.J. (1977) Decision-making in animals. Nature, 269: 15-21

McFarland, D.J. (1985) Animal behaviour. Pitman Books, London

McFarland, D.J. (1989) Problems of animal behaviour. Longmans, London.

McFarland, D.J. (1991) What it means for robot behaviour to be adaptive. From animals to animats. (eds.) J. Meyer and S. W. Wilson. MIT Press, Cambridge, Mass

McFarland, D.J. (1992b) Animals as cost-based robots. International Studies in the Philosophy of Science, 6, 133-153

McFarland, D.J. (1993) Animal behaviour. (second edition). Longman, London

McFarland, D.J. (1995) Autonomy and self-sufficiency in robots. In: Steels, L. and R. Brooks (1995) The Artificial Life Route to AI. Building Situated Autonomous Agents. Lawrence Erlbaum, New Haven. pp. 197-225

McFarland, D.J. and L'Angellier, A.B. (1966) Disinhibition of drinking during satiation of feeding behaviour in the Barbary dove. Animal Behaviour, 14, 463-467

McFarland, D.J. and Boesser, T. (1993) Intelligent behavior in animals and robots. MIT Press, Cambridge, Mass

McFarland, D.J. and Houston, A. (1981) Quantitative ethology - The state space approach. Pitman Books, London.

McFarland, D.J. and Sibly, R.M. (1975) The behavioural final common path. Philosophical Transactions of the Royal Society 270: 265-293

Sibly, R.M. and McCleery, R.H. (1985) Optimal decision rules for gulls. Animal Behaviour, 33, 449-465

Sibly, R.M. and McFarland, D.J. (1976) On the fitness of behaviour sequences. American naturalist, 110: 601-617

Stephens, D.W. and Krebs, J.R. (1986) Foraging theory. Princeton University Press, Princeton, N.J

Tinbergen, N. (1953) The herring gull's world. Collins, London.

On Quantitative Performance Measures of Robot Behaviour

Tim Smithers

Euskal Herriko Unibertsitatea, Informatika Fakultatea, 649 Postakutxa,
20080 Donostia, Spain

Universidad del Pais Vasco, Facultad de Informática, Apartado 649,
20080 San Sebastián, Spain

Abstract. This paper presents the background to and some initial results
of an attempt to develop the basis for quantitative performance measures of
robot behaviour. First, an example of a simple robot behaviour is used to motivate the need for a dynamical systems approach to the understanding and
investigation of robot behaviour. The background and initial theoretical developments necessary for defining some appropriate quantitative measures is
then presented. Finally, an example of the application of one of the techniques
proposed is presented, together with a discussion of the practical difficulties
involved and the future prospects of the presented approach.

Keywords. Autonomous systems, complex dynamical systems, correlation
dimension, quantitative measures, robotics

1 Introduction

Almost all reported work in intelligent robotics can be classified as making
some kind of technical or engineering contribution to our understanding of
robots and how to design and build them. So far, however, there has been
little or no attempt to develop and use quantitative measures of performance
with which to assess the behaviour of the robots built and tested. The performance of many of the techniques developed and used, particularly computational algorithms, are subjected to quantitative assessment, but these do
not constitute measures of robot behaviour, not directly at least. They do
not tell us (every qualitatively) how well the robot carries out its navigation
task, or how well it avoids obstacles or getting stuck, for example.

This is a somewhat surprising state of affairs, since without measurement
there is no science, and in particular, without quantitative measures it is hard
to develop and test theories. If the field of intelligent robotics and autonomous
systems is to develop beyond its current, essentially, empirical status, we need

to develop and use some effective quantitative measures with which we can assess robot behaviour, that is, the interactions of real (and complete) robots with their environments.

This paper presents the background and initial theoretical developments of a dynamical systems approach to understanding the behaviour of robots – situated embodied agents. This is then used as the basis for defining some effective quantitative measures of agent-environment interaction systems and their performance. An example of attempting to use one of the techniques proposed is then presented, together with a discussion of the practical difficulties involved and the future prospects of the approach. We start, however, with an example of some simple robot behaviour which illustrates our motives for attempting to develop a dynamical systems approach to understanding behaviour in agent-environment interaction systems.

2 Robot Behaviour and the Need for a Dynamical Systems Approach

The issue of devising and using quantitative measures of preformance for robot behaviour raises the problem of what to measure. In conventional systems controlled using feedback strategies, the behavioural performance of the controlled system can usually be adequately assessed by measuring aspects of the behaviour of the controller – the average deviation of control variables from their set point values, the sensitivity of the controller to noise, and the stability of the controller dynamics, for example. In mobile robots, however, this aproach is typically inadequate. Any set point values used do not specify the required behaviour so directly, and, in the case of more recent Behaviour-based approaches, [7], [25], to building robots, set point values may not even exist. In these approaches, robot behaviour is an emergent property of agent-environment interactions. We therefore need ways of measuring behaviour which do not assume there is a classical controller in side the robot.

In Behaviour-based approaches to building robots, there can be considerable variation in the form of sensor signals generated even when we have effective (good enough) robot behaviour. The form of the sensor signals does not, therefore, provide a good basis for quantitatively assessing performance. To illustate this an example of a simple robot behaviour is presented which is based on a real robot experiment conducted at the VUB AI Lab in Brussels. This particular experiment forms part of an ongoing investigation into the dynamics of agent-environment interactions, [41] and [42], during which the sampled eight bit signals, from the five infra-red (IR) sensors on the robot used, were recorded.

First the robot, experiment design and setup, and data recording, are described. The results are then presented to illustrate the kinds of normal signal variation that can occur in this quite simple agent-environment interaction system.

2.1 The Robot and Experimental Setup

The robot used is a second generation Lego vehicle[1] (LvII). It has a three-state bumper sensor (left bump, right-bump, and no-bump), and two thee-state whisker sensors (out-contact, inner-contact, no-contact) mounted at the front for contact detection, five eight-bit active IR sensors (all operating at the same IR and pulse frequencies) arranged in an arc from left to right across the front of the vehicle (with the centre one facing directly forward, inner side ones at 15° from forward, and outer side ones at 40° from forward), a revolution counter on the front wheel[2], and a lap counter (see below). It is powered by two Lego 9v motors supplied with 14v each controlled by a pulse-width modulated motor control channel. This allows more or less continuous variation of power setting of each motor, and thus effective speed control and direction of the vehicle. It has a third free wheel at the front. (See [14, figure 5] for an illustation of this arrangement.)

The LvII is programmed so that its motor state-space has a fixed point attactor corresponding to both motors forward, thus making the robot move forwards in a straight line. It uses its bumpers and whiskers for contact detection and IR sensors for object proximity detection in a 'don't get stuck' behaviour—do not get stuck in or trapped by things in the environment. The transfer functions, implemented by the program, between each sensor modality and motor states are dynamic (depending on recent sensor signal history), nonlinear, and independent of each other, i.e., there is no "sensor fusion" across sensor types done in the program. Thus, when sensor signals are non-zero, the motor state is perturbed (in different ways, depending on the sensor states) from the fixed-point and it intrinstic dynamics then takes back to its stable state, where the degree of perturbation depends upon recent sensor state history.

The experimental environment was formed by an enclosure and constructed to be a good "IR environment" so that the robot would normally not make contact with the sides of the enclosure or any of the objects within it, see figure 1. The parameters of the transfer functions, implemented by the control program, were adjusted so that, starting from the same nominal position and direction, the LvII would consistently take a route round the enclosure near the walls, passing under a cross-bar on each lap, thus triggering a sensor used to detect this event, see figure 3. Achieving this turned out to be significantly more difficult than it was for a first generation Lego vehicle (see [14]) which used only three one-bit IR sensors, which is why we decided to investigate more closely what was going on.

Each run of the LvII consisted of ten lap counts and lasted a little over two minutes. During each run the signals from the five IR-sensors were sampled every 20 milliseconds and recorded in memory, together with the time and revolution count at the start of each lap. These data were uploaded to a host computer and stored at the end of each run. A series of nine runs were done, each of nine complete laps. The data presented here is taken from one of

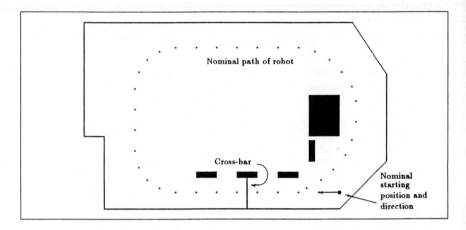

Figure 1. Schematic view of agent-environment setup used to record IR-sensor data as the LvII moves round the nominal path shown

these runs (selected at random), with the data from the first lap disgarded to avoid transients associated with the starting conditions, giving a total of eight laps.

2.2 IR-Sensor Data

The data sets of the eight laps are presented in figures 2 to 9.

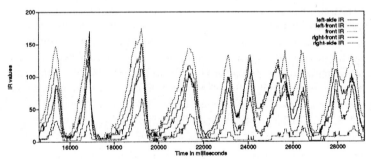

Figure 2. Lap 1 of 8, time=14,593ms (101.48% of average), count=219, giving a distance of 9.862m (101.45% of average), and an average speed of 0.676m/s.

As can be seen from figures 2 to 9, while there is an obvious common general form, there is considerable variation in detail of the five sensor signal profiles across each of the laps shown. What is important to note here is that

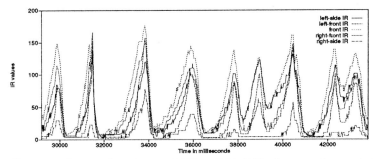

Figure 3. Lap 2 of 8, time=14,595ms (101.49% of average), count=219, giving a distance of 9.862m (101.48% of average), and an average speed of 0.676m/s.

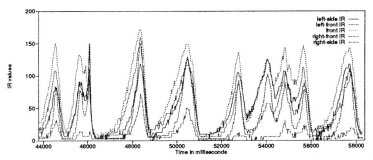

Figure 4. Lap 3 of 8, time=14,493ms (100.78% of average), count=219, giving a distance of 9.862m (101.45% of average), and an average speed of 0.676m/s.

this detailed variation is *not* due to noise. There is, of course, noise in these signals, but it can be shown to be at least an order of magnitude smaller than the variations that can be seen here. This variation is structural and arises as a direct result of the small variations in the actual path taken by the robot on successive laps. The lap times and distances travelled, and their percentage of the average values over the eight laps (see figure captions) indicate that this variation in path is small. However, it has a quite noticeable (structural)

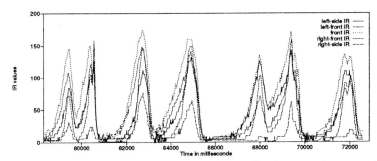

Figure 5. Lap 4 of 8, time=14,293ms (99.39% of average), count=215, giving a distance of 9.682m (99.60% of average), and an average speed of 0.677m/s.

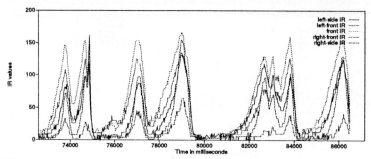

Figure 6. Lap 5 of 8, time=14,293ms (99.39% of average), count=210, giving a distance of 9.456m (97.28% of average), and an average speed of 0.662m/s.

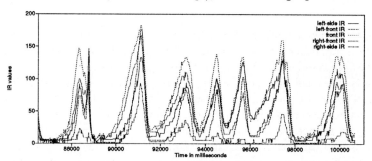

Figure 7. Lap 6 of 8, time=13,923ms (96.82% of average), count=212, giving a distance of 9.546m (97.89% of average), and an average speed of 0.686m/s.

effect on the actual sensor signals generated, as can be seen.

This kind of structural variation in sensor signal profiles is typical of other types of sensors too, ultra-sound, and vision, for example, when used on mobile robots. It can, of course, be reduced, and, at least sometimes, made to effectively go away, but only at the cost of controlling the motion of the robot so that the differences in the path taken become very small. This is both a difficult thing to do, and is not necessary for effective robot

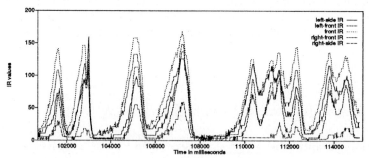

Figure 8. Lap 7 of 8, time=14,459ms (100.55% of average), count=216, giving a distance of 9.727m (100.06% of average), and an average speed of 0.673m/s.

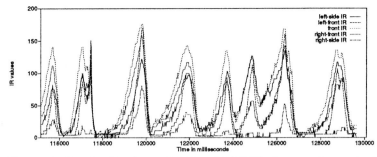

Figure 9. Lap 8 of 8, time=14,494ms (100.79% of average), count=217, giving a distance of 9.771m (100.20% of average), and an average speed of 0.674m/s.

behaviour—the behaviour achieved is quite stable and not overly sensitive to perturbations. In other words, to reduce the sensor signal variation to a level about the same as the level of noise in the signal, requires much higher degrees of motion control than is necessary for the robot to get around quite well enough. This is the case in the experiment described here: the actual path taken is sufficiently close to the nominal path on each lap and sufficiently robust. To try to reduce what variation there is, to make the sensor signal profiles more similar for each lap, would take us away from the real problem of achieving reliable and robust behaviour. In fact, any attempt to make the motion control more precise simply makes the robot more sensitive to any variations that do occur, as a result of environmental perturbations. Building a controller that can deal with the effects of such perturbations requires yet further machinery, again, none of which is actually of benefit in improving the performance of the robot - which was doing perfectly well in the first place, see [43] for more on this.

From this we conclude that the structural variation in sensor signal profiles is a natural and proper part of the robot-environment interaction. We cannot, therefore, seek to develop quantitative performance measures which presume that such variations are to be minimised or kept as small as possible in normal robot environment interactions. The attempt to reconsider this problem, and to develop a more appropriate basis for behavioural performance measurement, led to a radical reconsideration of the relationship between an agent and its environment, and, in particular, the question what is meant by terms like uncertain, variable, unpredictable, complex, and dynamic environments.

3   What is an Uncertain Environment Anyway?

In much work on adaptive behaviour and learning in agents, animats ([49]) or animals, environments are often described as uncertain, variable, unpredictable, dynamic, complex, and even threatening. For example, to list just a few from previous SAB (Simulation of Adaptive Behaviour) conference proceedings, from [26] we have, "In a changing, unpredictable, and more or less threatening environment the behavior of an animal is adaptive as long as the behavior allows the animal to survive.", from [49], "Environments differ enormously in their complexity, uncertainty, and degree of reinforcement.", from [13], "An autonomous robot which has to move in an uncertain environment has to deal with the problem of how to perform a goal-oriented behaviour.", and from [24], talking about Behavior-Based AI, we have, "... a general approach for building autonomous systems that have to deal with multiple, changing goals in a dynamic, unpredictable environment." There is, however, no clear agreement about the use of such descriptive terms, and the authors who use them seldom make clear what they are intended to mean. [49] is a rare exception, but one which is based upon an agent independent view.

The underlying assumption behind all these descriptions of environments, and others like them, is that the proper way to understand adaptive behaviour and learning in agents is in terms of systems of two separable components: environments and agents. This 'classical' decomposition (see figure 10) can be seen explicitly in [5], [6], [11], [46], and [45], for example, and is present, if only implicitly, in much other recent work besides. It presumes the agent to be an input-output information processing system taking information from the environment, via its sensing, processing this into selected actions (typically with the aim of satisfying some goal or goals), and then performing these actions on the environment. The uncertainty, unpredictability, complexity, etc., of the environment are properties of the environment component, and are seen as something the agent has to battle with, and overcome, in its sensing, acting, and information processing.

This classical view of agents and environments is not new. Essentially the same input-output related agent and environment combination can be seen in Cybernetics, see [2, 3], for example, and is also the model that has dominated work in Artificial Intelligence (AI) since its beginnings, see [28], for example.

All approaches to understanding behaviour that adopt this classical agent and environment decomposition characterise behaviour as a property of the agent expressed through its actions on the environment. They seek explanations of behaviour—why agents act the way they do—in information processing terms. The only real differences between most current work in adaptive behavior and learning in agents and most work in AI is in the kinds of behaviour investigated, and that the former takes a mostly bottom-up approach, while AI traditionally works top-down.

Much of our work has involved building real robots with which to investigate adaptive behaviour and learning, [29], [14], [30], [48], [31], [42], and [43], for example. This experience, and that of others who also build real robots, [15], [8], and [10], for example, has led us to become increasingly suspicious of the suitability of this classical view to support a proper understanding of

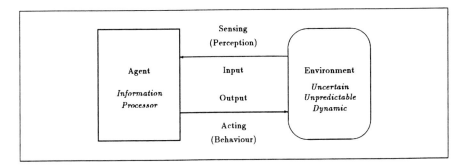

Figure 10. Classical input-output related agent and environment combination

real agent-environment systems, and of quantitative measures of behaviour performance of such systems, in particular. Behaviour is not a property of an agent, it is a dynamical process constituted of the interactions between an agent and its environment. The concept of information processing abstracts away from this essential dynamical aspect: a dynamics which is governed by physical laws which thus determine the nature and form of all agent-environment interactions, and which should therefore appear in any proper explanation of adaptive behaviour.

In this section we introduce an alternative view of agent-environment interaction systems, and present the beginings of a theory based upon it which we believe does less damage to the real nature of behaviour in the abstractions it uses. We then suggest how descriptions of variable, uncertain, unpredictable, dynamic, and complex environments can be more appropriately formulated as questions about the agent-environment interaction space, and propose some ways in which such questions may be operationalised as tests of variability and complexity of agent-environment interaction spaces, and quantitative measures of robot behaviour. We end with a brief review and discussion of some related work.

4 Situated Embodied Agents

First, some working characterisations which form important background to the view of agent-environment systems we propose here. (See [41] for an explanation of the role, nature, and use of working characterisations.) The concepts of situated, embodied, and agent, as used here, are informally captured in the following working characterisations:

- *Situated* means that the here and now counts all the time and everywhere, it always did, and it always will;
- *Embodied* means that physical properties such as inertial mass, shape and space occupancy, and the forces acting on them count, a lot, and *all* interactions involve energy transduction; and, to be an
- *Agent* means having the sufficient means to sustain and maintain a continuous identity through situated embodied interaction with the environment without (necessary) dependence upon external support or assistance.

The characterisation of situatedness is intended to capture the important fact that for a physical agent interacting with a real environment, the present—the current situation for the agent—is an integral and inseparable aspect of the dynamics of an irreversible interaction process that takes the agent-environment system from the past through the present into the future. Situated agents cannot 'step outside' or 'turn away from' their interactions with their environment, as they can in computational models of agent-environment systems, where, rather than "waiting for no man", time always waits for the agent, and for as long as it takes it to decide what to do next, see [5] and [11], for example. It is also intended to capture the further important fact that, as for almost all the things that go on in the physical world, in general, agent-environment interaction processes are irreversible, and thus ones in which history counts. Those processes that are reversible and in which history does not count (systems having Hamiltonian dynamics or are first order Markovian processes, for example) are exceptions, and, although they are important, we must not be misled by their relative simplicity into thinking that studying such degenerate forms will lead to a proper understanding of the general class of interaction processes.

The characterisation of embodiedness used here is intended also to capture the important fact that real world systems, and in particular agent-environment systems, are essentially dissipative dynamical systems whose behaviour is constrained by physical law. All agent-environment interactions involve energy transduction which are never perfect, i.e., there is always some dissipation of the energy involved. This dissipation results in (typically small) statistical variations in the details of repeated interactions, and a dissipative (i.e., non-Hamiltonian) dynamics.

Embodied agents cannot 'step outside' or 'turn away from' the consequences of the physical laws that govern their interactions with their environments, as they can, and do, in computational models of agent-environment systems. In real agent-environment systems the physics of the real world comes for free, and is forever unavoidable. In computational models, however, the effects of physical laws have to be added explicitly, but this is difficult to do and hardly ever done in practice. As a result, the agent-environment systems modelled computationally are of a different class of system to the class of real agent-environment systems, and the relationship between particular systems in these two classes is not, in general, obvious or simple. Understanding a computational agent-environment system does not, therefore, necessarily give us any grounds for claming we understand some real agent-environment system, no matter how similar the two systems are said to look.

The characterisation of agenthood used here is intended to pick out a certain kind of physical entity, a class of systems that are composed of two essential components. All physical entities interact with their environment, they cannot *not* do so. Hence, all physical entity-environment combinations have intrinsic behaviour governed by what we will call the intrinsic dynamics of the entity or agent—their inescapable or inevitable dynamics. Intrinsic behaviour is not, in general, sufficient to sustain and maintain identity through interaction. If the environmental forces inducing motion, for example, change or go to zero, the behaviour of the entity-environment system may change, and change in such a way that the original entity can no longer be picked out from the general goings on in its environment: its identity may be lost. To sustain and maintain identity through interaction with the environment requires the entity to have the means to introduce additional dynamical properties in the interactions with its environment. This requires the entity to have what we here call inherent dynamics—the dynamics it takes on. Having inherent dynamics, in addition to intrinsic dynamics, is what makes an entity an agent. Inherent dynamics do not replace the intrinsic dynamics, they complement and modify them.

This distinction between systems having only intrinsic dynamics and systems having both intrinsic and inherent dynamics can be seen in biology in the distinction between plants and animals[3]. The evolution of networks of motor-sensory and nervous systems and morphology in animals—mobile biological systems—can be seen as providing the inherent dynamics needed to sustain and maintain effective identity through agent-environment interactions. Plants, however, which do not move about, can depend well enough on the intrinsic dynamics of their structure and material properties. Having intrinsic and inherent dynamics can thus be seen as a necessary (but perhaps not sufficient) condition for agenthood.

5 Towards a Theory of Agent-Environment Systems

In this section we introduce the beginnings of a theory of agent-environment systems. (To call it a full theory would be premature and perhaps presumptious given that it has yet to be extensively tested in terms of the predictions and explanations it supports.) It is intended as the beginnings of a general theory of behaviour in agent-environment interaction systems, composed of what we here call situated embodied agents and real environments. It is based upon a further working out and development of ideas first presented in [41].

5.1 The Components of Agent-Environment Systems

First, the essential components of agent-environment systems and their relationships will be defined. For the purposes of this introduction and for simpicity, we will be concerned only with single agent systems. The extension to multi-agent cases follows naturally, though it becomes necessarily more complicated.

The world consists of an *environment*, E_a, in which there is an *agent*, A_e^4 An agent consists of a sets of intrinsic, d_t, and inherent, d_h, dynamical processes which interact with the dynamical processes in the rest of the environment via the agent's physical properties and motor-sensory interaction systems. In general, an agent will only interact with certain aspects of its environment, not with all of it. Thus, for any given agent-environment system, there will be a *set of all possible interactions*, S_i: the set of all possible agent-environment interactions. At any point in time, only a subset of these possible interactions will be occurring. We define a *situation*, s_t as the combination of possible interactions that can occur at the same time. The *set of possible situations*, S_s, is thus defined as the set of all such combinations, which can be written:

$$S_s = C_{is}(S_i) \tag{1}$$

where C_{is} is a function that takes the set of possible agent-environment interactions and returns the set of possible agent-environment situations. From this we have that for all situations s_t, $s_t \in S_s$.

For any given agent-environment system the elements of S_s will be related in the sense that not all sequences of situations will be possible. This further structural organisation over S_s, T_s, together with S_s defines a state space, Ψ_{ae}, for the agent-environment interaction process, P_i, which takes the agent-environment system through possible sequences of situations over time. We call this state space the *agent-environment interaction space*. It is a product of the interactions between a particular agent and its environment, and therefore *not* something which can be identified or specified independent of either the agent or its environment. The agent-environment interaction process can also be understood as what the agent experiences: the agent is not properly understood as experiencing (sensing or perceiving) its environment directly, the 'world out there', so to speak. The agent is thus directly involved in bringing about the 'world' it experiences, rather than being an external observer of it able to act on it, as it is in the classical view. This view of situated embodied agent-environment systems is shown in figure 11.

5.2 Behaviour and the Dynamics of Interaction Processes

In this view of agent-environment systems, when we talk about and describe behaviour we talk about and describe the form of P_i and properties of Ψ_{ae}. An explanation of any particular agent-environment system behaviour must

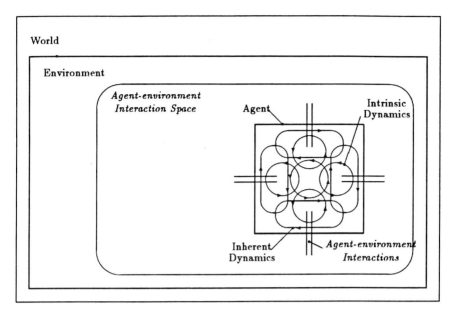

Figure 11. A situated embodied agent-environment system, where the agent has both intrinsic and inherent dynamics which combine in interactions with the environment to form an agent-environment interaction process that traces a trajectory through the agent-environment interaction space over time

therefore be made in terms of the dynamical nature and properties of P_i and Ψ_{ae}. To understand the behaviour of agents in environments, according to this view, means that we must investigate the agent-environment interaction process as a dynamical process. We therefore need to see what general form this can take and what kinds of dynamics we can expect in real agent-environment systems.

Each situation s_t in S_s can be fully characterised by a set of time varying variables, $\mathbf{v}_t(t)$, that represent the instantaneous level of energy tranduced by each type of agent-environment interaction forming the current situation. The set of all such variables thus define the dimensions of the interaction space Ψ_{ae}, and their values, at any point in time, combine to form the state vector, $\mathbf{V}$, of the agent-environment interaction process P_i. From the discussion, we can see that the time variation of $\mathbf{V}$ will be governed by the particular properties of the agent-environment system involved: the intrinsic and inherent dynamics of the agent and the dynamics of its environment. We collect these governing parameters together to form an *agent-environment system parameter vector* $\mathbf{Q}$. The time variation of $\mathbf{V}$ for any agent-environment system can therefore be expressed in the general form:

$$\dot{\mathbf{V}} = F_{ae}(\mathbf{V}; \mathbf{Q}, \mathbf{n}) \tag{2}$$

where $\mathbf{n}$ is a vector of the noise on each element of $\mathbf{V}$ arising from the dissipation in the agent-environment interactions, and where F_{ae} is (typically) a nonlinear function.

Equation (2), in general, thus defines a high dimensional nonlinear dynamical system. It is the properties of this dynamical system that we must investigate if we are to understand the behaviour of any particular situated embodied agent-environment system. We must also understand the relationship between these properties and the intrinsic and inherent dynamics of the agent, if we are to understand the processes and mechanisms necessary for adaptive behaviour in agent-environment systems.

5.3 The Nature of Agent-Environment Interaction Dynamics

In general it is not possible in practice either to pre-specify or empirically establish a complete description of S_i. We cannot, therefore, expect to be able to define C_{is} explicitly for any given agent-environment system, and thus determine S_s. Without a definition of S_s, and thus the possibility of specifiying the dimensions of Ψ_{ac} and elements of V, it would seem that we have little chance, in practice, of investigating agent-environment system dynamics as defined by Equation (2).

There are two reasons why this may not be the barrier it might at first seem. The first is that Equation (2) defines the form of many kinds of dynamical systems, not just agent-environment interaction systems, and these other kinds of complex nonlinear systems are much studied in physics, chemistry, and biology, see [33], [20], and [4], for example. The second, related, reason is to do with the kind of agent-environment system behaviour we are generally interested in. From the theory presented above we can see that adaptive behaviour is to do with an agent's ability to sustain and maintain its identity through its interactions with its environment. We can further see that this occurs when a high degree of spatio-temporal coherence is maintained in the interaction process trajectory—as opposed to taking a Brownian motion form. In other words, when T_s defines a strong organisation over S_s. In such conditions the behaviour of the agent-environment system can be well characterised by a much smaller number of characteristic variables than the dimension of its interaction space. The analogy of this condition in classical examples of far from equilibrium nonlinear systems, such as Bénard cells and Belousev-Zabotinski reaction difussion systems[5], for example, is called self-organisation, and it arises via the instabilities which exist at critical regions of the parameter space Q of the dynamical systems. In all such cases the behaviour of the systems, its dynamics, become dominated by just a few collective modes, called *order parameters*. They represent a compression of the degrees of freedom of the dynamical system, from the typically very high number in the state vector to a relatively small number. This type of compression has been called the *slaving principle* and has been given an exact mathematical form by Haken for a large class of systems, see [20] and [21] and references therein, for example.

What is therefore being hypothesised here is that, although agent- environment systems will in general give rise to very high dimensional nonlinear dissipative dynamical systems, the kind of behaviour of these systems that we are interested in occurs near critical regions in their parameter spaces, and, as with other examples of this kind of phenomena, leads to self-organised spatio-temporal structures that are fully described by a much smaller number of collective modes. See [22], and references therein, for a presentation of essentially similar ideas and their application to understanding dynamic pattern formation in animal movement.

5.4 Complexity and Variability of Interaction Processes

The interaction space of any particular agent-environment system can be used to characterises completely its nature and type. In particular, certain properties can be used to define measures of variation and complexity which can be used to compare and assess different agent-environment systems, particularly under experimental conditions—the same agent in different environments, or controlled modifications of agents in the same environment, for example.

For example, using techniques for estimating the underlying hidden Markov process of empirical time series data (collected from robot experiments) an estimate of T_s and the number of situations in S_s can be obtained. This gives us a transition matrix which specifies the number of different states and the transition probabilities between them. A measure of complexity, K_{ae}, of the agent-environment system can thus be defined as:

$$K_{ae} = \frac{1}{N} \sum_{i=1}^{N} \sum_{j=1}^{N} T_p(i,j) \tag{3}$$

where N is the number of estimated states, and $T_p(i,j)$ is the estimated transition probability from state i to state j. From the estimate of T_s we can construct a representative set of Markov chains, and from these we can plot the frequency distribution of the different states, and define the variability, J_{ae}, of the agent-environment system, as the area under this curve.

Other measures can be proposed as ways of usefully comparing and classifying agent-environment systems. These include: the number of attractors in Ψ_{ae}; the dimension of each attractor; and the number of negative Ljapunov exponents each attractor has. These can be estimated from an estimate of T_{ae}, and by using numerical techniques that can be applied to empirically collected times series data. (See, for example, [34] and references therein, for an introduction to algorithms for estimating the capacity dimension, information dimension, correlation dimension, and Ljapunov exponents, from time series data, and [1] for a review of the analysis of empirical data from chaotic physical systems.) These techniques, and effective procedures for applying them to robot-environment systems, are currently being developed and tested.

6   Related Work and Discussion

Concern for a proper treatment of the dynamics of agent and environment interactions has been expressed before, by [9] and [35], for example. Others have also suggested that agents are better characterised as dynamical systems, [17] and [12], but in both these cases agents and environments are considered as joined, but separable, dynamical systems. This work recapitulates the work of Ashby, [2], who propossed the *coupling* of agent and environment dynamics to form state determined systems, as a basis for analysing stability in machine and animal control. The concentration is on the intrinsic and (in particular) the inherent dynamics of agents, and not with explicitly characterising the agent-environment interaction process in and of itself. This work assumes, as Ashby did, that stability requires equilibrium conditions in the dynamics, whereas here stability (of behaviour) is associated with the kinds of self-organised structures that can occur in far from equilibrium nonlinear processes. Also, again following Ashby, Gallagher and Beer make no definitive distinction between what is agent and what is environment: it is an arbitrarily drawn boundary. In their work, the leg of a robot is defined as the environment of a controller within the body of the agent. The analysis they present is therefore of the performance of the controller, not of the behaviour of the agent interacting with its environment. This is controller design practice, but not analysis of agent-environment system behaviour.

Complex dynamics has previously been suggested as an appropriate framework for studying autonomous systems, [23], and as an alternative view to the traditional view of cognition as computational in Cognitive Science, [47]. A dynamical system approach to understanding pattern generation in animal movement, similar to the one presented here, is proposed in [22], see also [39], with an application to robot motion control in [40].

This as yet small but growing interest in dynamical systems approaches to behaviour and cognition, and the complex dissipative dynamical theory presented here, in particular, brings the investigation of the behavioural phenomena studied in work on adaptive behaviour closer to other work on far from equilibrium self-organising phenomena found in physics, chemistry, and biology. This proximity give us effective access to some powerful and fast developing theoretical and computational techniques in the field of complex dynamical systems, as well as forming a basis for relating the behaviour of agent-environment interaction systems to other complex phenomena of the physical world—something which has not been possible up to now. Behaviour, and, in particular, the kinds of intelligent behaviour traditionally investigated in AI, has never been related in any way to the phenomena studied in other sciences. A situation which is both odd, and, probably, an indication of a need for change.

7 An Example of Quantitative Measurement of Behaviour

In this section we present the results from, and some discussion of, an attempt to use one of the analysis techniques mentioned in the last section, on data collected from a real robot experiment. First we will briefly describe the experimental set up, then the particular analysis technique used, correlation dimension estimation, and then the results obtained.

7.1 Experimental Setup

This experiments involved a first generation Lego vehicle, equiped with binary bumper and whisker sensors and three one-bit active IR sensors, see [14] for details. The robot's default state was to move forward in a straight line and was programmed to avoid obstacles detected using any of its sensors. This program is similar to that described in section 2.1, in other words, there is a dynamically adaptive mapping from sensor signals to motor state changes for each of the three sensor modalities used, which depend on recent signal history.

An experimental environment similar to that presented in figure 1 was used and the robot program adjusted so that the robot would normally travel around the outside of the enclosure. In this experiment the robot would sometimes not detect obstacles with its IR sensors and so there was some bumper and whisker mediated interaction. Also, the path taken would not always keep the robot near the walls of the enclosure: sometimes it would break away and cross to the other side, where it would typically re-establish 'wall-following' behaviour.

During each run, lasting about 26 mins., the amount of time (in milliseconds) spent moving forward (rather than turning or reversing) in each one second interval was recorded. The idea behind collecting this kind of data was that it can be conveniently collected by the robot and it is representative of the behaviour of the robot interacting with this kind of environment. At the end of each run this data was uploaded to the host computer and stored as a time series. The data from nine runs were recorded in this way and used in the subsequent analysis, with each data set containing 1600 points.

7.2 Correlation Dimension Estimation

Correlation dimension analysis (see [18] and [32]) is one of three techniques that have been developed (and which are currently widely used) to estimate the dimensionality of the attactor underlying the dynamics of a complex (normally dissipative) dynamical system. The other two techniques are know as Capacity Dimension and Information Dimension. These three technqiues are based on similar assumptions and mathematical developments, but differ in detail, see [27], for a comparison, and an algorithm for estimating all three.

We use the Correlation Dimension Estimation techniques because it has been more widely used on empirical data (as opposed to synthetic data) and is generally regared as being more appropriate for the kind of dynamical system we are concerned with here.

We will not repeat the details of the technique here—they can be found in the above two references. However, one particular point will be noted here. What makes this technique attractive (as well as those for capacity and information dimension) is that is can be applied to one-dimensional time series data. Essentially the method involves synthesising, from the original one-dimensional time series, a series of phase space descriptions of increasing dimension. The implementation used here is by Goldeberg, [19], from which the following brief explanation is borrowed.

The correlation function $C(r)$ for an N dimensional space defined by N vectors

$$\mathbf{y}(n), n = 1, \ldots, N, \tag{4}$$

is defined by:

$$C(r) = \frac{2}{N(N-1)} \sum_{m=1}^{N} \sum_{n=m+1}^{N} \theta(r - \|\mathbf{y}(m) - \mathbf{y}(n)\|) \tag{5}$$

where $\theta(x)$ is the Heaviside step function and $\|\mathbf{x}\|$ is any well defined vector norm. The correlation dimension (d_c) is then formally defined as the following limit:

$$d_c = \lim_{r \to 0} \frac{\log(C(r))}{\log(r)} \tag{6}$$

In practice, d_c is obtained by plotting $C(r)$ versus r on a log-log scale and reading off the slope from the portion of the graph that displays the appropriate scaling behavior.

In the analysis of a scalar time series data set (equally spaced in time, $i.e.$ $x(n) = x(t_o + n\tau)$), the synthesised dimensions are given by:

$$\mathbf{y}(n) = [x(n), x(n+T), x(n+2T), \ldots, x(n+(d_E-1)T)] \tag{7}$$

for various values of d_E—the synthesised embedding space dimension. As d_E increases, the value of d_c should saturate to the appropriate correlation dimension of the system. If the value of d_c is non-integer it is an indication of a chaotic dynamics with a some kind of strange attactor. The minimum d_E at which this saturation occurs is known as the embedding dimension of the data set.

This 'trick' of synthesising phase spaces of higher dimension form the orginal one-dimensional time series works well if the orginal data is of a 'pertinent' variable of the overall dynamcial system. In other words, we can build a picture of the full phase space dynamics, typically involving many variables, if we chose to record one system variable which 'bears the hall marks' of all these system variables.

The resulting estimation of d_c gives us an estimate of the dimensionality of the underlying dynamical system attractor and dimension of its embedding space. We can use these two quantities to both compare different robot-environment interaction systems and to quantitatively investigate the effects of changes to the robot or environment (or both) on the behaviour of the system. It gives us a quantitative way of comparing the overall (qualitative) behaviour of the system without having to record many different variables, and have to remove detailed variations in recorded signals.

7.3 Empirical Results

In this initial investigation we have run the robot with the same program and setting in the same environment in order to see what kind and amount of normal variation we can expect, and as a first attempt to apply this technique to empirical data of this kind. Each run was done on a different day. So, although the conditions were nominally the same, there was some inevitable variation in the setup of the environment and robot: the environment was set up in a room used by other people, and changing humidity levels in the air changed the IR-reflection characteristics of the enclosure walls, thus slightly (but noticably) affecting the robot-environment interaction.

Before the Correlation Diemension analysis as preformed the data was put through a low-pass filter to remove a high frequency component that resulted from a sampling error produced by the particular implementation of the data collection mchanism. The filtered datasets were called 'ppld5'.

The nine Correlation Dimension plots (part a) and associated first-return plots (part b) are presented in figures 12 to 20. Each Correlation Dimensions plot shows the distance dependence of the correlation function for a series of synthesied phases spaces, form dimension 1 to 10. The upper limit of 10 was found to be the point at which the slope of these graphs remained essentially the same (within the amount of variation that can be seen in the graphs shown). The Correlation Dimension is thus estimated by taking the slope of the linear part of the graph for dimensions 8 to 10, and the embedding dimension estimate is taken to be between 8 and 10 in each case. The first-return plot are presented to indicate that these is some kind of structure present in the data, and to show that the delay interval of one used in the Correlation Dimension analysis is reasonable – it produces and 'open look' to the first-return plots.

7.4 Discussion

The correlation dimension, d_c, estimates for each of the nine data sets presented here are collected in table 1.

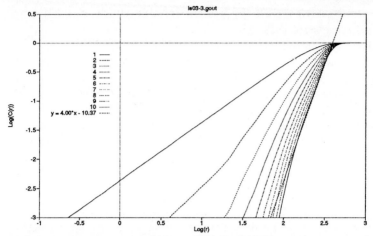

Figure 12a. Log-log plot (base 10) of correlation distance r against Correlation function $C(r)$ for data set ls03-3.ppld5. Estimated correlation dimension $d_c = 4.0$

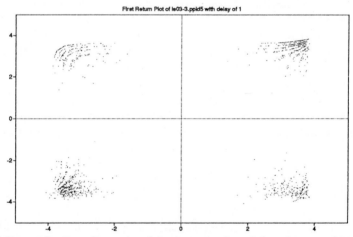

Figure 12b. Two dimensional first return plot of data ls03-3.ppld5 with a delay interval of 1

The first thing to say is that the interpretation of these results is not easy and must be subjected to a good deal of uncertainty. The practice of applying techniques such as this to empirical data is relatively new and not well understood in general.

The worst interpretation that can (quite possibly) be put on these results is that they tell us nothing more than there is a lot of noise in the data. We might be lead to this view by the fact that the correlation dimension estimates are quite high. Although it is generally regarded that anything over about ten should be treated as an indication of a purely noisy system, rather than one with a complex (possibly chaotic) dynamics, values greater than about

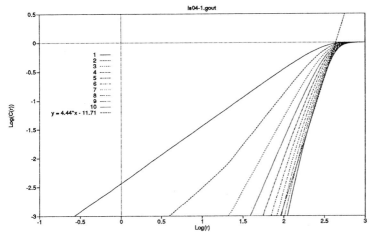

Figure 13a. Log-log plot (base 10) of correlation distance r against Correlation function $C(r)$ for data set ls04-1.ppld5. Estimated correlation dimension $d_c = 4.4$

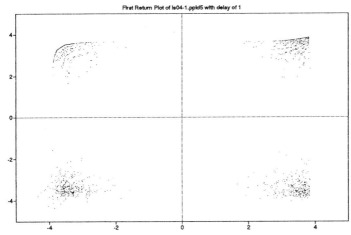

Figure 13b. Two dimensional first return plot of data ls04-1.ppld5 with a delay interval of 1

3 are difficult to verify for empirical data of this kind.

Another reason for being suspicious of these results is that they have been obtained from relatively small data sets. Currently, the question of how many points are required for a good correlation distance analysis, is the subject of argument and dispute. It is, however, generally agreed that the higher the dimensionality of the system being investigated the more data points are required. With estimated correlation dimensions greater than four to five, the 1600 points used here may well be grossly inadequate.

However, there are also some indications for why we might also be justified in being more positive (and optimistic) about interpreting these results. The

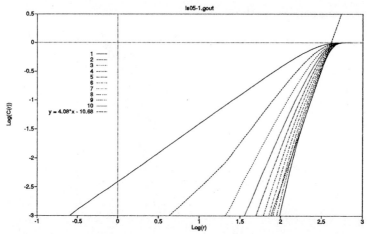

Figure 14a. Log-log plot (base 10) of correlation distance r against Correlation function $C(r)$ for data set ls05-1.ppld5. Estimated correlation dimension $d_c = 4.1$

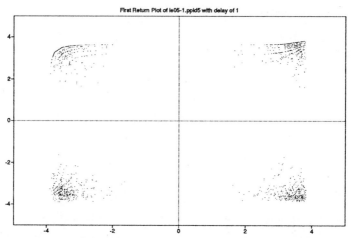

Figure 14b. Two dimensional first return plot of data ls05-1.ppld5 with a delay interval of 1

first is that all the graphs (figures 12 to 20) display the characteristic slope saturation phenomenon that we expect from this analysis, and all in the same range of dimensions, from $N = 8$ to 10. If the data contained large amounts of noise we would expect the slope of the $Log(C(r))/Log(r)$ plots to keep getting steeper as we increase the dimension of the synthesised embedding space. The second good sign is that there is some consistency in the results, with six (numbers 1, 3, 4, 5, 6, and 9) of the nine estimates for d_c being between 4.0 and 4.2. The third, and perhaps most encouraging sign, is that the estimates for d_c that diverge the most from A_{d6} (numbers 2, 7, and 8) correlate with runs which, in the case of numbers 2 and 7, involved more

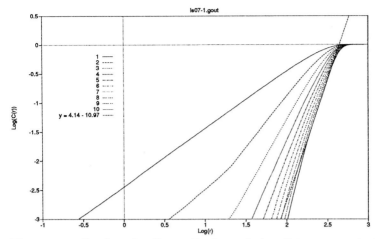

Figure 15a. Log-log plot (base 10) of correlation distance r against Correlation function $C(r)$ for data set ls07-1.ppld5. Estimated correlation dimension $d_c = 4.1$

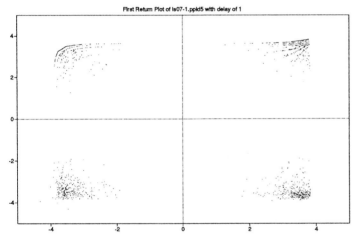

Figure 15b. Two dimensional first return plot of data ls07-1.ppld5 with a delay interval of 1

collisions with the walls of the enclosure and objects within it, resulting in the 'perimeter following' behaviour breaking down more often than usual, and, in the case of number 9, a run which was untypically free of any such behaviour – on this occasion the robot maintained a consistent path round the outside of the enclosure for the whole of the run.

The question then is, given this correlation between the estimates of d_c and differences in the runs, why should we expect the dimension to be higher in the cases of numbers 2 and 7 and lower in the case of number 8?

First we need to consider why the dimension estimates might really be as high as ten (and more) in the first place. A plausible reason for this, apart

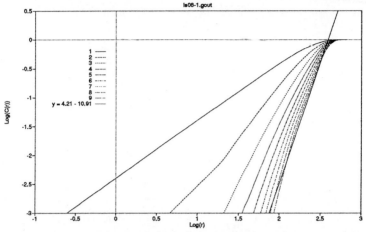

Figure 16a. Log-log plot (base 10) of correlation distance r against Correlation function $C(r)$ for data set ls08-1.ppld5. Estimated correlation dimension $d_c = 4.2$

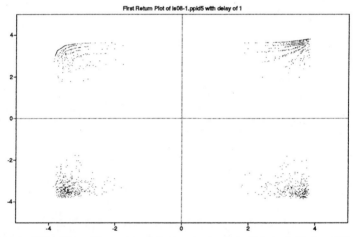

Figure 16b. Two dimensional first return plot of data ls08-1.ppld5 with a delay interval of 1

from noise in the data being dominant, is that we are here dealing with a dynamical system which has history in it. In other words, we do not have a (first order) Markovian process, where each subsequent state is completely determined by the current state. Instead, because of both the inertia of the robot in motion, and the nature of the control program (which uses sensor signal history to adapt the motor actions) we have a system in which the influence of the current state propagates forward over significant intervals of time compared to the data sampling rate (of one data point per second). It is well know that the dynamics of systems with history are typically more complex, and so require higher dimensional attactors to describe them.

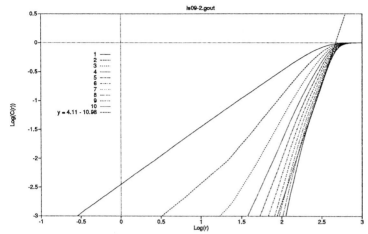

Figure 17a. Log-log plot (base 10) of correlation distance r against Correlation function $C(r)$ for data set ls09-2.ppld5. Estimated correlation dimension $d_c = 4.1$

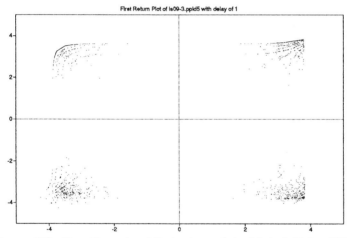

Figure 17b. Two dimensional first return plot of data ls09-2.ppld5 with a delay interval of 1

At the moment, not enough is known about the robot-environment interaction system investigated here, to be able to say what the relationship is between the amount of history in the system and the dimensionality of the underlying attractor. This is a question currently being investigated. However, what we can say is that, as more modes of interation become involved (mediated by bumpers, whiskers, and IR sensors), and as the amounts of these different modes of interaction increases, so the complexity of the interaction process increases, as does the amount of history in the system. (This is predicted by the agent-environment systems theory presented in the previous section.) So, we might seek to explain the higher estimates of d_c in the

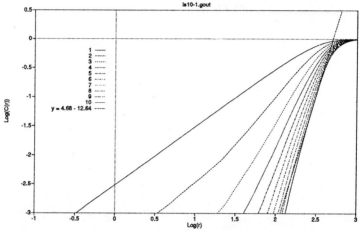

Figure 18a. Log-log plot (base 10) of correlation distance r against Correlation function $C(r)$ for data set ls10-1.ppld5. Estimated correlation dimension $d_c = 4.7$

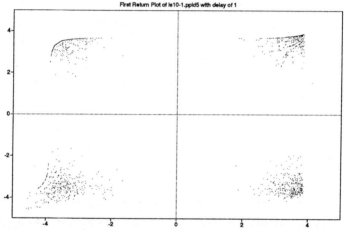

Figure 18b. Two dimensional first return plot of data ls10-1.ppld5 with a delay interval of 1

cases of runs 2 and 7, by the fact that in these runs there were more bumper and whisker mediated interaction than usual. And, in the case of run number 8, we might seek to explain the lower estimate of d_c by the fact that there was only IR-mediated interaction and also, that the form of the interaction was unusually constant – with no break downs of the perimeter following behaviour. If all this is a true, then what we have here, is quantitative estimates for the way in which the overall form of the dynamics of interaction changes as the nature of the robot-environment interaction changes.

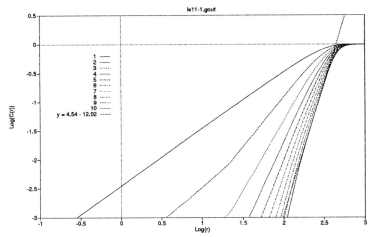

Figure 19a. Log-log plot (base 10) of correlation distance r against Correlation function $C(r)$ for data set ls11-1.ppld5. Estimated correlation dimension $d_c = 4.5$

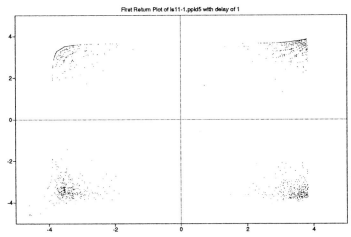

Figure 19b. Two dimensional first return plot of data ls11-1.ppld5 with a delay interval of 1

8 Conclusions

What we can conclude from all this is that we have either shown very little, apart from the inadequacy of this experiment to produce useful data, or we have some very preliminary, but encouraging, indications as to how we might use correlation dimension estimation technqiues to make quantitative estimates of overall robot behaviour. Taking the more positive view, we can therefore say, we have shown how quantitative measures of robot behaviour can be developed and used in practice, together with an appropriate theoretical framework for agent-environment interaction systems.

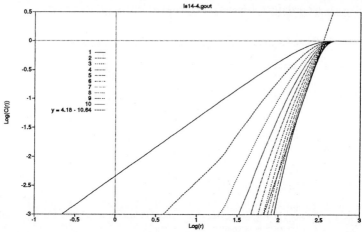

Figure 20a. Log-log plot (base 10) of correlation distance r against Correlation function $C(r)$ for data set ls14-4.ppld5. Estimated correlation dimension $d_c = 4.2$

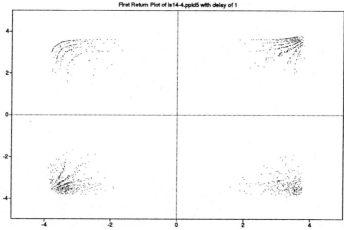

Figure 20b. Two dimensional first return plot of data ls14-4.ppld5 with a delay interval of 1

Work is currently under way to further test the ideas and approach presented here, and, in particular, to try to establish the proper interpretation of this kind of results. Power spectrum analysis is being used to estimate the amount and nature of the noise in the data, and to also provide indications for temporal structure. And further robot-environment experiments are being conducted to collect new data with larger numbers of data points, as well as for setups with known amounts of changes in them.

Table 1. Correlation dimension estimates for each data set

No.	Data set	Estimated d_c
1	ls03-3.ppld5	4.0
2	ls04-1.ppld5	4.4
3	ls05-1.ppld5	4.1
4	ls07-1.ppld5	4.1
5	ls08-1.ppld5	4.2
6	ls09-2.ppld5	4.1
7	ls10-1.ppld5	4.7
8	ls11-1.ppld5	4.5
9	ls14-4.ppld5	4.2

Acknowledgements

The work reported here was mostly carried out while the author held the SWIFT AI Chair at the VUB AI Lab, Brussels. Miles Pebody, Ian Porter, Piet Ruyssinck, and Danny Vereertbrugghen, were responsible for much of the development of the second generation Lego vehicle technology used. Anne Sjoström assisted in building, testing, and debugging the robot program used, and in carrying out the experiments from which the data were taken.

The development of the theoretical ideas presented here have benefited from discussions with numerous people. I am particularly grateful to Agnessa Babloyantz, Tony Bell, Grégoire Nicolis, Rolf Pfeifer, Cecilia Sarasola, Gregor Schöner, Luc Steels, and Javier Torreldea for their various suggestions, corrections, demands for clarification, support, and encouragement. The misconceptions and confusions that remain are all my own. I am also grateful to to Jess Goldger for help in using his program, and to Amaia Bernaras, who read and commented on earlier drafts of this paper. My sincere and grateful thanks to them all! Lastly, but least, I am happy to acknowledge the financial support of the University of the Basque Country for my current position, and to the Faculty of Informatics, in particular, for providing a very supportive academic home.

Notes

1. Second generation Lego vehicles are essentially the same as first generation, see [14], except that a two microprocessors architecture is used, one (a MC68HC11) to service the sensor and motor-control channels, [50], and the

second (a MC68340 with 0.5MByte RAM) to run the control program. It also
has 8 bit IR and light sensors (instead of 1 bit sensors), and a more efficient and
flexible time-slicing runtime kernel, plus a number of other advanced features.

2. This uses a Hall-effect switch and six magnets placed in the front wheel with
even spacing and alternating pole directions to produce three counts per complete revolution in one direction—changes in direction are detected in the program so that forward and reverse counts are maintained separately. The wheel
diameter is 43mm, giving a circumference of 135.089mm, and thus a distance
per wheel count of 45.03mm. Having the front wheel fixed (i.e., not a caster)
means that it is always in line with the direction of motion or tangential to it.
This means that, using this wheel counter, it is possible to reliably estimate
the distance travelled: essentially it integrates only the forward motion of the
robot, much as a planimeter does—an instrument used for measuring the area
of closed plane figures, such as ship hull sections, etc.

3. I am indebted to Francisco Varela for bringing this distinction to my attention. While it constitutes a very important and fundamental difference between
plants and animals, it does not form the basis for saying that animals can behave adaptively and plants cannot. Both the intrinsic and inherent dynamics
can be changed, but the nature of such changes is typically quite different.

4. Here, we already see an important difference from the classical view, in which
agents are a seperate entity from the environment they interact with, they are
not seen as being a part of the environment, as they are here. It is by identifying
the agent as a part of its environment that we capture the situatedness of real
agents.

5. See [33], [20], or [4] for details of these phenomena, or many other texts on
self-organisation in nonlinear far from equilibrium systems.

References

1. Henry D. I. Abarbanel, Reggie Brown, and Lev Sh. Tsimring, 1992. The analysis
of observed chaotic data in physical systems, Reviews of Modern Physics
2. W. Ross Ashby, 1956. An Introduction to cybernetics, Chapman & Hall
3. W. Ross Ashby, 1960. Design for a brain, second edn., Chapman & Hall
4. Agnessa Babloyantz, 1986. Molecules, dynamics and life, Wiley, New York
5. Randall D. Beer, 1990. Intelligence as adaptive behaviour, Academic Press
6. Lashon B. Booker, 1991. Instinct as an inductive bias for learning behavioural
sequences, in [36], 230
7. Rodney A. Brooks, 1985. Achieving artificial intelligence through building robot,
AI Memo 899, Artificial Intelligence Laboratory, Massachusetts Institute of Technology, May, 1986
8. Rodney A. Brooks, 1990. Challenges to complete creature architectures, AI
Memo, Artificial Intelligence Laboratory, Massachusetts Institute of Technology
9. Rodney A. Brooks, 1991. Intelligence without reason, IJCAI-91, Sydney, Australia, 569–595
10. Rodney A. Brooks, 1992. Artificial life and real robots, in [16], 3–10
11. Dave Cliff, 1991. The computational hoverfly; a study in computational neu-
roethology, in [36], 87–96

12. Dave Cliff, Inman Harvey, and Phil Husbands, 1993. Explorations in evolutionary robotics, Journal of Adaptive Behavior, Vol 2, no 1, 73–110

13. Holk Cruse, 1991. Coordination of leg movement in walking animals, in [36], 105

14. Jim Donnett and Tim Smithers, 1991. Lego Vehicles: A technology for studying intelligent systems, in [36], 540–549

15. Anita M. Flynn and Rodney A. Brooks, 1989. Battling reality, AI Memo 1148, MIT AI Laboratory, October 1989

16. Francisco J. Varela and Paul Bourgine (eds.) Towards a practice of autonomous systems, proc of the first European Conference on Artificial Life, held in Paris in 11–13 November 1991, MIT Press

17. John C. Gallagher and Randall D. Beer, 1993. A qualitative dynamical analysis of evolved locomotion controllers, in [37], 71–80

18. P. Grassberger and I. Procaccia, 1983. Physics Review Letters, vol 50, 346

19. Jesse Goldberg, 1993. Using the correlation dimension code: grass, program documentation, January, 1993

20. H. Haken, 1983. Synergetics, an introduction: Non-equilibrium phase transitions and self-organisation in physics, chemistry and biology, 3rd edn., Springer-Verlag, Berlin

21. H. Haken, 1983. Advanced synergetics: instability hierarchies of self-organisating systems and devices, Springer-Verlag, Berlin

22. J. A. Scott Kelso, M Ding, and Gregor Schöner, 1992. Dynamic pattern formation: A primer, in J.E. Mittenthal and A.B. Basin (eds.), Principles of Organization in Organisms, SFI Studies in the Sciences of Complexity, Vol XIII, Addison-Wesley

23. George Kiss, 1992. Autonomous agents, AI and chaos theory, in [36], 518–524

24. Pattie Maes, 1993. Behaviour-based artificial intelligence, in [37], 2–10

25. Chris Malcolm, Tim Smithers, and John Hallam, 1989. An emerging paradigm in robot architecture, in T. Kanade, F.C.O. Groen, and L.O. Hertzberger, (eds.), Intelligent Autonomous Systems 2, Amsterdam, 1989

26. Jean-Arcady Meyer and Agnès Guillot, 1991. Simulation of adaptive behavior in animats: review and prospect, in [36], 2–14

27. Francis C. Moon, 1987. Chaotic vibrations, Wiley

28. Allen Newell, 1981. The knowledge level, Journal of Artificial Intelligence, vol 18, no 1, 87–127

29. Ulrich Nehmzow and Tim Smithers, 1991. Mapbuilding using self-organising networks in "Really Useful Robots", in [36], 152–159

30. Ulrich Nehmzow and Tim Smithers, 1992. Using motor actions for location recognition, in [16], 96–104

31. Ullrich Nehmzow, Tim Smithers, and Brendan McGonigle, 1993. Increasing behavioural repertoire in a mobile robot, in [37], 291–297

32. Grégoire Nicolis, 1986. Dissipative systems, Rep. Prog. Phys. vol 49, 873–949, The Institute of Physics

33. Grégoire Nicolis and Ilya Prigogine, 1977. Self-organisation in nonequilibrium systems, Wiley, New York

34. Heinz-Otto Peitgen, Hartmut Jügens, and Dietmar Saupe, 1992. Chaos and fractals: New frontiers of science, Springer-Verlag, Berlin

35. Rolf Pfeifer and Paul Verschure, 1992. Distributed adaptive control: A paradigm for designing autonomous agents, in [16], 21–30

36. Jean-Arcady Meyer and Stewart W. Wilson (eds.), 1991. From animals to animats, proceedings of the first International Conference on Simulation of Adaptive Behavior, Paris, December 1990, MIT Press

37. Jean-Arcady Meyer, Herbert L. Roitblatt, and Stewart W. Wilson, 1993. From animals to animats 2, proceedings of the second International Conference on Simulation of Adaptive Behavior, Honolulu, December 1992, MIT Press

38. Dave Cliff, Philip Husbands, Jean-Arcady Meyer, and Stewart W. Wilson, (eds.), From animals to animats 3, proceedings of the third International Conference on Simulation of Adaptive Behavior, Brighton, August, 1994, MIT Press

39. Gregor Schöner and J. A. Scott Kelson, 1988. Dynamic pattern generation in behavioral and neural systems, Science, Vol 239, 1513–1520

40. Gregor Schöner and Michael Dose, 1993. A dynamical systems approach to task-level system integration used to plan and control autonomous vehicle motion, Journal of Robotics and Autonomous Systems, vol 10, 253–267

41. Tim Smithers, 1992. Taking eliminative materialism seriously: A methodology for autonomous systems research, in [16], 31–40

42. Tim Smithers, 1993. On behaviour as dissipative structures in agent-environment interactions processes, presented at the workshop prerational Intelligence: Phenomenology of Complexity Emerging in Systems of Agents Interacting Using Simple Rules, held at the Centre for Interdisciplinary Studies (ZiF), Bielefeld, November 22–26, 1993, as part of the Prerational intelligence Research Project

43. Tim Smithers, 1994. On why better robots make it harder, in [38], 64–72

44. Tim Smithers, 1995. Are autonomous agents information processing systems? in Luc Steels and Rodney A. Brooks, (eds.), The 'artificial life' route to 'artifical intelligence': building situated embodied agents, Lawrence Erlbaum Associates, forthcoming 1995

45. Richard S. Sutton, 1991. Reinforcement learning architectures for animats, in [36], 288–296

46. Toby Tyrrell, 1993. The use of hierarchies for action selection, Journal of Adaptive Behaviour, Vol 1, no 4, 387–420

47. Timothy van Gelder, 1992. What might cognition be if not computation, Technical Report, Department of Cognitive Science, Indiana University, Bloomington, IN

48. Barbara Webb and Tim Smithers, 1992. The connection between AI and biology in the Study of Behaviour, in [16], 421–429

49. Stewart W. Wilson, 1991. The Animate Path to AI, in [36], 15–21

50. Dany Vereertbrugghen, 1993. Design and Implementation of a sensor-motor control unit for mobile robots, Licentie Thesis, AI Laboratory, Vrije Universiteit Brussel.

Robot Adaptivity*

Carme Torras

Institut de Cibernètica (CSIC-UPC). Diagonal 647, 08028 Barcelona. Spain
e-mail: torras@ic.upc.es

1 Overview

These lecture notes review neural network techniques for achieving adaptivity
in autonomous agents. They are structured around three main topics:

- **Neural learning algorithms.** A distinction is made between the short-
  term and long-term levels of processing in neural networks. Long-term
  processing is what is referred to as "learning", although it should be more
  properly called "adaptation", and it amounts to modifying the connection
  weights. A classification of the learning tasks, learning rules and learning
  models that have appeared in the literature is presented. In particular,
  learning rules are grouped into three classes: correlational rules (Hebbian), error-minimization rules (perceptron, LMS, back-propagation) and
  reinforcement rules (associative search, associative reward-penalty).
- **Applications to robot control.** Within the field of control, neural
  networks have been applied to system identification and to the design of
  controllers. The two main approaches that have arisen are direct-inverse
  modelling and forward modelling. In the more specific field of robot control, efforts have been oriented to the learning of inverse models (inverse robot kinematics and dynamics) and of goal-oriented sensorimotor
  mappings (path finding, hand-eye coordination). Two systems are next
  described in detail: topology-conserving maps for learning visuomotor
  coordination of a robot arm [32], where a correlational rule is combined
  with an error-minimization rule; and the reinforcement-based path finder
  for mobile robots described in [26].
- **Limitations of neural control: a need for planning?** This block is
  intended to foster a discussion of what the advantages and limitations of
  subsymbolic and symbolic approaches are. Neural controllers obviate the
  programming phase by exploiting learning, but their generality and opac-
  ity make it impossible to take advantage of problem-specific information.
  This results in very long learning times. Motion planners, on the other
  hand, rely on geometric reasoning and heuristic search, thus allowing the

* The author acknowledges support from the Comisión Interministerial de Cien-
  cia y Tecnología (CICYT) under the project SUBSIM (TAP93-0451) and from
  the ESPRIT III Program of the European Community under contract No. 7274
  (project "B-LEARN II: Behavioural Learning: Combining Sensing and Action").

use of domain-specific knowledge to gain efficiency. However, they are hard to program and computationally expensive when high precision is required. A case is made out for the combination of a one-shot symbolic acquisition of knowledge (initial setting of the system) and a subsymbolic adaptation of skills through repetitive trial and error (subsequent tuning).

2 Biological Adaptivity

Adaptivity is the capability of *self-modification* that some agents have, which allows them to maintain a level of performance in front of environmental changes, or to improve it when confronted repeatedly with the same situation.

In the biological world, adaptivity occurs at several levels of an almost continuous scale, taking different forms at each level. Thus, at the cell level, adaptation is often *built-in* in the sense that, for instance, photoreceptors adapt to the ambient level of illumination providing always an image with roughly the same brightness. At the level of individuals, adaptation takes the form of an *association*, leading to the two well-known types of conditioning that will be described next. Finally, at the species level, adaptation is attained through mutation and *natural selection*.

Autonomous robots could be designed so as to incorporate all these forms of adaptivity. Adaptive thresholding, for example, is a way of building adaptivity into sensors. Perhaps genetic algorithms could be used to change robot morphology dynamically in order to tailor it to a given environment. In this paper we will concentrate on adaptivity at the individual level, e.g. both the robot morphology and its components are assumed to be fixed and what may change with experience is the functional relationship between sensors and actuators. Moreover, only approaches to adaptivity developed within the Neural Networks field will be considered, disregarding other more symbolic schemes originated in the domain of Artificial Intelligence.

The approaches to adaptivity pursued within the Neural Networks field have their roots in the learning paradigms developed in the domain of Behavioural Psychology (refer to Fig. 1). This is probably the reason why the rules to attain neural adaptivity are usually called learning rules. The role of the animal in the behavioural learning experiments is played here by the neuron.

The most basic learning paradigm is **classical conditioning**, as introduced by Pavlov [31], which consists of repeatedly presenting to an animal (e.g. a dog) an initially meaningless stimulus (e.g. the sound of a bell) together with an unconditioned stimulus (e.g. food) that triggers a reflex response (e.g. salivation). As a result of such paired presentations, the animal builds up an association so that, when presented with the conditioned stimulus (e.g. the sound of a bell) alone, it produces the same response as before (e.g. salivation). This type of learning is completely open-loop, in the sense

TYPE OF LEARNING	DEGREE OF FEEDBACK	DIAGRAM	KEYWORDS	NEURAL RULES

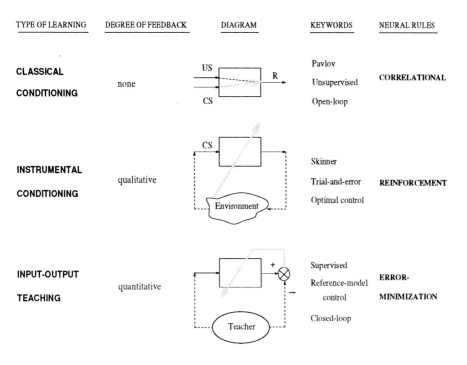

CLASSICAL CONDITIONING	none	US / CS / R	Pavlov / Unsupervised / Open-loop	**CORRELATIONAL**
INSTRUMENTAL CONDITIONING	qualitative	CS / Environment	Skinner / Trial-and-error / Optimal control	**REINFORCEMENT**
INPUT-OUTPUT TEACHING	quantitative	Teacher	Supervised / Reference-model control / Closed-loop	**ERROR- MINIMIZATION**

Fig. 1. Learning paradigms used in the field of Neural Networks

that it entails no feedback. The neural learning rules mimicking this type of conditioning are called *correlational rules*.

Instrumental conditioning was introduced by Skinner [36] and requires that the animal under experimentation performs an arbitrary action (e.g. pressing a lever, walking around) in the presence of an initially meaningless stimulus (e.g. a flickering light). If the action is "appropriate" to the given stimular situation, the animal receives a reward (e.g. food). Otherwise, it receives nothing or punishment, depending on the particular experimental design being applied. Thus, this learning paradigm strongly relies on providing the animal with a reinforcement signal dependent on the action performed. This can be conceptualized as a qualitative feedback. The neural learning rules implementing this type of conditioning at the neuron level are known as *reinforcement-based rules*.

Note that the natural progression in the degree of feedback supplied suggests the use of a quantitative error signal to guide learning. This is represented in the third row of Fig. 1 under the name of **input-output teaching**. Here, after presenting an input to the system and observing the emitted response, a teacher supplies the desired output whose difference with the emitted one provides the error signal which is fed back to the system. This is an entirely closed-loop learning process that requires perfect knowledge of the input-output pairs to be associated. The most widely used neural learning rules follow this scheme and we call them *error-minimization rules*.

3 Neural Levels of Processing

Neural networks are a type of connectionist model that has threshold elements as nodes and whose connectivity is plastic. Two levels of processing happening at different time scales can be distinguished. The *short-term level of processing* consists of the propagation of activity through the network according to the following expression:

$$x_j(t + \delta t) = f[\sum_{i \in I_j} w_{ij}(t)x_i(t)] \tag{1}$$

where x_j is the state of neuron j, I_j is the set of inputs to this neuron — which can come from the environment or be the outputs of other neurons, w_{ij} is the synaptic weight of the connection from neuron i to neuron j, t is the time instant, and f takes usually the form of either a deterministic or a stochastic threshold function. The short-term processing, or dynamics of node activation, has computational interest only in the case of recurrent networks: Hopfield model [17], Boltzmann machine [1].

The *long-term level of processing* is what is conceptualized as learning, and relies on the application of an unineuronal learning rule uniformly throughout the network. The majority of the learning rules postulated work by modifying the synaptic weights, according to the following generic expression:

$$w_{ij}(t + k\ \delta t) = g(w_{ij}(t), x_l(t)_{l \in I_j}, x_j(t), \overline{e}(t)), \tag{2}$$

where g is a polynomial function, $\overline{e}(t)$ is the training information supplied at time t, and k is a constant big enough to allow for the propagation of activity through the network before the weight update is carried out.

Several surveys of the various neural *learning models* proposed can be found in the literature [4, 10, 13, 16, 22, 38, 39]. In what follows we will concentrate on those that have been applied in robotics. First of all, we will give an overview of the different classes of *learning tasks* tackled, to next describe some of the *learning rules* proposed to solve these tasks. The triple distinction between learning tasks, learning rules and learning models is important, because the same learning rule can be incorporated into different network models in order to accomplish different tasks.

4 Learning Tasks

Without aiming at an exhaustive classification, we will consider four types of learning tasks: pattern reproduction, pattern association or classification, feature discovery or clustering, and reward maximization.

The task of **pattern reproduction** consists in retrieving one of a set of stimulus patterns that had been repeatedly shown to the system during training, upon presentation of a portion or a distorted version of the original pattern. When a portion of the original pattern is used as retrieval cue, the task is also referred to as *pattern completion*. Furthermore, Rumelhart and Zipser [34] call this task *auto-association*, interpreting that each pattern becomes associated with itself.

The most classical and widely studied learning task is that of **pattern association or classification**, which constitutes the main objective of the Pattern Recognition discipline. It consists in training the system with pairs of patterns —of which the second can be interpreted as the class to which the first belongs— so that when the first pattern of a pair is presented, the system produces the second. Although some authors consider association and classification to be two different tasks, we envisage the first as a limit case of the second in which each class consists of only one pattern.

Feature discovery or clustering has also been called *unsupervised learning* within the Pattern Recognition literature. Its goal is to find out statistically salient features of the population of stimulus patterns that permit establishing clusters of patterns with similar features.

Finally, the task of **reward maximization** requires that the system is given reward or punishment depending on the action it takes in response to each stimulus pattern. The system has to configure itself in a way that maximizes the amount of reward it receives.

It is clear that pattern association and reward maximization are closely related to input-output teaching and instrumental conditioning, respectively. We insist, however, that one thing is the task to be performed by the entire system and another different one is the learning rule applied at the single neuron level. One can, for instance, use a correlational rule (analog of classical conditioning) to make the whole network learn a pattern association task (input-output teaching).

5 Learning Rules

Since they are intended to constitute biologically plausible hypotheses, most neural learning rules postulate parameter changes on the basis of only local information available to a single neuron.

5.1 Correlational Rules

These rules adjust a connection weight according to the correlation between the states of the two neurons connected. They can all be considered variants of the classical **Hebbian learning rule** [12], whose expression in terms of the generic neuron model (1) is:

$$w_{ij}(t+1) = w_{ij}(t) + cx_i(t)x_j(t), \tag{3}$$

where c is a positive constant that determines the speed of learning. To prevent the unbounded growth of weights, different normalization procedures have been used, the most common one being that based on the euclidean metrics.

This rule can be viewed as a unineuronal analog of classical conditioning: one has only to pair repeatedly an unconditioned stimulus x_1 with a conditioned stimulus x_2, assigning initially to w_{1j} a value high enough to force the neuron to fire $(x_j = 1)$ when activating x_1; after some pairings, the neuron will discharge when only x_2 is presented.

The most extensively studied application of the Hebbian rule is the implementation of *associative memories* [15, 19], which are network learning models able to carry out pattern association tasks as described in the preceding section. The three most interesting aspects of this kind of networks: resistance to noise, addressing by content, and generalization capability, derive from the distributed way in which information is stored.

This rule has also been incorporated into *competitive learning models* [34, 41] able to accomplish feature discovery tasks. As shown in Fig. 2, these models consist of a set of hierarchically layered neurons, each neuron receiving excitatory input from the layer immediately below. Futhermore, the neurons in each layer are grouped into disjoint clusters, each neuron in a cluster inhibiting all other neurons within the cluster. The name "competitive learning" comes from the fact that the neurons within a cluster "compete" with one another to respond to the pattern appearing on the layer below; the more strongly any particular neuron responds, the more it shuts down the other members of its cluster, which therefore becomes a winner-take-all network as defined in [9]. A cluster containing n neurons can be considered an n-ary feature, every stimulus pattern being classified as having exactly one of the n possible values of this feature. It has been proved that, if the stimulus patterns naturally fall into classes, the system will find exactly these classes and the attained classification will be very stable. However, when presented with arbitrary input environments, competitive learning models can become very unstable and the need appears of stabilizing their response through the use of specialized mechanisms, leading to *adaptive resonance models* [11]. These are recurrent modular networks able to form a new cluster whenever they are presented with an input pattern that is very different from the patterns previously seen.

Following the same line of competitive learning, it has been proposed [20] to use *self-organizing feature maps* to construct mappings that preserve topography (i.e. neurons that are spatially close in the network are maximally activated by input vectors close according to the euclidean metrics). Essentially, this is realized through two-layer networks with intralayer lateral inhibition and interlayer plastic excitatory connections.

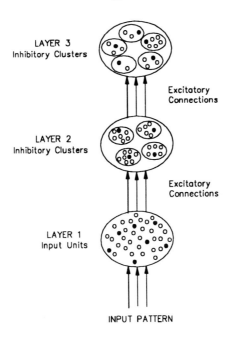

LAYER 3
Inhibitory Clusters

Excitatory
Connections

LAYER 2
Inhibitory Clusters

Excitatory
Connections

LAYER 1
Input Units

INPUT PATTERN

Fig. 2. Generic form of competitive learning models. Active neurons are represented by filled dots, inactive ones by open dots. The neurons within a cluster inhibit one another in such a way that only one neuron per cluster may be active (adapted from [34])

5.2 Error-Minimization Rules

These rules work by comparing the response to a given input pattern with the desired response and then modifying the weights in the direction of decreasing error. Depending on whether the desired response is specified at the single neuron or overall network levels, the gradient of the error with respect to each synaptic weight will necessarily be or may only be locally computable; moreover, the repertoire of learning tasks that can be carried out in the latter case is wider than that accomplishable in the former case.

The most classic error-correction rule that requires specification of the desired response for each single neuron is the **perceptron learning rule** [33]. Its expression in terms of the generic neuron model that we use as reference (1) is:

$$w_{ij}(t+1) = w_{ij}(t) + c(x_j^*(t) - x_j(t))x_i(t), \qquad (4)$$

where both the desired response $x_j^*(t)$ and the actual response $x_j(t)$ of the neuron j are binary, since the f in equation (1) is taken to be a threshold function. The same consideration about normalization made for the Hebbian rule applies also here.

The networks of neurons using the perceptron learning rule, called *"perceptrons"*, are especially well-suited to carry out pattern classification tasks. The simplest such network proposed [33] has three layers: sensory, preprocessor and actuator (see Fig. 3); only the weights of the connections between the second and the third layer are modifiable through learning. If $x_j^*(t)$ supplies the desired classification for each input pattern $X(t) = (x_1(t), \ldots, x_{d_j}(t))$ to the actuator neuron j, and the classes are linearly separable, then the synaptic weights will converge to a configuration that classifies all patterns correctly. The rule provides thus an iterative algorithm to find the solution of a system of linear inequalities.

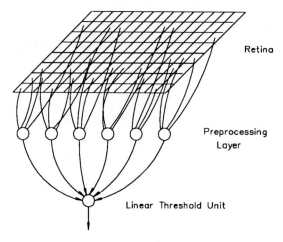

Retina

Preprocessing Layer

Linear Threshold Unit

Fig. 3. Perceptron with preprocessing layer (adapted from [3])

Nilsson [30] proved convergence theorems for different types of perceptron, always presupposing the existence of a solution, or in other words, of a configuration of weights that would give rise to the correct classification. Minsky and Papert [29] approached the complementary question, characterizing the limits of the discriminative ability for various types of perceptron. Essentially, they proved that topological features such as connectedness and symmetry cannot be discriminated by simple perceptrons —the work must be done in the preprocessing layer.

The **Widrow-Hoff learning rule** [42] is expressed by the same equation (4), but considering $x_j^*(t)$ and $x_j(t)$ to take real values and the f in equation (1) to be the identity.

Through this rule, the repeated presentation of pairs $(X_l, z_l), l = 1, \ldots k$, with $X_1, \ldots X_k$ linearly independent, to the network shown in Fig. 3, causes the convergence of the synaptic weights toward the proper configuration for the response to each stimulus X_l to be the desired real number z_l. The rule thus provides an iterative procedure to find the solution of a system of linear equations. If the patterns $X_1, \ldots X_k$ are not linearly independent, the rule can be slightly modified (converting the parameter c into a variable that goes to zero with time) so that the convergence of the weights minimizes the quadratic error between the actual output and the desired one. This slightly modified version of the rule computes a linear regression iteratively.

The implementation of *associative memories* through networks of neurons that incorporate the Widrow-Hoff rule has been extensively studied [21, 2]. While, as we saw previously, the Hebbian rule only provides a perfect association if the stimulation patterns are orthogonal, the Widrow-Hoff rule relaxes the restriction to their being linearly independent. As could be foreseen, if c approaches zero with time, optimal associations are formed according to the criterion of quadratic minimization, even when the stimulation patterns are linearly dependent.

An extension of the error-minimization rules so far described to the case where the desired response is specified only for a subset of neurons (those whose outputs constitute the output of the network) is **back-propagation** [24, 35]. As its name indicates, it proceeds by propagating error signals from the output neurons back to the sensory neurons, through all intermediate layers, so that appropriate corrections can be applied to all connection weights. Note that this procedure works only for layered networks with unidirectional interlayer connections and no intralayer connections.

Back-propagation generalizes the perceptron rule by minimizing the mean squared error E between the actual and the desired responses to all input patterns, through the repeated application of the rule:

$$w_{ij}(t+1) = w_{ij}(t) - c\frac{\partial E}{\partial w_{ij}}. \tag{5}$$

Observe that, the Widrow-Hoff learning rule is a particular instance of this generic rule, since for the former the f in equation (1) is the identity and thus $\partial x_j / \partial w_{ij} = x_i$, leading to:

$$\frac{\partial E}{\partial w_{ij}} = \frac{\partial E}{\partial x_j} \cdot \frac{\partial x_j}{\partial w_{ij}} = (x_j^* - x_j)x_i. \tag{6}$$

By incorporating this result into (5), equation (4) is obtained.

Rumelhart et al. [35] have applied the generic rule (5) to multilayered networks of neurons having a sigmoidal input-output function (i.e., the f in equation (1) takes the form $f(x) = 1/(1 + e^{-x})$). In this case, $\partial x_j / \partial w_{ij} = x_i x_j (1 - x_j)$ and the factor $\partial E / \partial x_j$ has to be calculated from the activity levels x_k of the neurons in the next layer. Hence, starting with the neurons in the output layer, for which $\partial E / \partial x_j = (x_j^* - x_j)$, the computation proceeds backwards:

$$\frac{\partial E}{\partial x_j} = \sum_k \frac{\partial E}{\partial x_k} \cdot \frac{\partial x_k}{\partial x_j} = \sum_k \frac{\partial E}{\partial x_k} w_{jk} x_k (1 - x_k) \tag{7}$$

and therefore:

$$\frac{\partial E}{\partial w_{ij}} = \frac{\partial E}{\partial x_j} \cdot \frac{\partial x_j}{\partial w_{ij}} = x_i x_j (1 - x_j) \sum_k \frac{\partial E}{\partial x_k} w_{jk} x_k (1 - x_k). \qquad (8)$$

Back-propagation is especially well-suited for solving feature discovery tasks; features relevant for discrimination between input patterns get progressively encoded in the activity of the neurons belonging to intermediate layers. This learning procedure has been applied to a variety of tasks; in particular, to the detection of symmetry [35] and to the translation invariant recognition of patterns [14]. It has the drawbacks of all gradient descent techniques, namely the possibility of getting stuck in local minima and a slow convergence rate.

5.3   Reinforcement Rules

These rules do not require being supplied with the desired responses, either at the single neuron or at the overall network levels, but instead a measure of the adequacy of the emitted responses suffices. This measure is reinforcement, which is used to guide a random search process to maximize reward. Hence, reinforcement rules can be considered neuronal analogs of instrumental conditioning in that the neuronal spontaneous responses are favored or weakened through the application of certain reinforcement schemes. Depending on whether the reinforcement signal is provided at the overall network level or is particularized for each single neuron, the structural credit-assignment problem does or does not arise. This is the problem of correctly assigning credit or blame to the action of each neuron that contributed to the overall evaluation received (refer to [4] for details).

The two reinforcement learning rules that we are going to describe in this section incorporate the required source of randomness in the input-output function of the neuron model they use, i.e. a noise with gaussian distribution is added to the weighted sum of inputs in equation (1).

The **associative search learning rule** [6] is the reinforcement-based counterpart of the correlational rules (equation 3). Its simplest expression is:

$$w_{ij}(t + 1) = w_{ij}(t) + c x_i(t) x_j(t) r(t) \qquad (9)$$

where $r(t)$ is the reinforcement signal.

A neuron model equipped with this rule learns to maximize $r(t)$ for each stimulus situation. If $r(t)$ is a random variable, its mathematical expectation is instead maximized. The neural networks that incorporate the above rule are called Associative Search Networks (ASN) and, if certain conditions are satisfied, they learn to respond to each stimulus situation $X_l = (x_{l1}, \ldots x_{ln})$ of a set $\{X_1, \ldots X_k\}$ repeatedly presented, with the vector $Y = (y_1, \ldots y_m)$ that maximizes the reinforcement function r. The conditions that have to be satisfied are: (a) the function r has to be unimodal, and (b) for each neuron, the subset of stimulus situations in which the optimum response is 0 has to be linearly separable from the corresponding subset in which the optimum response is 1. To the advantages already described for classical associative memories —resistance to noise, generalization capacity, and addressing by content— the above network adds two more: not needing the explicit presentation of the desired response for each stimulus situation, and using a single reinforcement signal common to all neurons.

The rule just described has quite limited pattern discrimination abilities and is not able to accomplish reward maximization tasks that require the discovery of complex features in the stimulus patterns and in the reinforcement contingencies. Since hidden neurons are needed to represent such features, this amounts to saying that this rule is not well-suited for networks incorporating neurons that do not contribute directly to the response that is being reinforced. Thus, the associative search learning rule suffers from the structural credit-assignment problem.

To overcome this problem, Barto and Anandan [5] have proposed the **associative reward-penalty learning rule** which, being the reinforcement-based counterpart of the error-minimization rules (equation 4), has the following expression:

$$w_{ij}(t+1) = w_{ij}(t) + c_{r(t)}[r(t)x_j(t) - E\{x_j(t) \mid \sum_{i=1}^{d_j} w_{ij}(t)x_i(t)\}]x_i(t), \quad (10)$$

where the parameter $c_{r(t)}$ makes the rule asymmetric by adopting different values for reward $(r(t) = +1)$ and penalty $(r(t) = -1)$, and $E\{\cdot\}$ is the expected value for the output given the weighted sum of inputs.

Convergence of this rule under broad conditions has been proved for a single neuron [5] and several network configurations including hidden neurons behaving according to this rule have been shown through simulation to be able to accomplish difficult nonlinear associative reinforcement learning tasks [4]. Provided each input pattern is presented with a given frequency, and since neurons are here stochastic components, the reward maximization attained is in fact a maximization of the probability of reward.

The models described so far have addressed a structural credit-assignment issue, which is sufficient for dealing with static pattern classification tasks. However, when dynamical situations need to be considered, because what is of interest is a temporal sequence of events, then a temporal credit-assignment problem also arises. Instead of assigning credit or blame to the action of each neuron in the network, credit or blame must here be assigned to each action in a sequence. In [37] it has been proposed to use *temporal-difference methods* for this purpose. Methods of this type have been embodied, for example, in "critic modules" used in conjunction with reinforcement learning approaches. The goal of these modules in this setting is to produce an heuristic reinforcement signal which, by predicting future outcomes, is more informed than that directly supplied by the environment.

6 Neurocontrol Approaches

Generally, a system to be controlled can be characterized by a transition function f and an output function g. The control signal $\mathbf{u}(t)$ in conjunction with the current state of the system $\mathbf{x}(t)$ determines the next state $\mathbf{x}(t+1)$:

$$\mathbf{x}(t+1) = f(\mathbf{u}(t), \mathbf{x}(t)). \tag{11}$$

In each state $\mathbf{x}(t)$, the system produces an output $\mathbf{y}(t)$:

$$\mathbf{y}(t) = g(\mathbf{x}(t)). \tag{12}$$

A controller can be thought of as an *inverse model* of the system in that, given a desired output $\mathbf{y}^*(t)$ and the current state, the controller has to generate the control signal that will produce that output.

The most straightforward neural control approach, named **direct inverse modelling**, uses the system itself to generate input-output pairs and trains the inverse model directly by reversing the roles of inputs and outputs (Fig. 4). The applicability of this approach is restricted to systems characterized by one-to-one mappings (otherwise, the inverse is a one-to-many mapping) and its success depends on the quality of the sampling (the inputs have to be selected so that the induced outputs cover adequately the output space). A detailed treatment of these issues can be found in [18].

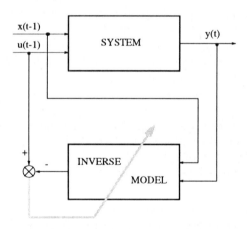

Fig. 4. Direct inverse modelling approach

The **forward modelling** approach proceeds in two stages. In the first stage a forward model of the system is learned from input-output pairs. The second stage consists of composing the obtained forward model with another network and training the composition of the two to approximate the identity mapping (Fig. 5). The weights of the forward model are held fixed in this second stage, while the weights of the controller network undergo adaptation. In the case that the forward mapping is many-to-one, this approach can be biased to find a particular inverse with certain desired properties.

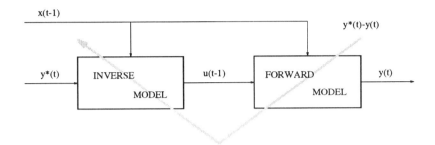

Fig. 5. Forward modelling approach.

The first stage can be obviated if the Jacobian matrix of the system is known. This is the matrix of partial derivatives, i.e. the multidimensional form of the derivative. The Jacobian can be approximated by slightly changing the input to the system and measuring the changes in the output. By a straightforward application of the chain rule, the Jacobian can be used to derive the input errors as a function of the output errors [22]. This is precisely the purpose of the forward model in the forward modelling approach and, therefore, its first stage is no longer needed.

Note also that, if instead of the desired outputs, only a reinforcement signal (i.e. a measure of how well the system is performing) is available, then this approach can still be applied. In this case, the forward model encompasses both the system and the reinforcement function, and it is used in the second stage to derive the suitable weight modifications to be performed in the controller network. Alternatively, the first modelling stage can be obviated and the reinforcement signal can be used directly to determine weight modifications in the controller network.

The usage neural networks have received in the control domain is described in [8, 28].

7 Two Representative Applications in Robot Control

Several compilations of applications of neural networks in robotics have been published [7, 23, 40]. Here we will only describe two of them to illustrate the use of the three types of learning rules described in Section 5 in a robot control setting. Thus, the first application chosen combines a correlational rule with an error-minimization one, while the second uses a reinforcement-based rule. Moreover, the former is intended for arm robots, while the second is appropriate for mobile robots.

7.1 Self-organizing Topologic Maps for Learning Eye-Hand Coordination

The use of neural networks to learn eye-hand coordination is of particular interest in the case of flexible robot arms or when, due to the operation conditions of the robots (in space, underwater, etc.), it is hardly possible to recalibrate them.

Ritter et al. [32] have used a Kohonen self-organizing map together with the LMS rule to learn the visuomotor mapping of a robot arm with three degrees of freedom in 3D space. A *direct inverse modelling approach* under a completely unsupervised training mode is used. The target position of the end-effector is defined as a spot registered by two cameras looking at the workspace from two different vantage points.

Neurons are arranged in a 3D lattice to match the dimensionality of physical space. It is expected that learning will make this lattice converge to a discrete representation of the workspace. Each neuron i has associated a four-dimensional vector $\mathbf{w}_i$ representing the retinal coordinates of a point of the workspace. The response of the network to a given input $\mathbf{u}$ is the vector of joint angles θ_k and the 3×4 Jacobian matrix $\mathbf{A}_k$ associated with the winning neuron k, i.e. that satisfying:

$$\mathbf{w}_k \mathbf{u} \geq \mathbf{w}_i \mathbf{u}, \ \forall i. \tag{13}$$

The joint angles produced for this particular input are then obtained with the expression:

$$\theta(\mathbf{u}) = \theta_k + \mathbf{A}_k(\mathbf{u} - \mathbf{w}_k). \tag{14}$$

A learning cycle consists of the following four steps:

1. First, the classical Kohonen rule is applied to the weights:

$$\mathbf{w}_i^{new} = \mathbf{w}_i^{old} + c\ h_k(i)\ (\mathbf{u}(t) - \mathbf{w}_k(t)), \tag{15}$$

where c is the learning rate and $h_k(.)$ is a Gaussian function centered at k used to modulate the adaptation steps as a function of the distance to the winning neuron.

2. By applying $\theta(\mathbf{u})$ to the real robot, the end-effector moves to position $\mathbf{u}'$ in camera coordinates. The difference between the desired position $\mathbf{u}$ and the attained one $\mathbf{u}'$ constitutes an error signal that permits applying an error-correction rule, in this case the LMS rule:

$$\theta^* = \theta_k + \Delta\theta = \theta_k + \mathbf{A}_k(\mathbf{u} - \mathbf{u}'). \tag{16}$$

3. By applying the correction increment $\mathbf{A}_k(\mathbf{u} - \mathbf{u}')$ to the joints of the real robot, a refined position $\mathbf{u}''$ in camera coordinates is obtained. Now, the LMS rule can be applied to the Jacobian matrix by using $\Delta\mathbf{u} = (\mathbf{u} - \mathbf{u}'')$ as the error signal:

$$\mathbf{A}^* = \mathbf{A}_k + (\Delta\theta - \mathbf{A}_k \Delta\mathbf{u})\frac{\Delta\mathbf{u}^T}{||\Delta\mathbf{u}||^2}. \tag{17}$$

4. Finally, the Kohonen rule is applied to the joint angles:

$$\theta_i^{new} = \theta_i^{old} + c' \; h_k'(i) \; (\theta *_i -\theta_k(t)), \qquad (18)$$

and the Jacobian matrix:

$$\mathbf{A}_i^{new} = \mathbf{A}_i^{old} + c' \; h_k'(i) \; (\mathbf{A}_i^* - \mathbf{A}_k(t)), \qquad (19)$$

where again c' is the learning rate and $h_k'(.)$ is a Gaussian function centered at k used to modulate the adaptation steps as a function of the distance to the winning neuron.

Extensive experimentation by the authors and other groups has shown that the network self-organizes into a reasonable representation of the workspace in about 30.000 learning cycles. This should be taken as an experimental demonstration of the powerful learning capabilities of this approach, because the conditions in which it is made to operate are the worst possible ones: no a priori knowledge of the robot model, random weight initialization, and random sampling of the workspace during training.

7.2   Reinforcement Learning for Mobile Robot Navigation

Millán and Torras [26] have developed a *reinforcement connectionist learning* system able to find safe paths for a mobile robot in a non-maze-like 2D environment. By saying "non-maze-like" it is stressed that finding a path in such an environment does not require sophisticated planning capabilities, but mainly obstacle-avoidance skills.

The input to the system consists of an attraction force exerted by the goal and 4 repulsion forces exerted by the obstacles. The output of the system represents a robot action, i.e. a step length and orientation. A reinforcement signal assesses how good is the robot move in response to the given sensorial situation, according to the optimization goal pursued, which is a compromise between minimizing path length and maximizing clearance. All signals are coded as real numbers.

The system consists of 2 networks: a step generator and a critic. The critic is a 3-layered backpropagation network, while the step generator is built incrementally starting from a 2-layered network with 2 stochastic real-valued output units. After learning a rough association between inputs and outputs, the weights are frozen and a hidden layer of units is added to the network. Adaptation of the new weights permits fine tuning the association map.

After 75.000 learning steps, the system had learned a reasonable sensorimotor map, leading to the generation of safe paths from almost all initial placements in the workspace. The sensorimotor map exhibited a remarkable noise tolerance (it was almost insensitive to 20% white noise), good generalization abilities in front of goal changes and new obstacles, and appropriate handling of dynamic goals and obstacles. The learning time can be considerably reduced by using a modular network and appropriately initializing it through a local search process [25].

Although the approach is reminiscent of the use of potential fields for path planning, there are some fundamental differences between them. The most important one is that the approach here described is *adaptive* and thus the sensorimotor map (which can be thought of as a sort of potential field) is learned through the interaction of the robot with the environment, instead of being predefined by the programmer. Moreover, the kinds of maps that can arise from this interaction are much more varied than those defined by a weighted sum of the potential fields originated by the different obstacles, since a 3-layered network with sigmoidal units is used to encode such maps. Another minor difference is that stochasticity is embedded in the approach itself, while it has instead to be added to the potential fields to avoid local minima. Finally, here the map is not defined in an absolute reference framework, but instead it is a function of sensory input. This implies that the map changes *automatically* when the obstacles or the goal move, leading to the remarkable generalization and dynamic abilities mentioned above.

Recently, the navigation system has been considerably enhanced and installed in a wheeled cylindrical platform of the NOMAD 200 family [27]. The codification of inputs has been adapted to the 2 rings of infrarred and ultrasound sensors mounted on the robot, and the learning time has been further reduced by incorporating a set of basic reflexes into the robot.

8 Limitations of Neural Control: A Need for Planning?

Neural learning procedures seem specially well-suited for implementing adaptivity in autonomous robots facing changing environments. However, they suffer from an extremely time-consuming learning phase and they have no way to encode a priori knowledge about the environment to gain efficiency. I believe that these limitations can be overcome with the concourse of symbolic techniques.

To be more precise, let us exemplify this claim in the context of robot navigation. If the environment is roughly known a priori (i.e. some sort of map is available), then this information should be exploited by a *planning* method to obtain a tentative path in physical space. This path would ease considerably the task of a navigator such as the one described in the preceding section, especially in the case of a "maze-like" environment. Note that then the navigator will have to refine this path in physical space into an obstacle-avoinding trajectory in configuration space (i.e. the space of degrees of freedom of the robot).

Even if no map is available, it looks useful to build such a symbolic representation when the robot is to carry out several tasks in the same environment. The map building process would be done symbolically at a central level through exploration of the environment, while the learning of skills would be carried out simultaneously at a neural peripheral level.

The key issue here is *how to integrate symbolic and neural techniques*. An initial idea for carrying out such integration in a hybrid system is the following. The symbolic planner would supply to the neural navigator a path in physical space together with a set of landmarks along this path. Then, a reinforcement learning algorithm, such as that presented in the preceding section, could be used to navigate between two consecutive landmarks. The subpath provided by the path planner could be used as the initial guess at each step, from which random deviations would be explored.

Thus, in this setting, we envisage two levels of concourse of symbolic techniques. The first is in the generation of the path and the landmarks, and the second is in the initialization of the navigator. The latter level thus leads to a combination of a one-shot symbolic acquisition of knowledge (initial setting of the neural system) and a neural adaptation of skills through repetitive trial and error (subsequent tuning).

Once this simple coupling of the two schemes has been achieved, one could aim at a further degree of integration by providing the symbolic planner with knowledge about the capabilities of the neural navigator. Thus, communication would be bidirectional and not just one-directional. The way this could be achieved is by designing a library of neural skills that the symbolic planner could include in the plans it develops.

We foresee that, as learning proceeds, a progressive competence flow from the symbolic to the neural part should take place, so that previously planned activities would progressively turn into automatic reactions.

Assuming that it is crucial to devise interfaces between symbolic and neural systems, the contents of this presentation may raise some speculative questions to foster a discussion:

- Is it possible to delineate, in biological systems, how plans are represented and transformed into distributed control commands? This could provide some inspiration for designing the interfaces above.
- In the Machine Learning domain, may neural skills be considered as building-blocks for symbolic learning systems?
- Does it help to distinguish between a one-shot symbolic acquisition of knowledge and a neural learning of skills through repetitive trial and error?
- How does an initially symbolic learning (e.g. learning to drive) become progressively an acquisition of automatic skills? Does it somehow relate to the distinction between the initial setting of the system and subsequent tuning?

References

1. Aarts, E.H.L., Korst, J.H.M.: Boltzmann machines and their applications. In: de Bakker, J.W., Nijman, A.J., Treleaven, P.C. (eds.): Proceedings of Parallel Achitectures and Languages Europe (PARLE). Lecture Notes in Computer Science **258**, Berlin: Springer-Verlag 1987, pp. 34–50

2. Amari, S.: Neural theory of association and concept-formation. Biological Cybernetics **26**, 175–185 (1977)

3. Arbib, M.A., Kilmer, W.L., Spinelli, D.N.: Neural models of memory. In: Rozenzweig, M.R., Bennet, E.L.: Neural Mechanisms of Learning and Memory. Cambridge, MA: MIT Press 1976

4. Barto, A.G.: Learning by statistical cooperation of self-interested neuron-like computing elements. Human Neurobiology **4**, 229–256 (1985)

5. Barto, A.G., Anandan, P.: Pattern-recognizing stochastic learning automata. IEEE Transactions on Systems, Man, and Cybernetics **15(3)**, 360–375 (1985)

6. Barto, A.G., Sutton, R.S., Brouwer, P.S.: Associative Search Network: A reinforcement learning associative memory'. Biological Cybernetics **40**, 201–211 (1981)

7. Bekey, G.A., Goldberg, K.Y.: Neural Networks in Robotics. Kluwer Academic Publishers 1993

8. Cembrano, G., Wells, G.: Neural networks for control. In: Artificial Intelligence in Process Control. Pergamon Press 1992

9. Feldman, J.A., Ballard, D.H.: Connectionist models and their properties. Cognitive Science **6**, 205–254 (1982)

10. Fogelman-Soulié, F.: Le connexionnisme. Support de cours MARI 87 - COGNITIVA 87, Paris, May 1987

11. Grossberg, S.: Competitive learning: from interactive activation to adaptive resonance. Cognitive Science **11**, 23–63 (1987)

12. Hebb, D.O.: The Organization of Behavior. New York: Wiley 1949

13. Hecht-Nielsen, R.: Neurocomputing. Reading, MA: Addison-Wesley 1990

14. Hinton, G.E.: Learning translation invariant recognition in massive parallel networks. In: de Bakker, J.W., Nijman, A.J., Treleaven, P.C. (eds.): Proceedings of Parallel Achitectures and Languages Europe (PARLE). Lecture Notes in Computer Science **258**, Berlin: Springer-Verlag 1987, pp. 1–13

15. Hinton, G.E., Anderson, J.A.: Parallel Models of Associative Memory. Hillsdale, NJ: Erlbaum 1981

16. Hinton, G.E.: Connectionist learning procedures. Artificial Intelligence **40**, 185–234 (1989)

17. Hopfield, J.J.: Neural networks and physical systems with emergent collective computational abilities. Proceedings National Academy of Sciences USA **79**, 2554–2558 (1982)

18. Jordan, M.I., Rumelhart, D.E.: Forward models: Supervised learning with a distal teacher. Cognitive Science **16**, 307–354 (1992)

19. Kohonen, T.: Associative Memory: A System Theoretic Approach. Berlin: Springer-Verlag 1977

20. Kohonen, T.: Self-Organization and Associative Memory (second edition). Berlin: Springer-Verlag 1988

21. Kohonen, T., Oja, E.: Fast adaptative formation of orthogonalizing filters and associative memory in recurrent networks of neuron-like elements. Biological Cybernetics **21**, 85–95 (1976)

22. Kröse, B.J.A., van der Smagt, P.P.: An Introduction to Neural Networks, 5th edition. University of Amsterdam 1993

23. Kung, S-Y., Hwang, J-N.: Neural network architectures for robotic applications. IEEE Transactions on Robotics and Automation **5(5)**, 641–657 (1989)

24. LeCun, Y.: Une procedure d'aprentissage pour reseau au seuil assymetrique. Proceedings of COGNITIVA, 1985, pp. 599–604

25. Millán, J. del R.: Building reactive path-finders through reinforcement connectionist learning: Three issues and an architecture. Proc. 10th European Conf. on Artificial Intelligence (ECAI'92), 1992, pp. 661–665

26. Millán, J. del R., Torras, C.: A reinforcement connectionist approach to robot path finding in non-maze-like environments. Machine Learning **8(3/4)**, 363–395 (1992)

27. Millán, J. del R., Torras, C.: Efficient reinforcement learning of navigation strategies in an autonomous robot. Proc. Intl. Conf. on Intelligent Robotics Systems (IROS'94), September 1994

28. Miller, W.T., Sutton, R.S., Werbos, P.J.: Neural Networks for Control. Cambridge, MA: MIT Press 1990

29. Minsky, M., Papert, S.: Perceptrons: An Introduction to Computational Geometry. Cambridge, MA: MIT Press 1969

30. Nilsson, N.J.: Learning Machines. McGraw-Hill 1965

31. Pavlov, I.P.: Conditioned Reflexes. Oxford University Press 1927

32. Ritter, H., Martinetz, T., Schulten, K.: Neural Computation and Self-Organizing Maps. New York: Addison-Wesley 1992

33. Rosenblatt, F.: Principles of Neurodynamics. Spartan Books 1962

34. Rumelhart, D.E., Zipser, D.: Feature discovery by competitive learning. Cognitive Science **9**, 75–112 (1985)

35. Rumelhart, D.E., Hinton, G.E., Williams, R.J.: Learning representations by back-propagating errors. Letters to Nature **323**, 533–535 (1986)

36. Skinner, B.F.: The Behavior of Organisms: An Experimental Analysis. Appleton Century 1938

37. Sutton, R.S.: Learning to predict by the methods of temporal differences. Machine Learning **3**, 9–44 (1988)

38. Torras, C.: Temporal-Pattern Learning in Neural Models. Lecture Notes in Biomathematics No. 63, Berlin: Springer-Verlag 1985

39. Torras, C.: Relaxation and neural learning: points of convergence and divergence. Journal of Parallel and Distributed Computing **6**, 217–244 (1989)

40. Torras, C.: From geometric motion planning to neural motor control in robotics. AI Communications **6(1)**, 3–17 (1993)

41. von der Malsburg, C.: Self-organization of orientation sensitive cells in the striate cortex. Kybernetik **14**, 80–100 (1973)

42. Widrow, B., Hoff, M.E.: Adaptive switching capatibility and its relation to the mechanisms of association. Kybernetik **12**, 204–215 (1960)

Intelligence - Dynamics and Representations

Luc Steels

Artificial Intelligence Laboratory, Vrije Universiteit Brussel
Pleinlaan 2, B-1050 Brussels, Belgium
E-mail: steels@arti.vub.ac.be

Abstract. The paper explores a biologically inspired definition of intelligence. Intelligence is related to whether behavior of a system contributes to its self-maintenance. Behavior becomes more intelligent (or copes with more ecological pressures) when it is capable to create and use representations. The notion of representation should not be restricted to formal expressions with a truth-theoretic semantics. The dynamics at various levels of intelligent systems plays an essential role in forming representations. An example is given how behavioral diversity spontaneously emerges in a globally coupled network of agents.
Keywords. Intelligence, self-organisation, representation, complex dynamical systems

1 Introduction

Artificial intelligence research is concerned with an investigation into the phenomenon of intelligence using the methods of the artificial [24]. This means that systems are built which exhibit intelligent behavior and that this is seen as a way to progressively derive and test a theory of intelligence. After three decades of research, nobody denies that AI has resulted in many spinoffs for computer science, such as list processing, declarative programming, search algorithms, etc. Nobody denies that a whole range of programs have been written that exhibit features of (human) intelligence. For example, chess programs now compete at grandmaster level, expert systems have demonstrated human-level performance in difficult problems like scheduling, diagnosis, or design, natural language programs of high complexity have been built for parsing and producing natural language, and some machine learning programs have been capable to extract compact representations from examples. But substantial progress is still possible.

First of all, it seems that research efforts so far have not resulted in a coherent, widely accepted theory of intelligence. There is a body of engineering methods, techniques, and intuitive insights, which are usually taught using case studies [30]. This absence of theory is undoubtely due to the engineering bias of AI and the push from society to produce useful artefacts as opposed to theories. In addition, the bits of theory that do exist (for example Newell's cognitive architecture [19]) have no connection with theories of the physical and biological world, so that there is a wide gap between AI and other natural sciences.

Second, there is a continuing criticism that the achievements of AI systems rest completely on the intelligence of the AI programmers: They extract and formalise the knowledge from experts, they set the conceptual framework and determine the set of good and bad examples for machine learning algorithms, they synthetise the grammars going into the natural language systems. The problem of how structures for knowledge and behavior may develop is largely unresolved. This is particularly a bottleneck in the area of sensory-motor intelligence and common sense, where the task of analysis, formalisation, and explicit programming is so formidable that little success has come from using the classical AI approach.

Current research in intelligent autonomous agents promises to tackle these two gaps in a fundamental way: It is seeking a theory of intelligence compatible with the basic laws of physics and biology and a theory which explains how intelligence may come from non-intelligent, material processes. Obviously we are far from achieving these goals. Only the contours of the theory are visible and at the moment the artefacts that can be built are promising with respect to sensory-motor intelligence but still weak with respect to 'higher level' cognitive tasks. But a new methodological track has been opened in which solid work can proceed.

The main pillars of the new approach are as follows:

+ A biologically oriented definition of intelligence is the starting point of the investigation. Intelligence is defined with respect to the capability of an autonomous system to maintain itself. This gives an objective criterion, as opposed to a subjective criterion based on judgement of performance or the ascription of knowledge and reasoning. This definition is refined by considering the functionalities used to increase the chances of survival: representation, specialisation, cooperation, communication, reflection, etc.

+ A theory of intelligence must be compatible with the basic laws of physics and biology and it must be a universal theory, i.e. independent of a particular embodiment (wetware or silicon) or system level (brain component, individual agent, society). This universality can be achieved by using complex dynamical systems theory as a foundation. Intelligence then is seen as the result of a set of non-linear processes which exhibit properties also found in other physical systems. Phenomena like behavioral coherence, cooperation, or the emergence of diversity between agents can be explained using bifurcation theory, chaos, self-organisation, etc.

The goal of this paper is to discuss these two directions of research in more detail. First we focus on the biologically oriented definition of intelligence (section 2), further refining it with the notion of representation (section 3). Then we discuss in which way dynamical systems theory can act as the theoretical foundation of a theory of intelligence.

2 Defining intelligence

AI has wrestled since the beginning with the question of what intelligence is, which explains the controversies around the achievements of AI. Let us first look at some common definitions and then turn to a biologically oriented definition.

The first set of definitions is in terms of comparative performance with respect to human intelligence. The most famous instance of such a definition is the Turing test. Turing imagined interaction with either a human or an intelligent computer program through a terminal. When the program managed to trick the experimenter into believing that it was human often enough, it would qualify as artificial intelligence.

If we consider more restricted versions of the Turing test, for example compare performance of chess programs with human performance, then an honest observer must by now agree that computer programs have reached levels of competence comparable to human intelligence. The problem is that it seems possible (given enough technological effort) to build highly complex programs which are indistinguishable in performance from human intelligence for a specific area, but these programs do not capture the evolution, nor the embedded (contextual) nature of intelligence. As a consequence 'intelligent' programs are often qualified as being no longer intelligent as soon as the person inspecting the program figures out how the problem has been solved. For example, chess programs carry out relatively deep searches in the search space and the impressive performance is therefore no longer thought to be due to intelligence. To find a firmer foundation it seems necessary to look for a definition of intelligence which is not related to subjective judgement.

The second set of definitions is in terms of knowledge and intensionality. For example, Newell has worked out the notion of a knowledge level description [19]. Such a description can be made of a system if its behavior is most coherently described in terms of the possession of knowledge and the application of this knowledge (principle of rationality). A system is defined to be intelligent if a knowledge-level description can be made of it and if it maximally uses the knowledge that it has in a given situation. It follows that artificial intelligence is (almost by definition) concerned with the extraction of knowledge and the formalisation and encoding in computer systems. This approach appears problematic from two points of view. First of all knowledge level descriptions can be made of many objects (such as thermostats) where the label 'intelligence' does not naturally apply. Second, the approach assumes a sharp discontinuum between intelligent and non-intelligent systems and hence does not help to explain how intelligence may have arisen in physical systems nor how knowledge and reasoning relates to neurophysiology.

There are still other definitions, which however are not used within AI itself. For example, several authors, most notably Roger Penrose, claim that intelligence is intimately tied up with consciousness and self-consciousness [21]. This in turn is defined as the capability to intuit mathematical truths or perform esthetic judgements. The topic of consciousness is so far not at the center of discussion in AI and no claims have ever been made that artificial intelligence systems exhibit consciousness (although see the discussion in [29]). Whether this means, as Penrose suggests, that consciousness falls outside the scope of artificial systems, is another matter. In any case it seems that the coupling of intelligence with consciousness unnecessarily restricts the scope of intelligent systems.

Let me now introduce an alternative to these definitions which finds its roots in biology (see for example [17], [6]). We start from the observation that intelligence is a property of living systems. Living systems are defined as systems which actively maintain themselves, using essentially two mechanisms:

+ They continuously replace their components and that way secure existence in the face of unreliable or short-lived components. The individual components of the system therefore do not matter, only the roles they play.
+ The system as a whole adapts/evolves to remain viable even if the environment changes, which is bound to happen.

The drive towards self-maintenance is found at many different levels:

The genetic level. This is the level which maintains the survivability of the species. Mechanisms of copying, mutation, and recombination together with selection pressures operating on the organisms carrying the genes, are the main mechanisms in which a coherent gene pool maintains itself and adapts itself to changing circumstances. At the moment we do not normally use such a level for artificial robotic agents, although we could say that the building plans, the design principles, and the initial structures of one type of agent when it starts its operation correspond to a kind of genetic level.

The structural level. This is the level of the components and processes making up the individual agents: cells, cell assemblies, organs, etc. Each of these components has its own defense mechanisms, renewal mechanisms, and adaptive processes. In the case of the brain, there are neurons, networks of neurons, neural assemblies, regions with particular functions, etc. What appropriate functional units are and how they coherently operate together is the main topic of interest in the study of intelligent autonomous agents. In artificial systems, they involve internal quantities, electronic and computational processes, behavior systems regulating relations between sensory states and actuator states, etc.

The individual level. This is the level of the individual agent which has to maintain itself by behaving appropriately in a given environment. In many biological systems (for example bacteria or ant colonies) individuals have no or little self-interest. But it is clear that the individual becomes gradually more important as evolution proceeded its path towards more complexity, and conflicts arise between genetic pressures, group pressures, and the tendency of the individual to maintain itself. Greater individuality seems to be linked tightly with the development of intelligence. In the case of artificial systems, the individual level corresponds to the level of the robotic agent as a whole which has to survive within its ecological niche.

The group level. This is the level where groups of individuals together form a coherent whole and maintain themselves as a group. This may include defense mechanisms, social differentiation according to the needs of the group, etc. In the case of artificial systems, the group level becomes relevant when there are groups of robotic agents which have to cooperate in order to survive within a particular ecosystem and accomplish tasks together.

Obviously there is a continuum for living systems in terms of the power they have to determine their own destiny, in other words the degrees of freedom or the available choices. A simple bacteria has very little control over its surroundings. It can at best move towards food sources and away from danger. For the rest it is at the mercy of environmental factors. Bacterial genes delineating the species are nevertheless still very successful mostly due to the rate of copying and the multitude and range of environments in which the individuals can survive. The more degrees of freedom living systems have, the more their chances of survival will increase but also the more structures and processes must be dedicated to making appropriate choices.

The qualification 'intelligent' must be seen against this general background. A biologically inspired definition of intelligence focuses on the interaction between a living system (at whatever level) and the environment, which includes other living systems. Such an interaction is typically called a behavior. The main requirement for a behavior to be intelligent is that it contributes to the continued survival of the system, directly or indirectly. Thus when an animal which is starving from hunger does not go to a food source but performs some other action such as fight another animal, then we would say that this behavior is non-intelligent. Obviously intelligent behavior depends strongly on the environment and a particular system will only behave intelligently within a certain environmental niche to which it is adapted. Because environments tend to evolve, a necessary aspect of intelligence is that it is adaptive.

The advantage of this definition is that it can be precisely quantified. It is possible to identify the characteristic pressures in an ecosystem (for example the availability of resources, the presence of dangers, etc.) and to measure behavior and its impact on the viability [16]. Well worked out theories already exist for example at the population dynamics level or at the genetic level. At this point all this is still a somewhat academic exercise for robotic agents because there are no autonomous agents yet in the world that are viable for a sufficiently long time in real world environments. But once the technology matures, it will be common place to take this perspective.

3 Representations

Many researchers would find a definition of intelligence in terms of survivability not strong enough. They would argue, rightfully, that the appropriate metabolism, a powerful immune system, etc., are also critical to the survival of organisms (in the case of artificial systems the equivalent is the life time of the batteries, the reliability of microprocessors, the physical robustness of the body). They would also argue that many biological systems (like fungi) would then be more intelligent than humans because they manage to survive for much longer periods of time. So we need to sharpen the definition of intelligence by considering what kind of functionalities intelligent systems use to achieve viability.

Here we quickly arrive at the notion of representation. The term representation is used in its broadest possible sense here. Representations are physical structures (for example electro-chemical states) which have correlations with aspects of the environment and thus have a predictive power for the system. These correlations are maintained by processes which are themselves quite complex and indirect, for example sensors or actuators which act as transducers of energy of one form into energy of another form. Representations support processes that in turn influence behavior. What makes representations unique is that processes operating over representations can have their own dynamics independently of the dynamics of the world that they represent.

Although it seems obvious that the ability to handle representations is the most distinguishing characteristic of intelligent systems, this has lately become a controversial point. Autonomous agents researchers have been arguing 'against representations'. For example, Brooks [3] has claimed that intelligence can be realised without representations. Researchers in situated cognition [4], [22] and in 'constructivist' cognitive science [14] have argued that representations do not play the important role that is traditionally assigned to them. Researchers in neural networks in general reject 'symbolic representations' in favor of subsymbolic or non-symbolic processing [26]. All this is resulting in a strong debate of representationalists vs. non-representationalists [8]. Let me attempt to clarify the issues.

In classical AI, physical structures acting as representations are usually called symbols and the processes operating over them are called symbol processing operations. In addition the symbol processing is subjected to strong constraints: Symbols need to be defined using a formal system and symbolic expressions need to have a strict correspondence to the objects they represent in the sense of Tarskian truth-theoretic semantics. The operations that can be performed to obtain predictive power must be truth-preserving.

These restrictions on representations are obviously too narrow. States in dynamical systems [11] may also behave as representations. Representations should not be restricted to those amenable to formal semantics nor should processing be restricted to logically justified inferences. The relation between representations and reality can and usually is very undisciplined, partly due to the problem of maintaining strict correspondence between the environment and the representation. For example, it is known that the signals received by sonar sensors are only for 20 percent effectively due to reflection from objects. Sonar sensors therefore do not function directly as object detectors and they do not produce a 'clean representation' of whether there is an object or not in the environment. Rather they establish a (weak) correlation between external states (the presence of obstacles in the environment) and internal states (hypothesised positions of obstacles in an analogical map) which may be usefully exploited by the behavioral models.

Second, classical AI restricts itself mostly to *explicit representations*. A representation in general is a structure which has an influence on behavior. Explicit representations enact this influence by categorising concepts of the reality concerned and by deriving descriptions of future states of reality. An implicit (or emergent) representation occurs when an agent has a particular behavior which is appropriate with respect to the motivations and action patterns of other agents and the environment but there is no model. The appropriate behavior is for example due to an historical evolution which has selected for the behavior. The implicit representation is still grounded in explicit representations but they are at a different level.

Indeed, representations can be postulated at all levels of intelligent systems:

- *Genetic level*: The DNA molecules are the physical structures which act as explicit representations at the genetic level. More specifically, they represent directly the presence of particular structures in the organism using a code based on the position of the individual molecules. Of course DNA does not 'represent' in a Tarskian-style truth-theoretic sense. The relation between the DNA molecules and the resulting structure is complex and what structure is found in the organism is to a large extent determined by the environment as well. At the same time, the genes represent implicitly a set of environments in which the resulting organism can survive.
- *Structural level*: The neural structures causally responsible for behavior have their own explicit representations although there is no consensus on what they are: levels of activation, patterns of spikes, mass behavior of neurons. Also in artificial systems explicit internal representations in the form of electro-magnetic states are common and computer technology has made it possible to create and manipulate millions of representational states in very short time periods.
- *Individual level*: It seems appropriate to make a distinction between representations at the structural level and at the level of the individual. These are the representations that we are most familiar with. They take the form of drawings, models, conventions, internal languages, etc. The individual's representations are implemented by the representations and processes at the structural level.

– *Group level*: Groups of agents may use representations at the group level to maintain coherence and increase their chances of survival, from the pheromone trails deposited by ant societies to languages and literature used in human societies.

For a long time, science has made progress by reducing the complexity at one level by looking at the underlying components. Behavior at a particular level is explained by clarifying the behavior of the components at the next level down. For example, properties of chemical reactions are explained (and thus predicted) by the properties of the molecules engaged in the reactions, the properties of the molecules are explained in terms of atoms, the properties of atoms in terms of elementary particles, etc. Also in the case of intelligence, we see that many researchers hope that an understanding of intelligence will come from understanding the behavior of the underlying components. For example, most neurophysiologists believe that a theory of intelligence will result from understanding the behavior of neural networks in the brain. Some physicists go even so far as to claim that only a reduction of the biochemical structures and processes in the brain to the quantum level will provide an explanation of intelligence ([21]).

At the moment there is however a strong opposing tendency to take a wholistic point of view, also in the basic sciences [5]. This means that it is now understood that there are properties at each level which cannot be reduced to the level below, but follow from the dynamics at that level, and from interactions (resonances) between the dynamics of the different levels ([20]). In the case of intelligence, this means that it will not be possible to understand intelligence by only focusing on the structures and processes causally determining observable behavior. Part of the explanation of intelligence will come from the dynamics in interaction with the structures and processes in the environment, and the coupling between the different levels.

This viewpoint is adopted in the strongest possible sense in our own work, and is common in current research on intelligent autonomous agents. One implication is for example, that it makes less sense from a methodological point of view to build a particular robot which executes a particular task. In our laboratory, we have created a complete robotic ecosystem (figure 3) which involves an environment with different pressures for the robots (e.g. the need to collect energy and ensure that it is available), different robotic agents which have to cooperate but are also in competition with each other, and a growing repertoire of adaptive structural components (called behavior systems) which are causally responsible for behavior. (see [27], [17]).

Such an integrated experimental environment ensures that all the different levels (genetic, structural, individual, group) are present at the same time, each with strong interactions to the environment. This way a wholistic approach to the study of intelligence is possible.

A wholistic position must also be adopted for intelligent behavior and for representations. When a particular behavior is observed, it is not at all

Fig. 1. Robotic ecosystem constructed at the VUB AI laboratory. There is a charging station which robots can use to recharge their batteries. There are also 'parasites' in the form of lamps which take energy away from the charging station. Robots temporarily kill off parasites by pushing against the boxes.

clear at which level the representation causally co-responsible for it, should be located. Classical AI is too much focused on representations at the individual level and assumes that there is a direct correspondence between the individual and the structural level. But many representations of the world could be implicit, i.e. they are assumed by particular behavior patters due to a historical selection process.

So 'intelligence' is no longer viewed here as restricted to a unique capability that only (conscious) humans have. Intelligence occurs at all levels, although one can say that the most developed forms are at the level of humans and their cultures; most complex in the sense that at all levels we find the most complex representations, the most versatile behavior creation, and strong and elaborate forms of cooperation and communication.

Nor is intelligence viewed here as an all-or-none phenomenon. In each case, it is possible to trace an evolutionary path (with many co-evolutionary couplings re-enforcing the build up of complexity) in which the first signs of a particular functionality (e.g. communication) are becoming apparent and the functionality then gradually develops under the ecological pressures into the complex forms that we can observe in humans. Understanding such evolutionary paths is one of the main challenges of autonomous agents research.

Science proceeds by formulating abstract mathematical theories with which it is possible to describe a wide range of natural phenomena. The mathematical deduction or calculation based on the theory can then be mapped onto predictions of reality in order to check whether the theoretical description is valid. At the moment the most worked out formal theory used in theories of intelligence is based on logic (e.g. [9]). But the search for a theory of intelligence which is compatible with physics and biology and which sees intelligence as a universal phenomenon present at many different levels of biological systems, pushes us into another direction. Most theories of complex natural phenomena are phrased in terms of the recently developed theory of complex dynamical systems, which includes theories of chaos and self-organisation. It is therefore no surprise that several researchers in the field of intelligent autonomous agents have been seeking a foundation in the same direction ([7], [25], [27], [13]).

4 Emergent diversification in agent behavior

To illustrate many points of the paper, I will now develop a concrete example in the context of the robotic experiments mentioned earlier. We will be interested in the question how in a group of agents, individual differences between agents may arise.

Three aspects need to be described: (i) the contextual setting of the experiment, (ii) the experimental results, and (iii) the underlying theory.

(i) Contextual setting.

The experiment involves a group (at least 2) of robotic agents. Each agent has 30 sensors (for infrared, touch, visible light, battery level, forward movement, sound) mounted on a body which also houses batteries and motors connected to a left and right wheel (figure 2).

The robots have a limited behavioral repertoire described in more detail in [28]: It includes forward movement, touch-based obstacle avoidance, infrared-based (smooth) obstacle avoidance, phototaxis towards a light mounted on a charging station, phototaxis towards lamps mounted in boxes, etc. Each behavior is a dynamical process implemented in a dedicated programming language PDL implemented on top of C.

The processes directly relate sensory states (which are continuously varying) to actuator parameters. They typically take the form of feedback control processes with three components:

1. A cue c, for example the amount of infrared light perceived. In the case of infra-red based obstacle avoidance, this infrared light needs to be minimised.

2. A rate r, which determines the slope with which the desired state is approached.

3. An intensity i, which is determined by motivational processes.

These components exert an influence on an actuator parameter a based on the formula:

$$a \leftarrow i\, r\, c$$

Fig. 2. Robotic agents used in the experiments. Each robot is autonomous from the viewpoint of energy (thanks to batteries) and processing (thanks to onboard dedicated electronics, microcontrollers located on a sensory-motor unit, and the main processing unit).

The PDL system freezes at time step t all quantities, executes all processes, sums the influences and enacts them on the quantities. It then sends the final values to actuators, reads new sensory values and proceeds to the next time step (t+1). Such a cycle takes place 100 times per second (for a repertoire of up to 500 processes).

Motivational processes monitor dimensions related to viability of the agent (for example presence of enough energy in the batteries). A motivational quantity (such as EnergyNeed) increases as its monitored quantity (Energy in battery) decreases. This quantity is then coupled to the behavioral intensity of those behaviors that may contribute to decreasing the motivational quantity. A motivational process has also a cue (for example the difference between the maximum battery level and the current battery level) and a rate, which is the rate at which the motivational quantity should increase w.r.t. the cue.

$$m \leftarrow r\,c$$

A further important characteristic is that all processes (both behavioral and motivational) are adaptive in the sense that their rate r changes depending on the resulting behavior. For example, if the impact of the behavioral processes monitoring infrared does not cause the robot to turn away from obstacles and collisions occur, then the rate with which there is an influence on the actuators must be increased. These adaptations are carried out by a third set of processes (the adaptation processes) which monitor the performance of behaviors and couple it to the increase or decrease of the rate parameters.

Although the basic control mechanisms are linear, there are two sources of non-linearity in the overall system. First of all, each behavioral process is adaptive. As a consequence its rate may increase or decrease, thus making the process non-linear. Second, the cue strength typically increases non-linearly as a particular behavior is enacted. For example, the amount of light received during phototaxis will increase non-linearly as the robot approaches the light.

It is impossible within the limited space available here to explain in full detail how such a collection of distributed adaptive processes may give rise to coherent behavior, particularly because there is no central action selection mechanism that decides what the most appropriate behavior is at any point in time. Instead, all behavior systems are active at all times and they cooperate to give the globally coherent behavior. But the most important point for the further discussion is that the interactions between the different behaviors and the environment causes an activity cycle to emerge which has a characteristic cycle time. The activities in the cycle are:

- seek and destroy competitors
- seek charging station
- recharge

McFarland [17] has worked out the optimality conditions for such a cycle and also the possible evolutions towards more optimality.

Each agent individually goes through these characteristic activity cycles spending varying amounts of time depending on the parameter settings. For example, if the motivation to recharge overtakes the motivation to do more work because the battery is getting low, there is a switch from the 'seek parasites' activity to the 'seek charging station' activity, which means that the various behaviors involved with these activities have a stronger tendency to occur. But the individual cycles are coupled: to the environment, because the ability to recharge will depend on the availability of energy in the charging station, and to the other agents, because if one agent is in the charging station, another agent cannot have access.

The question now is whether the different agents could smoothly cooperate in exploiting the available resources. This means that the right setting must be found of the various parameters in the different motivational processes. From an analytic point of view this problem is extremely difficult (see [18]). And the question is whether the robots would be able to discover themselves the right parameter settings using adaptation processes. The answer turns out to be yes, and there are some surprising side effects to be explained in the next paragraphs.

We focus on one parameter w, which could be called 'the amount of work that a robot believes it has to do before going to the charging station'. This parameter has an impact on the process that relates the units U of work already done to the motivation M to do more work. One unit of work is equal

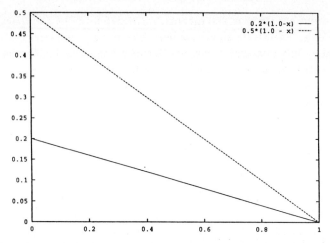

Fig. 3. Function relating the amount of work done with the motivation to do more work. When the parameter w is high (e.g. w = 0.5, left figure) the robot 'believes' that it has to do more work than when the parameter is low (e.g. w = 0.2, right figure)

to one push against the box that houses a lamp, i.e. a parasite. This process is defined as follows:

$M \leftarrow w(Max - U)$

Max is equal to the maximum amount of work.

The adaptation process is as follows:

1. When the robot arrives in the charging station and is forced to leave the charging station before its battery is full due to the parasites which take away too much energy, then the parameter w is increased. The rationale behind this is that not enough work (i.e. killing of parasites) was done, so next time more needs to be done.

2. When the robot has been able to recharge itself fully and there is still energy left in the charging station, then the parameter w is decreased. The rationale here is that too much work was done.

(ii) Experimental results

Simulations in the context of the complete ecosystem for a period of 4 hours with 2 robots show that after a while three different types of situations occur:

1. There is a situation, illustrated in fig. 4, in which the robots are both viable and oscillate around the same nearly optimal values of the parameters.

2. There is a situation in which one of the robots is no longer viable. This is partly due to the fact that in this experiment no communication exists between the robots. One robot locks the other one out of going into the charging station.

3. There is a third situation, which is the most interesting and also the most common one, in which two types of robots emerge (fig. 5). The first

robot has a high rate for w which means that it will do a lot more work than the second robot. Effectively there has been a diversification between a 'hard working' and a 'less working' behavior. Because the second robot has more free time, it will have an opportunity to develop other behaviors.

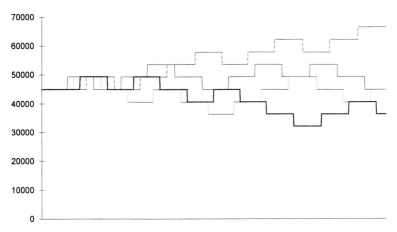

Fig. 4. Situation in which both robots oscillate around the same nearly optimal parameter value for w.

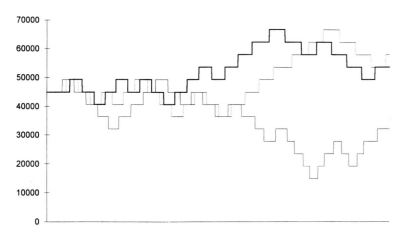

Fig. 5. Situation in which diversification between different behaviors emerges. There is now a hard working robot with high value of w and a less working robot with low value of w. An equilibrium situation has however developed making both robots viable.

It is interesting to raise in this context the question what kind of representations are used. Clearly each invididual agent does not have an explicit model of the energy left in the charging station, what other agents do, how much energy it will need to find back the charging station, etc. Consequently the agents can not do any planning (in the traditional sense). An agent only has explicit representations of the rates relating the amount of work done with motivation to work and the amount of energy left in the battery with motivation to go recharging. The agent has implicit representations of the aspects of the environment relevant for their decision making. Nevertheless the agents behave optimally with respect to an exploitation of the energy available in the overall environment.

(iii) Underlying theory

These experimental results are fascinating but do not yet make up a scientific theory. For that we need to go one step further and find abstract mathematical objects which exhibit the same properties. Then we know that the behavior is due to the dynamics itself (and not to some artefact of the robotic ecosystem or other intervening factors). We will also have a predictive theory and possibly learn about other possible situations which we have not been able to investigate yet experimentally. If the particular dynamics also explains other natural phenomena, particularly in biology, then this is further evidence that aspects of intelligence can be explained using concepts from the natural sciences.

It turns out that there is a theory which is appropriate here. This is the theory of coupled map lattices, developed by Kaneko [12] and various coworkers. Kaneko has used this theory for quite different biological phenomena, such as cell differentiation or the maintenance of diversity in population dynamics.

Globally coupled maps consist of a network of elements which each have a certain (possibly chaotic) dynamics. An example is the following:

$$x_{n+1}(i) = (1 - \epsilon)f(x_n(i)) + \frac{\epsilon}{N}\sum_{j=1}^{N} f(x_n(j))$$

where n is a discrete time step, i is the index of an element and $f(x) = 1 - ax^2$. Synchronous oscillation occurs due to the interaction between the different elements, whereas the chaotic instability inherent in f introduces the potential for destruction of the coherence. A cluster is defined as a set of elements for which x(i) is the same. k is the number of clusters in the total network.

Kaneko [12] has shown that depending on the degree of non-linearity (captured by the parameter a) four phases occur:

1. There is a coherent phase in which all elements oscillate in harmony (k = 1).
2. There is ordered phase in which few clusters are observed.
3. There is partially ordered phase with a coexistence of attractors with many and few clusters.

4. All attractors have N clusters, where N is the number of elements in the network.

Kaneko has also observed phenomena of intermittency, where the self-organisation towards coherent structure is seen in cascade with the occurrence of high-dimensional disordered motion.

The situation being investigated maps onto this model as follows. The individual cycles of the agents map onto the function f which is present in each of the elements. The elements are globally coupled through the charging station and the parasites. The degree of non-linearity is related to the rate r with which the rate w is adapted, because adaptation causes the linear functions to become non-linear. In the experiments, an adaptation rate r was apparently chosen so that the second regime was observed (an ordered phase with few clusters). The theory predicts that under other parameter selections high-dimensional disordered motion may occur or the other two situations. These predictions have not yet been tested experimentally.

The example illustrates the following points:

- It shows how behavioral diversity (and thus diversity between different agents) can emerge spontaneously even though every agent starts with the same initial structure, and even though every agent has the same self-interest to survive in optimal circumstances.
- It shows that the diversity is not due to structure inside each agent, but through the coupling between the different agents through the environment (and particularly the global constraints imposed by the charging station and the parasites). This illustrates that behavior (at one level) cannot be understood in isolation but that a wholistic point of view must be taken.
- It shows that the diversity is due to the dynamics. In other words, no new principle or explanation is required because universal properties of abstract dynamical systems (in this case globally coupled maps) already exhibit these phenomena. More generally, the clustering phenomena shown here are a special case of the theory in which order arises through fluctuations causing bifurcation and thus evolution in systems ([20]).
- The agents have no explicit models of the strategies of other agents, of the time it takes to recharge, the average time to find the charging station, etc. Their behavior could nevertheless be called intelligent from the viewpoint of optimal use of resources within the given ecological constraints. This is an example where implicit representations dominate.

5 Conclusions

The paper discussed approaches towards a theory of intelligence which are grounded in biological theory and the theory of complex dynamical systems. The approach starts from the idea that intelligence centers around the ability of a system to maintain itself through the creation and use of representations. Intelligence is seen at many different levels and is partly due to the coupling between the different levels. Representations are not necessarily explicit but may be implicit in distributed behavior. This has been illustrated with a concrete example of how behavioral diversity may originate from adaptive processes.

Acknowledgement

The viewpoints discussed in this paper have been shaped and greatly enhanced by discussions with many people, including discussions at the Trento NATO ASI. Thanks are due in particular to Thomas Christaller, David McFarland, Rolf Pfeifer, Tim Smithers, and Walter Van de Velde. Danny Vereertbrugghe, Peter Stuer, and Filip Vertommen have constructed the physical ecosystem and the robots. Johan Myny has contributed in the construction and investigation of the robotic ecosystem. This research was partially sponsored by the Esprit basic research project SUBSYM and the DPWB concerted action (IUAP) CONSTRUCT of the Belgian government.

References

1. Arkin, R. (1989) Motor schema based mobile robot navigation. Int. Journal of Robotics Research. Vol 8, 4 92-112
2. Brooks, R. 1991. Intelligence without reason, IJCAI-91, Sydney, Australia, 569–595
3. Brooks, R. 1991. Intelligence without representation, Artificial Intelligence, 47, 1991, 139-160
4. Clancey, W.J. (1993) Situated action: A neuropsychological interpretation. Cognitive Science 17(1), 87-116
5. Cohen, J. and I. Stewart (1994) The collapse of chaos. Discovering simplicity in a complex world. Viking Books. London
6. Dawkins, R. (1976) The Selfish Gene. Oxford University Press. Oxford
7. Engels, C. and G. Schoener (1994) Dynamic fields endow behavior-based robots with representations. Robotics and Autonomous Systems
8. Ford, K. and P. Hayes (1994) On babies and bathwater: a cautionary tale. AIMagazine, Vol 15, 4, 14–26
9. Genesereth, M. and N. Nilsson (1987) Logical foundations of Artificial Intelligence. Morgan Kaufmann, Los Altos
10. Haken, H. (1983). Advanced synergetics: instability hierarchies of self-organisating systems and devices, Springer-Verlag, Berlin
11. Jaeger, H. (1994) Dynamic symbol systems, Ph.D. thesis. Faculty of Technology. Bielefeld
12. Kaneko, K. (1994) Relevance of dynamic clustering to biological networks. Physica D 75, 55-73

13. Kiss, G. (1993) Autonomous agents, AI and chaos theory. In: Meyer, J.A., et al. (eds.) From Animals to Animats 2 Proceedings of the Second Int. Conference on Simulation of Adaptive Behavior. MIT Press, Cambridge, MA, 518-524

14. Maturana, H.R. and F.J. Varela (1987) The tree of knowledge: The biological roots of human understanding. Shamhala Press, Boston

15. McFarland, D. (1990) Animal behaviour. Oxford University Press, Oxford.

16. McFarland, D. and T. Boesser (1994) Intelligent behavior in animals and robots. MIT Press/Bradford Books, Cambridge, MA

17. McFarland, D. (1994) Towards Robot Cooperation. Proceedings of the Simulation of Adaptive Behavior Conference. Brighton. MIT Press.

18. McFarland, D., E. Spier, and P. Stuer (1995) Environmental constraints on the behavioural stability of autonomous robots, Adaptive behavior, to be submitted

19. Newell,A. (1981). The knowledge level, Journal of Artificial Intelligence, 18 (1), 87-127

20. Nicolis, G. and I. Prigogine (1985) Exploring Complexity. Piper, München

21. Penrose, R. (1990) The Emperor's new mind. Oxford University Press. Oxford

22. Pfeifer, R. and P. Verschure (1992) Distributed adaptive control: A paradigm for designing autonomous agents. In: Varela, F.J. and P. Bourgine (eds.) (1992) Toward a practice of autonomous systems. Proceedings of the First European Conference on Artificial Life. MIT Press/Bradford Books, Cambridge MA 21-30

23. Schöner, G. and M. Dose (1993) A dynamical systems approach to task-level system integration used to plan and control autonomous vehicle motion, Journal of Robotics and Autonomous Systems, vol 10, 253-267

24. Simon, H. (1969) The Sciences of the artificial. MIT Press, Cambridge MA

25. Smithers, T. (1994) Are autonomous agents information processing systems? In: Steels, L. and R. Brooks (eds.), The 'artificial life' route to 'artifical intelligence': building situated embodied agents, Lawrence Erlbaum Associates, New Haven

26. Smolensky, P. (1986) Information processing in dynamical systems. Foundations of Harmony Theory. In: Rumelhart, D.E., J.L. McClelland (eds.), Parallel distributed processing. Explorations in the microstructure of cognition. Vol 1. MIT Press, Cambridge, MA. 194-281

27. Steels, L. (1994a) The artificial life roots of artificial intelligence. Artificial Life Journal, Vol 1,1. MIT Press, Cambridge, MA

28. Steels, L. (1994b) A case study in the behavior-oriented design of autonomous agents. Proceedings of the Simulation of Adaptive Behavior Conference. Brighton. MIT Press. Cambridge, MA

29. Trautteur, G. (ed.) (1994) Approaches to consciousness. Kluwer Academic Publishing. Amsterdam

30. Winston, P. (1992) Artificial Intelligence. Addison-Wesley, Reading MA

An Introduction to Reinforcement Learning

Leslie Pack Kaelbling*   Michael L. Littman*   Andrew W. Moore[†]

*Computer Science Department, Brown University, Box 1910, Providence, RI 02912 USA {lpk,mll}@cs.brown.edu

[†]Carnegie Mellon University, School of Computer Science and Robotics Institute, 5000 Forbes Ave, Pittsburgh, PA 15213 USA awm@cs.cmu.edu

Abstract This paper surveys the historical basis of reinforcement learning and some of the current work from a computer scientist's point of view. It is an outgrowth of a number of talks given by the authors, including a NATO Advanced Study Institute and tutorials at AAAI '94 and Machine Learning '94. Reinforcement learning is a popular model of the learning problems that are encountered by an agent that learns behavior through trial-and-error interactions with a dynamic environment. It has a strong family resemblance to work in psychology, but differs considerably in the details and in the use of the word "reinforcement." It is appropriately thought of as a class of problems, rather than as a set of techniques. The paper addresses a variety of subproblems in reinforcement learning, including exploration vs. exploitation, learning from delayed reinforcement, learning and using models, generalization and hierarchy, and hidden state. It concludes with a survey of some practical systems and an assessment of the practical utility of current reinforcement-learning systems.

Keywords reinforcement learning, robot learning, Markov decision processes

1 Introduction

The field of reinforcement learning dates back to the early days of cybernetics, and work in statistics, psychology, neuroscience, and computer science. In the last five to ten years, it has attracted rapidly increasing interest in the machine learning and artificial intelligence communities.

This paper surveys the historical basis of reinforcement learning and some of the current work from a computer scientist's point of view. It is an outgrowth of a number of talks given by the authors, including a NATO Advanced Study Institute and tutorials at AAAI '94 and ML '94. We have tried to give a high-level overview and a taste of some specific approaches. It is, of course, impossible to mention all of the important work in the field; this should not be taken to be by any means an exhaustive account.

Reinforcement learning is a popular model of an agent that learns behavior through trial-and-error interactions with a dynamic environment. It has a strong family resemblance to eponymous work in psychology, but differs considerably in the details and in the use of the word "reinforcement." It is appropriately thought of as a class of problems, rather than as a set of techniques.

The rest of this section is devoted to establishing notation and describing the basic reinforcement-learning model. Section 2 explains the trade-off between exploration and exploitation and presents some solutions to the most basic of reinforcement-learning problems. Section 3 considers the more general problem in which payoffs can be delayed in time from the actions that were crucial to gaining them. Section 4 demonstrates a continuum of algorithms that are sensitive to the amount of computation an agent can perform between steps of action in the world. Section 5 explores the problems of generalization as they apply to reinforcement learning. Section 6 considers the problems that arise when the agent does not have complete perceptual access to the state of the world. Section 7 catalogues some of reinforcement-learning's successful applications. Finally, Section 8 concludes with some speculations about important open problems and the future of reinforcement learning.

1.1 Reinforcement-Learning Model

In the standard reinforcement-learning model, an agent is connected to its environment via perception and action. On each step of interaction the agent receives as input some indication of the current state of the environment; the agent then chooses an action to generate as output. The action changes the state of the environment, and the value of this new state is reflected to the agent in a special scalar *reinforcement input*. The agent should choose actions that tend to increase the long-run sum of values of the reinforcement signal; it can learn to do this over time by systematic trial and error and a wide variety of algorithms that are the subject of later sections of this paper. This interaction model is depicted in Figure 1.

Formally, the model consists of

- a discrete set of environment states, $\mathcal{S}$
- a discrete set of agent actions, $\mathcal{A}$
- a scalar set of reinforcement values; typically $\{0, 1\}$, or the real numbers

This figure also includes an input function I, which determines how the agent views the environment state; we will assume that it is the identity function (that is, the agent correctly perceives the exact state of the environment) until we consider partial observability in Section 6.

An intuitive way to understand the relation between the agent and its environment is with the following example dialogue.

Environment: You are in state 65. You have 4 possible actions.
Agent: I'll take action 2.
Environment: You received a reinforcement of 7 units. You are now in state 15. You have 2 possible actions.
Agent: I'll take action 1.
Environment: You received a reinforcement of -4 units. You are now in state 65. You have 4 possible actions.
Agent: I'll take action 2.
Environment: You received a reinforcement of 5 units. You are now in state 44. You have 5 possible actions.

$\vdots$ $\vdots$

The agent's job is to find a policy $\pi : \mathcal{S} \to \mathcal{A}$ that maximizes some long-run measure of reinforcement. We expect, in general, that the environment will be non-deterministic; that is, that taking the same action in the same state on two different occasions may result in different next states and/or different reinforcement values. This happens in our example above: from state 65, applying action 2 produces differing reinforcements and differing states on two occasions. However, we assume the environment is stationary; that is, that the probabilities of making state transitions or receiving reinforcement values do not change over time.

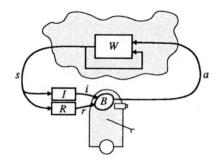

Fig. 1. The standard reinforcement learning model

Reinforcement learning differs from supervised learning for several reasons. Firstly, there is no presentation of input/output pairs. Instead, after choosing an action the agent is told the immediate reward and the subsequent state, but is *not* told which action would have been in its best long-term interests. Instead, it is necessary for the system to actively gather useful experience about the possible system states, actions, transitions and rewards. Another difference from supervised learning is that on-line performance is important: in many practical situations such as mobile robotics, juggling, or robot motion control, an optimally-chosen action which takes a significant time to compute is worthless.

Some aspects of reinforcement learning involve issues that are closely related to search and planning issues in artificial intelligence. AI search algorithms generate a satisfactory trajectory through a graph of states. Planning operates in a similar manner, but typically within a construct with more complexity than a graph, in which states are represented by compositions of logical expressions instead of atomic symbols. These AI algorithms are less general than the reinforcement learning methods in that they require a predefined model of state transitions, and with a few exceptions assume determinism. On the other hand, reinforcement learning, at least in the kind of discrete cases for which theory has been developed, assumes that the entire state space can be stored in memory—an assumption to which conventional search algorithms are not tied.

1.2 Models of Optimal Behavior

Before we can start thinking about algorithms for learning to behave optimally, we have to decide what our model of optimality will be. In particular, we have to specify how the agent should take the future into account in the decisions it makes about how to behave now. There are three models that have been the subject of the majority of the work.

The *finite horizon* model is the easiest to think about; at a given moment in time, the agent should optimize its reward for the next k steps:

$$E(\sum_{t=0}^{k} r_t) \; ;$$

it need not worry about what will happen after that. This model can be used in two ways. In the first, the agent will have a non-stationary policy; that is, one that changes over time. On its first step it will take a k-step optimal action (the best action available given that it has k steps remaining in which to act and gain reinforcement), on the next a $k - 1$-step optimal action, and so on, until it finally takes a 1-step optimal action and terminates. In the second, the agent does *receding horizon control*, in which it always takes the k-step optimal action. The agent always acts according to the same policy, but the value of k limits how far ahead it looks in choosing its actions. The finite-horizon model is not always appropriate. It can be more computationally expensive than other models of optimality, and we do not usually know exactly how much time the agent has left in which to act.

In the average case model, the agent is supposed to take actions that optimize its long-run average reward:

$$lim_{k\to\infty} E(\frac{1}{k} \sum_{t=0}^{k} r_t) \; .$$

One problem with this basic criterion is that there is no way to distinguish between two policies, one of which gains a large amount of reward in the initial phases and the other of which does not. Reward gained on any initial prefix of the agent's life is overwhelmed by the long-run average performance. It is possible to generalize this model so that it takes into account both the long run average and the amount of extra reward than can be gained. In the generalized model, we prefer policies with the maximum value of the long-run average, but if there is a tie, it is broken by the initial extra reward.

The infinite-horizon discounted model takes the long-run reward of the agent into account, but rewards that are received in the future are exponentially discounted according to discount factor γ, which is between 0 and 1:

$$E(\sum_{t=1}^{\infty} \gamma^t r_t) \ .$$

This model is similar to receding horizon control, but the discounted model is much more mathematically tractable than the finite horizon model. For this reason, we will predominately concentrate on this model.

1.3 Measuring Learning Performance

Once we have selected a measure of optimality, we have criteria for assessing the quality of a learning algorithm. Unfortunately, there are several measures of performance, and they are not necessarily consistent.

- **Eventual convergence to optimal.** Many algorithms come with a provable guarantee of asymptotic convergence to optimal. This is not completely satisfactory for examining on-line performance. An agent that learns quickly and cheaply but plateaus at 99% of optimality may, in many practical situations, be preferable to an agent which has a guarantee of eventual optimality but a sluggish early learning rate.
- **Speed of convergence to optimality.** Optimality is usually an asymptotic result, and so convergence speed is an ill-defined measure. More practical is the *speed of convergence to near-optimality*. This measure begs the definition of how near to optimality is sufficient. A related measure is *level of performance after a given time*, which similarly requires that someone defines the given time.
  It should be noted that, again, we have a difference between reinforcement learning and conventional machine learning. In the latter, expected future predictive accuracy or statistical efficiency are the prime concerns. For example, in the well-known PAC framework, there is a learning period during which mistakes do not count, then a performance period during which they do. That is usually an inappropriate view for an agent with a long existence in a complex world.
  Measures related to speed of learning have an additional weakness. An algorithm which merely tries to achieve optimality as fast as possible

may incur unnecessarily large penalties during the learning period. A less aggressive strategy that took longer to achieve optimality, but gained greater total reinforcement during its learning might be preferable.

- **Regret.** A more appropriate measure, then, is the expected decrease in reward gained due to executing the learning algorithm instead of behaving optimally from the very beginning. This measure is known as *regret*, and is related to the idea of *mistake bounds*. It penalizes mistakes wherever they occur during the run.

2 Exploitation versus Exploration: The Two-Armed Bandit

Let's consider the following simple problem, which has been the subject of a great deal of study in the statistics literature. The agent is in a room with two gambling machines (called "one-armed bandits" in colloquial English). These machines do not require a deposit to play; when arm i is pulled, machine i pays off 1 or 0, according to some underlying probability parameter p_i. The arm pulls are independent events and the p_i are unknown. What should the agent's strategy be? That is, what arm should it pull on step t given the history of arm-pulls and payoffs from the previous steps?

This problem illustrates the fundamental tradeoff between exploitation and exploration. The agent might believe that a particular arm has a fairly high payoff probability; should it choose that arm all the time, or should it choose another one that it has less information about, but seems to be worse? Answers to these questions depends on how long the agent is expected to play the game; the longer the game lasts, the worse the consequences of prematurely converging on a sub-optimal arm, and the more the agent should explore.

There is a wide variety of solutions to this problem. We will consider a representative selection of them. A number of important theoretical results are included in a survey by Berry and Fristedt (Berry & Fristedt 1985).

2.1 Bayesian Approach

If the agent is going to be in the casino for a total of h steps, we can use basic Bayesian reasoning to solve for an optimal strategy as follows. We will use a dynamic programming approach that computes a mapping from *belief states* to actions. Start with a set of prior probability distributions D_1 and D_2 on the underlying payoff parameters p_1 and p_2.

Choose the best last action for every possible sequence of $h - 1$ actions and observations, by using Bayesian updating of the belief distribution. Given that, compute the best action for step $h - 1$ assuming that that the agent will act according the policy just determined for step h. Iterate until the current state is reached.

The computational expense is quartic in the number of steps h. In a generalization to more than two arms, it is exponential in the number of arms.

2.2 Minimax Approach

Rather than assuming some prior beliefs about the arms, we could adopt a model in which we try to do as well as we can, assuming that the casino management is going to make the most difficult possible assignment of probabilities to arms. This does not mean that they will merely provide low or hard to distinguish arms, but, having been allowed to inspect the algorithm we will use, they chose arm probabilities to maximize the expected regret of our algorithm. This is known as the *minimax model* and a very simple algorithm can be shown to be optimal in terms of regret.

For finite horizon h, let $m = \sqrt{h}$. Choose arms alternately until arm i has m more successes than arm j. Pull arm i forever more. This algorithm has regret $O(\sqrt{h})$, which is optimal in the minimax sense. For infinite-horizon discounted optimality, set $m = \sqrt{\frac{1}{1-\gamma}}$.

2.3 Greedy Strategies

As we move away from the two-armed bandit model, we now use the term "action" instead of "arm", and we consider the case of more than two available actions.

The first strategy which comes to mind is to always choose the action with the best estimated reward. The flaw with always being greedy in this manner is easily spotted. If the true optimal action is initially unlucky its estimated reward may be below the actual reward obtained from another suboptimal action. The suboptimal action will always be picked, the true optimal action will be starved of data and so its superiority will never be discovered. Exploration is needed.

A useful heuristic is *optimism in the face of uncertainty* in which actions are chosen greedily, but strongly optimistic prior beliefs are put on their rewards so that considerable negative evidence is needed to eliminate an action from experimentation. This still has a finite danger of starving an optimal but unlucky action, but the danger can be made extremely small. Techniques like this have been used in several reinforcement learning algorithms including the Interval Exploration method (Kaelbling 1993 b) (described shortly), the *exploration bonus* idea in Dyna (Sutton 1990) and the exploration mechanism in Prioritized Sweeping (Moore & Atkeson 1993).

2.4 Ad Hoc Randomized Strategies

In reinforcement-learning practice, some simple, *ad hoc* strategies have been popular. They are rarely, if ever, the best choice for the models of optimality we have used, but they may be viewed as reasonable, computationally tractable, heuristics. Thrun (Thrun 1992) has surveyed a variety of these techniques.

The simplest exploration strategy is to take the action with the best estimated expected reward by default. But with probability p, choose an action at random. Some versions of this strategy start with large values of p to encourage initial exploration, then slowly decrease them.

An objection to the simple strategy is that when it experiments with a non-greedy action it is no more likely to try a promising alternative than a clearly hopeless alternative. A slightly more sophisticated strategy is *Boltzmann exploration*. In this case, we estimate the expected reward for taking action a, $ER(a)$ and choose an action probabilistically according to the distribution

$$P(a) = \frac{e^{ER(a)/T}}{\sum_{a' \in A} e^{ER(a')/T}} \ .$$

The *temperature* parameter T can be decreased over time to decrease exploration. This method works well if the best action is well separated from the others, but suffers when the values of the actions are close. It may also converge unnecessarily slowly unless the temperature schedule is manually tuned with great care.

2.5 Learning Automata

There is a whole branch of the theory of adaptive control devoted to *learning automata* (see Narendra and Thathachar for a survey (Narendra & Thathachar 1989)). They were originally described explicitly as finite state automata as in the *Tsetlin automaton* shown in Figure 2.

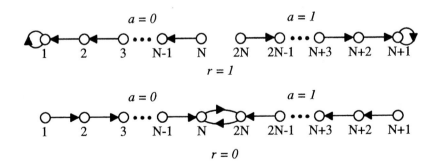

Fig. 2. The Tsetlin automaton

It is inconvenient to describe algorithms in this way, so a move was made to describe the internal state as a probability distribution according to which actions would be chosen. The probabilities of taking different actions would be adjusted according to their previous successes and failures.

An example, which was also independently developed in the mathematical psychology literature, is the *linear reward-punishment* algorithm. When action a_i succeeds,

$$p_i := p_i + \alpha(1 - p_i)$$
$$p_j := p_j - \alpha p_j \text{ for } j \neq i$$

When action a_i fails,

$$p_i := p_i - \beta p_i$$
$$p_j := p_j + \beta(\frac{1}{n-1} - p_j) \text{ for } j \neq i$$

This algorithm performs poorly unless β is set to 0, yielding a version called *linear reward-inaction*, which converges with probability 1 to a vector containing a single 1 and the rest 0's (choosing a particular action with probability 1). Unfortunately, it does not always converge to the correct action; but the probability that it converges to the wrong one can be made arbitrarily small by making α small. There is no understanding of the regret of this algorithm.

2.6 Interval-based Techniques

Exploration is often more efficient when it is based on second-order information about the certainty or variance of the estimated values of actions. Kaelbling's *interval estimation* algorithm (Kaelbling 1993 b) stores statistics for each action a_i: s_i is the number of successes and n_i the number of trials. An action is chosen by computing the upper bound of a $(1 - \alpha)\%$ confidence interval on the success probability of each action and choosing the action with the highest upper bound. The method works very well in empirical trials. It is also related to a certain class of statistical techniques known as *experiment design* methods, which are used for comparing multiple treatments (for example, fertilizers or drugs) to determine which treatment (if any) is best in as small a set of experiments as possible.

2.7 Extensions

Most of the above methods can be fairly straightforwardly extended to work with real-valued reinforcement. In many cases, they can also be extended in *ad hoc* ways to work in *non-stationary* environments, in which there may be slow changes or occasional changes in the underlying probabilities of reinforcement. Non-stationarity makes learning hard because it is never safe to permanently commit to a single action. There are very few theoretical analyses of algorithms for non-stationary environments because it is hard to characterize reasonable kinds of non-stationarity.

2.8 Dealing with Multiple States

There is a trivial extension of this class of problems, in which there are multiple states, but the agent is expected to take actions only to optimize its next reward (that is, with finite horizon 1). These problems can be thought of as bandit problems in which the agent is in a very large casino, each room of which presents a basic k-armed bandit problem. After choosing an arm, the agent is randomly placed in another room. They can also be seen as a problem from the class described in the next section, with the discount factor set to 0.

Such problems can be trivially solved by making a copy of any of the algorithms discussed above for each state. Such algorithms exhibit no generalization behavior, though, and are for the most part uninteresting. See Section 5 for a suite of techniques that could be applied to these problems.

3 Delayed Reinforcement

In the general case of the reinforcement learning problem, the agent's actions determine its immediate reward, but also (at least probabilistically) determine the next state of the environment. Such environments can be thought of as networks of bandit problems. But the agent must take into account the next state as well as the immediate reward when it decides which action to take. The model of long-run optimality the agent is using determines exactly how it should take the value of the future into account. The agent will have to be able to learn from delayed reinforcement: it may take a long sequence of actions, receiving no reinforcement, then finally arrive at a state with high reinforcement. The agent must be able to learn which of its actions are desirable based on reward that can take place arbitrarily far in the future.

3.1 Markov Decision Processes

Problems with delayed reinforcement are well modeled as *Markov decision processes* (MDPs). An MDP consists of

- a finite set of states S
- a finite set of actions A
- a reward function $R : S \times A \rightarrow \Re$
- a state transition function $T : S \times A \rightarrow \Pi(S)$, where $\Pi(S)$ is the set of probability distributions over the set S.

The state transition function probabilistically specifies the next state of the environment as a function of its current state and the agent's action. It is *Markov* if the state transitions are independent of any previous environment states or agent actions. There are many good references to MDP models; see, for instance, (Bellman 1957, Bertsekas 1987, Howard 1960).

3.2 Finding a Policy Given a Model

Before we consider algorithms for learning to behave in MDP environments, we will explore techniques for determining the optimal policy given a correct model. These techniques will serve as the foundation and inspiration for the learning algorithms to follow. We will restrict our attention mainly to finding optimal policies for the infinite-horizon discounted model, but most of these algorithms have analogues for the finite horizon and average-case models as well.

We will speak of the optimal *value* of a state; that is the expected infinite discounted sum of reward that the agent will gain if it starts in that state and executes the optimal policy. It is written

$$V^*(s) = \max_\pi E(\sum_{t=0}^\infty \gamma^t r_t) \ .$$

This optimal value function is unique and can be defined as the solution to the recursive equations

$$V^*(s) = \max_a (R(s,a) + \gamma \sum_{s' \in \mathcal{S}} T(s,a,s')V^*(s')) \ ,$$

which assert that the value of a state s is the expected instantaneous reward plus the discounted expected value of the next state. Given the optimal value function, we can specify the optimal policy as

$$\pi^*(s) = \arg\max_a (R(s,a) + \gamma \sum_{s' \in \mathcal{S}} T(s,a,s')V^*(s')) \ .$$

Value Iteration One way, then, to find an optimal policy is to find the optimal value function. It can be determined by a very simple iterative algorithm called *value iteration*.

```
initialize V(s) arbitrarily
loop
    loop for s ∈ S
        V(s) := maxₐ(R(s,a) + γ∑ₛ'∈S T(s,a,s')V(s'))
    end loop
end loop
```

An approximately optimal value function yields an approximately good policy:

$$|V^* - \hat{V}|_{\sup} \leq \epsilon \Rightarrow |V^* - V_{\hat{\pi}}|_{\sup} \leq 2\epsilon \frac{\gamma}{1-\gamma} \ .$$

That is, if the maximum difference between the optimal value function V^* and an approximate value function $\hat{V}$ at any state is less than ϵ, then the maximum difference between the optimal expected value and the expected value derived from executing policy $\hat{\pi}$, which was derived from $\hat{V}$, will be less than $2\epsilon\frac{\gamma}{1-\gamma}$.

It is not obvious when to stop the value iteration algorithm. There is a theorem about *Bellman residual* that tells us if the maximum difference between two successive value functions is less than ϵ, then the maximum difference between one of those value functions and the optimal value function is $2\epsilon\gamma/(1-\gamma)$. This provides an effective stopping criterion for the algorithm.

Value iteration is very flexible. The assignments to V need not be done in strict order as shown above, but instead can occur asynchronously in parallel provided that they all get updated infinitely often on an infinite run. These issues are treated extensively in (Bertsekas & Tsitsiklis 1989), where convergence results are also proved.

The computational complexity of the algorithm is polynomial in the number of states. The convergence in terms of number of iterations is first order.

Policy Iteration The *policy iteration* algorithm manipulates the policy directly, rather than finding it indirectly via the optimal value function. It operates as follows:

```
choose an arbitrary policy π
loop
      compute the value function of policy π:
```
$$V_\pi(s) = R(s, \pi(s)) + \gamma \sum_{s' \in \mathcal{S}} T(s, \pi(s), s')V_\pi(s')$$
```
      improve the policy at each state:
```
$$\pi'(s) := \arg\max_a (R(s, a) + \gamma \sum_{s' \in \mathcal{S}} T(s, a, s')V_\pi(s'))$$
$$\pi := \pi'$$
```
until no further improvement is possible
```

The value function of a policy is just the expected infinite discounted reward that will be gained, at each state, by executing that policy. It can be determined by solving a set of linear equations. Once we know the value of each state under the current policy, we consider whether the value could be improved by changing the first action taken. If it can, we change the policy to take the new action whenever it is in that situation. This step is guaranteed to strictly improve the performance of the policy. When no improvements are possible, then the policy is guaranteed to be optimal.

3.3 Learning an Optimal Policy with No Initial Model

In the previous section we reviewed methods for obtaining an optimal policy assuming that we already had a model. The model consists of knowledge of the state transition probability function $T(s, a, s')$ and the reinforcement function $R(s, a)$. Reinforcement learning is primarily concerned with how to obtain the optimal policy when such a model is not known in advance. Instead, the agent must interact with the real world to obtain information which, by means of an appropriate algorithm, can be processed to produce an optimal policy.

At this point, there are two ways to proceed.

- **Model-based:** Learn a model, and use it to derive a controller.
- **Model-free:** Learn a controller without learning a model.

Which approach is better? This is a matter of some debate in the reinforcement learning community. A number of algorithms have been proposed on both sides. This question also appears in other fields, such as adaptive control, where the dichotomy is between *direct* and *indirect* adaptive control.

Perhaps the most conceptually straightforward idea is a model-based algorithm: First learn the T and R functions by exploring the environment and keeping statistics about the results of each action. Then compute an optimal policy using one of the methods of the previous section. There are some serious objections to this method. It makes an arbitrary division between the learning phase and the acting phase. Worse, how should it gather data about the environment initially? Random exploration might be dangerous, and in some domains is an immensely inefficient method of gathering data, requiring exponentially more data than a system which interleaves experience gathering with policy-building more tightly. Another objection to the naive model-based strategy concerns possible changes in the environment. Breaking up an agent's life into a pure learning and a pure acting phase has a considerable risk that the optimal controller based on early life becomes, without detection, a suboptimal controller if the world changes.

Model-free algorithms can avoid these problems, and we will examine them first. In Section 4, we will return to model-based methods, and the techniques they use to avoid the same problems.

3.4 Model-Free Learning of an Optimal Policy

The biggest problem facing us is *temporal credit assignment.* How do we know whether the action we just took is a good one, when it might have far reaching effects? One strategy is to wait until the "end" and reward the actions we took if the result was good and punish them if the result was bad. In ongoing tasks, it is difficult to know what the "end" is, and this might require a great deal of memory. Instead, we will use insights from value iteration to adjust the value of a state based on the immediate reward and the value of the next state. This class of algorithms is called *temporal difference methods* (Sutton 1988). We will consider two different temporal-difference learning strategies for the discounted infinite-horizon model.

AHC and TD The *adaptive heuristic critic* algorithm is a learning analogue of policy iteration (Barto, Sutton & Anderson 1983). A block diagram is shown in Figure 3. It consists of two components, a critic and a reinforcement-learning component. The reinforcement-learning component can be an instance of any of the k-armed bandit algorithms, modified to deal with multiple states. But instead of acting to maximize instantaneous reward, it will be acting to maximize the heuristic value, v, that is computed by the critic. The critic uses the real external reinforcement signal to learn to map situations to their expected discounted values given that the policy being executed is the one currently instantiated in the RL component.

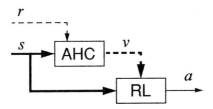

Fig. 3. Adaptive heuristic critic architecture

We can see the analogy with policy iteration if we imagine these components working in alternation. The policy π implemented by RL is fixed and the critic learns the value function V_π for that policy. Now we fix the critic and let RL component learn a new policy π' that maximizes the new value function, and so on. In most implementations, however, both components operate simultaneously.

It remains to explain how the critic can learn the value of a policy. It uses Sutton's TD algorithm (Sutton 1988) with the weight update rule

$$V(s) := (1 - \alpha)V(s) + \alpha(r + \gamma V(s')) \ .$$

Whenever a state s is visited, its value is updated to be closer to $r + \gamma V(s')$, where r is the instantaneous reward received and $V(s')$ is the value of the actually occurring next state. The motivating idea is that $r + \gamma V(s')$ is a sample of the value of $V(s)$, and it is more likely to be correct because it incorporates the real r. If the learning rate α is adjusted properly (it must be slowly decreased), TD is guaranteed to converge to the optimal value function.

The TD rule as presented above is really an instance of a more general class of algorithms called $TD(\lambda)$, with $\lambda = 0$. $TD(0)$ is only able to look one step ahead when adjusting value estimates; although it will eventually arrive at the correct answer, it can take quite a while to do so. The TD(0) rule above can be algebraically rewritten to

$$V(s) := V(s) + \alpha(r + \gamma V(s') - V(s)) \ ;$$

the general $TD(\lambda)$ rule is similar,

$$V(s) := V(s) + \alpha(r + \gamma V(s') - V(s))e(s) ,$$

but it is applied to *every state* according to its eligibility $e(s)$, rather than just to the immediately previous state. The eligibility trace is

$$e(s) = \sum_{k=1}^{t}(\lambda\gamma)^{t-k}I(s = s_k) ,$$

where I is an indicator function. It indicates the degree to which state s has been visited in the recent past; when a reinforcement is received, it is used to update all the states that have been recently visited, according to their eligibility. When $\lambda = 0$ this is equivalent to TD(0). When $\lambda = 1$, it is roughly equivalent to rewarding all the states according to their value at the end of a run. It is computationally more expensive to execute the general $TD(\lambda)$, though it often converges considerably faster for large λ. See (Dayan 1992, Dayan & Sejnowski 1994) for theoretical convergence results.

Q-learning The work of the two components of AHC can be accomplished in a unified manner by Watkins' Q-learning (Watkins 1989, Watkins & Dayan 1992). Q-learning is typically easier to implement in a general-purpose computer. In order to understand Q-learning, we have to develop some additional notation. Let $Q^*(s, a)$ be the expected discounted reinforcement of taking action a in situation s, then continuing by choosing actions to maximize Q^* (i.e., optimally). It can be written recursively as

$$Q^*(s, a) = R(s, a) + \gamma \sum_{s' \in S} T(s, a, s') \max_{a'} Q^*(s', a') .$$

Watkins has shown that $V * (s) = max_a Q^*(s, a)$ and thus that $\pi^*(s) = \arg\max_a Q^*(s, a)$ is the optimal policy.

Because the Q function makes the action explicit, we can estimate the Q values on-line using a method essentially the same as TD, but also use them to define the policy, because an action can be chosen just by taking the one with the maximum Q value for the current situation.

The Q-learning rule is

$$Q(s, a) := (1 - \alpha)Q(s, a) + \alpha(r + \gamma \max_{a'} Q(s', a')) ;$$

it is applied whenever action a is taken in situation s with resulting immediate reward r and next state s'. If each action is executed in each state an infinite number of times on an infinite run and α is decayed appropriately, the Q values will converge with probability 1 to Q^*. Q-learning can also be extended to update states that occurred more than one step previously, as in $TD(\lambda)$ (Peng & Williams 1994).

Q-learning is the most popular algorithm for learning from delayed reinforcement. It does not, however, address any of the issues involved in generalizing over large state and/or action spaces. In addition, it may converge quite slowly to a good policy. Both of these shortcomings are addressed in following sections.

4 Learning Optimal Policies by Learning Models

We have already briefly considered two different strategies for learning to behave in MDP environments. In the *model-free* strategy, we update our policy incrementally, based on the results of each action. In the *model-based* strategy, we can estimate the transition probabilities and the reward function by gathering statistics as we take actions in the world. We must then use this information to derive an optimal policy. The simplest approach would be to wait until sufficient data is gathered and then, once and for all, solve to find the optimal policy given the estimated model. But this begs the question of how to explore during data gathering and artificially divides learning into a pure exploring and pure exploiting phase. Model free methods such as TD and Q-learning suffers from neither of these drawbacks.

Another alternative for a model-based method is to re-solve for the optimal policy on every step using either policy or value iteration. This approach makes maximal use of information from interaction with the world; it learns very efficiently as a function of the number of steps it takes in the environment. But the computational costs are enormous because it has to recompute the entire optimal policy on every step. TD and Q-learning do very little computation per interaction with the world, so are fast, but they do not fully exploit the information it gets from the world, requiring it to visit some parts of the environment repeatedly before it can determine the optimal policy.

4.1 Dyna

Sutton's Dyna architecture (Sutton 1990, Sutton 1991) allows the middle ground to be exploited, yielding strategies that are both more effective than model-free learning and more computationally efficient than the naive model-based approach. It simultaneously uses experience to build a model, uses experience to adjust the policy, and uses the model to adjust the policy.

Dyna operates in a loop of interaction with the environment. Given a tuple $\langle s, a, s', r \rangle$ where s is the previous state, a is the action taken in s, r is the immediate reward received and s' is the current state, it behaves as follows:

- Update the model, incrementing statistics for the transition from s to s' on action a and for receiving reward r for taking action a in state s. The updated models are $\hat{T}$ and $\hat{R}$.

- Update the policy at state s based on the newly updated model using the rule

$$Q(s,a) := \hat{R}(s,a) + \gamma \sum_{s'} \hat{T}(s,a,s') \max_{a'} Q(s',a') \; ,$$

which is a version of the value iteration update for Q values.
- Perform k additional updates. Choose k state-action pairs at random and update them according to the same rule as before:

$$Q(s_k, a_k) := \hat{R}(s_k, a_k) + \gamma \sum_{s'} \hat{T}(s_k, a_k, s') \max_{a'} Q(s',a') \; .$$

- Choose an action a' to perform in state s', based on the Q values but perhaps modified by an exploration strategy.

The Dyna algorithm requires about k times the computation of Q-learning per instance, but this is typically vastly less than for the naive model-based method. In a sample gridworld navigation task, Dyna requires an order of magnitude fewer steps of experience than does Q-learning to arrive at an optimal policy. Dyna requires about six times more computational effort, however. A reasonable value of k can be determined based on the relative speeds of computation and of taking action in the world.

Models can accelerate convergence to the correct value function and are especially useful during initial exploration.

4.2 Prioritized Sweeping / Queue-Dyna

Although Dyna is a great improvement on previous methods, it suffers from being relatively undirected. It is particularly unhelpful when the goal has just been reached or when the agent is stuck in a dead end; it continues to update random state-action pairs, rather than concentrating on the "interesting" parts of the state space. These problems are addressed by Prioritized Sweeping (Moore & Atkeson 1993) and Queue-Dyna (Peng & Williams 1993), which are two independently-developed but very similar techniques. We will describe prioritized sweeping in some detail.

The algorithm is very similar to Dyna, except in the choice of which state-action pairs to update. To make appropriate choices, we must store additional information in the model. Each state remembers its *predecessors*: the states that have a non-zero transition probability to it under some action. In addition, each state has a *priority*, initially set to zero.

Instead of backing up k random state-action pairs, prioritized sweeping backs up k state-action pairs for states with the highest priority. For each state s and action a,

- Remember the current value of the state: $V_{old} = \max_a Q(s,a)$.

– Update the state-action pair's value

$$Q(s, a) := \hat{R}(s, a) + \gamma \sum_{s'} \hat{T}(s, a, s') \max_{a'} Q(s', a') \ .$$

– Set the state's priority back to 0.
– Compute the value change $\Delta = |V_{old} - \max_a Q(s, a)|$.
– Use Δ to possibly change the priorities of some of the predecessors of s.

If we have done a backup on state s' and it has changed by amount Δ, then the immediate predecessors of s' are informed of this event. Any state s for which there exists an action a such that $\hat{T}(s, a, s') \neq 0$ has its priority increased by $\Delta\hat{T}(s, a, s')$.

The global behavior of this algorithm is that when a real-world transition is "surprising" (the agent happens upon a goal state, for instance), then lots of computation is directed to propagate this new information back to relevant predecessor states. When the real-world transition is "boring" (the actual result is very similar to the predicted result), then computation continues in the most deserving part of the space.

Running prioritized sweeping on the problem discussed in the previous section, we see a large improvement over Dyna. The optimal policy is reached in about half the number of steps and one-third the computation as Dyna required.

4.3 Other Model-Based Methods

RTDP (real-time dynamic programming) (Barto, Bradtke & Singh 1993) is another model-based method that uses Q-learning to concentrate computational effort on the areas of the state-space in which the agent is most likely to be. It is specific to problems in which the agent is trying to achieve a particular goal state and the reward everywhere else is 0. By taking into account the start state, it can find the optimal policy along a path from the start to the goal, without necessarily visiting the rest of the state space.

The Plexus planning system (Dean, Kaelbling, Kirman & Nicholson 1993) exploits a similar intuition. It starts by making an approximate version of the MDP which is much smaller than the original one. The approximate MDP contains a set of states, called the ENVELOPE, that includes the agent's current state and the goal state, if there is one. States that are not in the envelope are summarized by a single "out" state. The planning process is an alternation between finding an optimal policy on the approximate MDP and adding useful states to the envelope. Action may take place in parallel with planning, in which case irrelevant states are also pruned out of the envelope.

5 Generalization

All of the previous discussion has tacitly assumed that it is possible to enumerate the state and action spaces and store tables of values over them.

Except in very small domains, this means impractical memory requirements. It is also inefficient. In a large, smooth, state space we expect similar states to generally have similar values and similar policies. Surely, therefore, there should be some more compact representation than a table? Most problems will have continuous or large discrete state spaces; some will have large or continuous action spaces. The problem of learning in large spaces is addressed through generalization techniques, which allow compact storage of learned information and transfer of knowledge between "similar" states and actions.

The large literature of generalization techniques from inductive concept learning can by applied to reinforcement learning. However, they often need to be tailored to specific details of the problem. In the following sections, we explore the application of standard function approximation techniques, adaptive resolution models, and hierarchical methods to the problem of reinforcement learning.

5.1   Using Function Approximation Techniques

The reinforcement learning architectures and algorithms discussed above have included the storage of a variety of mappings, including $\mathcal{S} \to \mathcal{A}$, $\mathcal{S} \to V$, $\mathcal{S} \times \mathcal{A} \to Q$, $\mathcal{S} \times \mathcal{A} \to \mathcal{S}$, and $\mathcal{S} \times \mathcal{A} \to R$. Some of these are straightforward supervised learning problems, and can be solved using any of the wide variety of function approximation techniques for supervised learning. Popular methods include various neural network methods and local memory-based methods, such as generalizations of nearest neighbor methods. Other mappings, especially the policy mapping from $\mathcal{S} \to \mathcal{A}$ typically need specialized algorithms.

CRBP   The complementary reinforcement backpropagation algorithm (Ackley & Littman 1989) (CRBP) addresses the problem of learning from immediate reinforcement when reinforcement is boolean and when the action is represented as a vector of boolean values. As shown in Figure 4, it consists of a feed-forward network from an encoding of the state to an encoding of the action. The action is determined probabilistically from the activation of the output units. If output unit i has activation y_i, then bit i of the action has value 1 with probability y_i, and 0 otherwise. Any backpropagation training procedure can be used. If the result of generating action a is $r = 1$, then train the network with input-output pair $\langle s, a \rangle$ to be more likely to reproduce this action. If the result is $r = 0$, then train the network with input-output pair $\langle s, \bar{a} \rangle$, where $\bar{a} = (1 - a_1, \ldots, 1 - a_n)$, to take an opposing action.

The idea behind this training rule is that whenever an action seems not to be working, we will try to generate an action that is different from the one that does not work. Although it seems like the algorithm might oscillate between an action and its complement, that does not happen. One step of training a network will only change the action a little bit; it will not immediately change

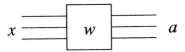

Fig. 4. Schematic for the CRBP learning algorithm

completely. Since the probabilities will tend to move toward 0.5, this makes the action more random and increases search. The hope is that the random distribution will generate an action that works better, and then that action will be reinforced.

ARC The associative reinforcement comparison (ARC) algorithm (Sutton 1984) is an instance of the AHC architecture for the case of boolean actions, consisting of two feed-forward networks. One learns the value of situations, the other learns a policy. These can be simple linear networks or can have hidden units.

In the simplest case, the entire system learns only to optimize instantaneous reward. First, let us consider the behavior of the network that learns the policy. If the output unit has activation y, then the action generated will be 1 if $y + \nu > 0$, where ν is normal noise, and 0 otherwise. The error for the output unit is most simply computed as

$$e = r(a - 1/2) \ ,$$

where the first term is the reward received for taking the most recent action and the second encodes which action was taken. The actions are encoded as 0 and 1, so $a - 1/2$ always has the same magnitude; if the reward and the action have the same sign, then action 1 will be made more likely, otherwise action 0 will be.

This method is not quite right, because it performs poorly in situations that have all positive rewards (or all negative rewards), but with different magnitudes. Rather than depending on the raw sign of the reward received, we should compare the reward to some baseline. We will change the error to

$$e = (r - p)(a - 1/2) \ ,$$

where p is the output of the second network. The second network is trained in a standard supervised mode to estimate r as a function of s. This is the *reinforcement comparison* learning rule (an instance of the REINFORCE framework (Williams 1992)).

One further generalization of this scheme is to use the TD rule to allow the prediction network to predict the long-run value of a state, rather than just the expected immediate reinforcement. In this case, we train the critic (TD) network on input s_t with error

$$e_v = (r + \gamma v_{t+1} - v_t) \ ,$$

and the policy network with error

$$e_p = (r + \gamma v_{t+1} - v_t)(a - 1/2) \ .$$

This network structure has been used in a variety of applications (Anderson 1986, Barto et al. 1983, Lin 1993 b, Sutton 1984).

Generalizing the Value Function Another popular generalizing algorithm is to use a function approximator to represent the value function. Moore and Boyan have used local memory-based methods in conjunction with value iteration; Lin has used back-propagation networks for Q-learning. An action is chosen as that which gives the largest predicted output.

There are unfortunate interactions between function approximation and the learning rules. In discrete domains there is a guarantee that any operation that backs up the value function (such as Bellman's equation, or the Q-learning rule) can only reduce the error between the current value function and the optimal value function. This guarantee no longer holds if generalization is present. These issues are discussed by Moore and Boyan, who give some simple examples of value function errors growing arbitrarily large when generalization is used with value iteration. Their solution to this, applicable only to certain classes of problem, prevents such divergence by only permitting backups whose estimated values agree with a direct value estimate obtained by Monte-Carlo experiments.

5.2 Continuous Action Spaces

When the action space is continuous, we cannot simply use a network for the Q values associated with each action. One strategy is to use a single network with both the state and action as input and Q value as the output. The associative training problem is not particularly difficult, but using the network is. When a new state is received as input, an action must be chosen. One method might be to do a local gradient-ascent search on the action in order to find one with high value. In addition, the exploration strategy is important.

Gullapalli (Gullapalli 1990, Gullapalli 1992) has developed a "neural" reinforcement-learning unit for use in continuous action spaces. The unit generates actions with a normal distribution; it adjusts the mean and variance based on previous experience. When the chosen actions are not performing well, the variance is high, resulting in exploration of the range of outputs. When an action performs well, the mean is moved in that direction and the variance decreased, resulting in a tendency to generate more action values near the successful one. This method was successfully employed to learn to control a robot arm with many continuous degrees of freedom.

5.3 Logic-Based methods

Another strategy for generalization in reinforcement learning is to reduce the learning problem to an associative problem of learning boolean functions. A boolean function has a vector of boolean inputs and a single boolean output. Taking inspiration from mainstream machine learning work, Kaelbling developed two algorithms for learning boolean functions from reinforcement: one uses the bias of k-DNF to drive the generalization process (Kaelbling 1994 b); the other searches the space of syntactic descriptions of functions using a simple generate-and-test method (Kaelbling 1994 a).

The restriction to a single boolean output makes the basic techniques impractical. In very benign learning situations, it is possible to use a collection of learners to learn independently the individual bits that make up a complex output. In general, however, that approach suffers from the problem of very unreliable reinforcement: if a single learner generates an inappropriate output bit, all of the learners receive a low reinforcement value. The CASCADE method (Kaelbling 1993 b) allows a collection of learners to be trained collectively to generate appropriate joint outputs; it is considerably more reliable, but can require additional computational effort.

5.4 Adaptive Resolution Models

In many cases, what we would like to do is partition the domain into regions of states that can be considered the same for the purposes of learning and generating actions. Without detailed prior knowledge of the domain, it is very difficult to know what granularity or placement of partitions is appropriate. This problem is overcome in methods that use adaptive resolution; during the course of learning, a partition is constructed that is appropriate to the domain.

Decision Trees In domains that are characterized by a set of boolean or discrete-valued variables, it is possible to learn compact decision trees for representing Q values. The *G-learning* algorithm (Chapman & Kaelbling 1991), works as follows. It starts by assuming that no partitioning is necessary and tries to learn Q values for the entire domain as if it were one state. In parallel with this process, it gathers statistics based in individual input bits; it asks the question whether there is some bit b such that the Q values for states in which $b = 1$ are significantly different from Q values for states in which $b = 0$. If such a bit is found, it is used to put a split in the decision tree. Then, the process is repeated in each of the leaves. This method was able to learn very small representations of the Q function in the presence of an overwhelming number of irrelevant, noisy state attributes. It outperformed Q-learning with backpropagation in a simple video-game domain. It cannot, however, acquire partitions in which attributes are only significant in combination (such as those needed to solve parity problems).

Variable Resolution Dynamic Programming The VRDP algorithm (Moore 1991) enables conventional dynamic programming to be performed in real-valued multivariate state-spaces where straightforward discretization would fall prey to the curse of dimensionality. A kd-tree (similar to a decision tree) is used to partition state space into coarse regions. The coarse regions are refined into detailed regions, but only in parts of the state space which are predicted to be important. This notion of importance is obtained by running "mental trajectories" through state space. This algorithm proved effective on a number of problems for which full high-resolution arrays would have been impractical. It has the disadvantage of requiring a guess at an initially valid trajectory through state-space.

PartiGame Algorithm Moore's PartiGame algorithm (Moore 1994) is another solution to the problem of learning to achieve goal configurations in deterministic high-dimensional continuous spaces by learning an adaptive-resolution model. It also divides the domain into cells; but in each cell, the actions available consist of aiming at the neighboring cells. The graph of cell transitions is solved for shortest paths in an online incremental manner, but a minimax criterion is used to detect when a group of cell is too coarse to prevent movement between obstacles or to avoid limit cycles. The offending cells are split to higher resolution. Eventually, the domain is divided up just enough to choose appropriate actions for achieving the goal, but no unnecessary distinctions are made. An important feature is that as well as reducing memory and computational requirements, it also structures exploration of state space in a multi-resolution manner. Given a failure, the agent will initially try something very different to rectify the failure, and only resort to small local changes when all the qualitatively different strategies have been exhausted.

Figure 5 shows a two-dimensional continuous maze. Figure 6 shows the performance of the robot during the very first trial. Figure 7 shows the second trial, started from a slightly different position.

This is a very fast algorithm, learning policies in spaces of up to nine dimensions in less than a minute. The restriction of the current implementation to deterministic environments limits its applicability, however.

5.5 Hierarchical Methods

Another strategy for dealing with large state spaces is to treat them as a hierarchy of learning problems. In many cases, hierarchical solutions introduce slight sub-optimality in performance, but gain a good deal of efficiency in execution time, learning time, and space.

Gated Behaviors Hierarchical learners are commonly structured as *gated behaviors*. There is a collection of *behaviors* that map environment states

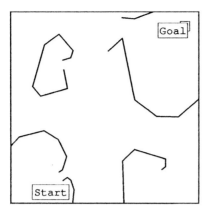

Fig. 5. A two-dimensional maze problem. The point robot must find a path from start to goal without crossing any of the barrier lines.

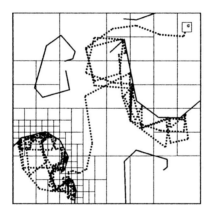

Fig. 6. The path taken by PartiGame during the entire first trial. It begins with intense exploration to find a route out of the almost entirely enclosed start region. Having eventually reached a sufficiently high resolution, it discovers the gap and proceeds greedily towards the goal, only to be temporarily blocked by the goal's barrier region.

into low-level actions and a *gating function* that decides, based on the state of the environment, what behavior's actions should be switched through and actually executed. Maes and Brooks (Maes & Brooks 1990) used a version of this architecture in which the individual behaviors were fixed *a priori* and the gating function was learned from reinforcement. Mahadevan and Connell (Mahadevan & Connell 1991 b) used the dual approach: they fixed the gating function, and supplied reinforcement functions for the individual behaviors, which were learned. Lin (Lin 1993 a) used this approach, first training the behaviors and then training the gating function. Many of the

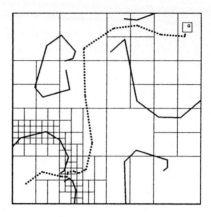

Fig. 7. The second trial

other hierarchical learning methods can be cast in this framework.

Feudal Q Feudal Q learning (Dayan & Hinton 1993, Watkins 1989) involves a hierarchy of learning modules. In the simplest case, there is a high-level master and a low-level slave. The master receives reinforcement form the outside world. Its actions consist of command that it can give to the low-level learner. When the master generates a particular command to the slave, it must reward the slave for taking actions that satisfy the command, even if they do not result in external reinforcement. The master, then, learns a mapping from world states to commands. The slave learns a mapping from commands and world states to real external actions. The set of "commands" and their associated reinforcement functions are established in advance of the learning.

This is really an instance of the general "gated behaviors" approach, in which the slave can execute any of the behaviors depending on its command. The reinforcement functions for the individual behaviors (commands) are given, but learning takes place simultaneously at both the high and low levels.

Compositional Q Singh's compositional learning (Singh 1992 b, Singh 1992 a) consists of a hierarchy based on the temporal sequencing of subgoals. The *elemental tasks* are behaviors that achieve some recognizable condition. The high-level goal of the system is to achieve some set of conditions in sequential order. The achievement of the conditions provides reinforcement for the elemental tasks, which are trained first to achieve individual subgoals. Then, the gating function learns to switch the elemental tasks in order to achieve the appropriate high-level sequential goal.

Hierarchical Distance to Goal Especially if we consider reinforcement learning modules to be part of larger agent architectures, it is important to consider problems in which goals are dynamically input to the learner. Kaelbling's HDG algorithm (Kaelbling 1993 a) uses a hierarchical approach to solving problems when goals of achievement (the agent should get to a particular state as quickly as possible) are given to an agent dynamically.

The HDG algorithm works by analogy with navigation in a harbor. The domain is partitioned (*a priori*, but more recent work addresses the case of learning the partition) into a set of regions whose centers are known as "landmarks." If the agent is currently in the same region as the goal, then it uses low-level actions to move to the goal. If not, then high-level information is used to determine the next landmark on the shortest path from the landmark to which the agent is closest to the landmark to which the goal is closest. Then, the agent uses low-level information to aim toward that next landmark. If errors in action cause deviations in the path, there is no problem; the best aiming point is recomputed on every step.

This algorithm results in improved computation time and space, but introduces suboptimality into the agent's actions. It is somewhat more difficult to describe in the "gated behaviors" framework. We can take the individual behaviors to be instructions to "aim" at a particular landmark; they are no different from the goals given to the high level, except there is a restricted set. Because the behaviors need to learn to achieve a goal, the "reinforcement" is implicit. The gating function maps the external goal to the next aiming point. The analogy breaks down when the current state and goal are within the same region, but gives insight into the commonalities of the hierarchical methods.

6   Partially Observable Environments

In any kind of realistic domain, it will not be possible for the agent to have perfect and complete perception of the state of the environment. Unfortunately, that is what is necessary for learning methods based on MDPs to be appropriate. In this section, we will consider the case in which the agent makes *observations* of the state of the environment, but these observations may be noisy and provide incomplete information. In the case of a robot, for instance, it might observe whether it is in a corridor, an open room, a T-junction, etc., and those observation might be error prone. This problem is also referred to as "incomplete perception" or "perceptual aliasing."

In this section, we will consider extensions to the basic MDP framework for solving partially observable problems.

6.1   State-Free Deterministic Policies

The most naive strategy for dealing with partial observability is to ignore it. That is, to treat the observations as if they were the states of the world and try to learn to behave. Consider a simple domain in which the agent is attempting to get "home" from the store. If it moves from the store, there is a good chance that the agent will end up in one of two places that look like "woods", but that require different actions for getting home. If we consider these states to be the same, then the agent cannot possibly behave optimally. But how well can it do?

The resulting problem is not Markovian, and Q-learning cannot be guaranteed to converge. Small breaches of the Markov requirement are well handled by Q learning, but it is possible to construct simple environments that cause Q learning to oscillate. It is possible to use a model-based approach, however; act according to some policy and gather statistics about the transitions between observations, then solve for the optimal policy based on those observations. Unfortunately, the transition probabilities depend on the policy being executed (when the environment is not Markovian), so executing this new policy will induce a new set of transition probabilities. This approach may yield plausible results in some cases, but again, there are no guarantees.

It is reasonable, though, to ask what the optimal deterministic policy (mapping from observations to actions, in this case) is. It is NP-hard (Littman 1994) to find this mapping, and it often has very poor performance. In the case of our agent trying to get home, for instance, the optimal deterministic policy will be to go right when in the woods. But if the agent happens to be in the woods on the right side of the graph, it will never get home.

6.2   State-Free Stochastic Policies

Some improvement can be gained by considering probabilistic policies; these are mappings from observations to probability distributions over actions. If there is randomness in the agent's actions, it will not get stuck in the woods forever. Jordan, Jaakola, and Singh (Singh, Jaakkola & Jordan 1994) have developed an algorithm for finding locally optimal stochastic policies, but finding a globally optimal policy is still NP hard. In our example, it turns out that the optimal stochastic policy is for the agent, when in the woods, to go right with probability $2 - \sqrt{2} \approx 0.6$ and left with probability $\sqrt{2} - 1 \approx 0.4$.

6.3   Policies with Internal State

The only way to behave truly effectively is to use memory of previous actions and observations to disambiguate the current state. There are a variety of approaches to learning policies with internal state.

Recurrent Q   One intuitively simple approach is to use a recurrent neural network to learn Q values. The network can be trained using backpropagation through time (or some other suitable technique) and learns to retain "history features" to predict value. This approach has been used by a number of researchers (Meeden, McGraw & Blank 1993, Lin & Mitchell 1992, Schmid-huber 1991). It seems to work effectively on simple problems, but can suffer from convergence to local optima on more complex problems.

POMDP approach Another strategy consists of using hidden Markov model (HMM) techniques to learn a model of the environment, including the hidden state, then to use that model to construct a *perfect memory* controller (Lovejoy 1991, Monahan 1982). Figure 8 illustrates the basic structure. The component on the left is the *state estimator*, which computes the agent's *belief state* as a function of the old belief state, the last action, and the current observation. A belief state is a probability distribution over states of the environment, indicating the likelihood, given the agent's past experience, that the environment is actually in each of those states. The state estimator can be constructed straightforwardly using the world model and Bayes' rule.

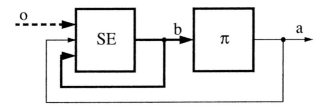

Fig. 8. Structure of a POMDP agent

Now we are left with the problem of finding a policy to map belief states into action. This problem can be formulated as an MDP, but it is difficult to solve because the input space is infinite. Chrisman's approach (Chrisman 1992) is myopic (does not take into account future uncertainty), but yields a policy without a large amount of computation. The standard approach in operations research is to solve for the optimal policy (or a close approximation thereof) based on its representation as a piecewise-linear and convex function on the belief space. This method is computationally intractable, but may serve as inspiration for methods that make further approximation (Cassandra, Kaelbling & Littman 1994).

7 Practial Reinforcement Learning

One reason that reinforcement learning is popular is that is serves as a theoretical tool for studying the principles of agents learning to act. But it is unsurprising that it has also been used by a number of researchers as a practical computational tool for constructing autonomous systems that improve themselves with experience. These applications have ranged from robotics, to industrial manufacturing, to combinatorial search problems such as computer game playing.

Practical applications provide a test of the efficacy and usefulness of learning algorithms. They are also an inspiration for deciding which components of the reinforcement learning framework are of practical importance. For example, a researcher with a real robotic task can provide a datapoint to questions such as:

- How important is optimal exploration? Can we break the learning period into exploration phases and exploitation phases?
- What is the best form of the long term reward function: Finite horizon? Discounted? Infinite horizon?
- How much computation is available between agent decisions and how should it be used?
- What prior knowledge can we build into the system, and which algorithms are capable of using that knowledge?

Let us examine a set of practical applications of reinforcement learning, whilst bearing these questions in mind.

7.1 Game Playing

Game playing has dominated the Artificial intelligence world as a problem domain ever since the field was born. Reinforcement learning applications are common here too. One application, spectacularly far ahead of its time, was Samuel's checkers playing system (Samuel 1959). This learned a value function represented by a linear function approximator, and employed a training scheme similar to the Bellman-equation backups of value iteration, temporal differencing and Q-learning.

More recently Tesauro (Tesauro 1992, Tesauro To appear) applied the temporal differencing algorithm to Backgammon. Backgammon has approximately 10^{20} states, making table-based reinforcement learning virtually impossible. Instead, Tesauro used a back-propagation-based three-layer neural network as a function approximator for the value function

Board Position $\rightarrow$ Probability of victory for current player.

Two versions of the learning algorithm were used. The first, which we will call Basic TD-Gammon, used very little pre-defined knowledge of the game, and the representation of a board position was a virtually raw encoding, sufficiently powerful only to permit the neural network to distinguish between conceptually different positions. The second, Regular TD-Gammon, was provided with the same raw state information supplemented by a number of hand crafted features of backgammon board positions. Providing hand crafted features in this manner is a good example of how inductive biases from human knowledge of the task can be supplied to a learning algorithm.

The training of both learning algorithms required several months of computer time, and was achieved by constant self-play. No exploration strategy was used—the system always greedily chose the move with the largest expected probability of victory. This naive exploration strategy proved entirely adequate for this domain, which is perhaps surprising given the considerable work in the reinforcement learning literature which has produced numerous counter-examples to show that greedy exploration can lead to poor learning performance. Backgammon, however, has two important properties. Firstly, whatever policy is followed, every game is guaranteed to end in finite time, meaning that useful reward information is obtained fairly frequently. Secondly, the state transitions are sufficiently stochastic that independent of the policy, all states will occasionally be visited—a wrong initial value function has little danger of starving us from visiting a critical part of state space from which important information could be obtained.

The results (Table 1) of TD-Gammon are impressive. It has competed at the very top level of international human play. Basic TD-Gammon played respectably, but not at a professional standard. When hand-crafted features were used, TD-Gammon's performance was remarkable.

Table 1. TD-Gammon's performance in games against the top human professional players. A backgammon tournament involves playing a series of games for points until one player reaches a set target. TD-Gammon won none of these tournaments but came sufficiently close that it is now considered one of the best few players in the world.

	Training Games	Hidden Units	Results
Basic			Poor
TD 1.0	300,000	80	Lost by 13 points in 51 games
TD 2.0	800,000	40	Lost by 7 points in 38 games
TD 2.1	1,500,000	80	Lost by 1 point in 40 games

Although experiments with other games have in some cases produced interesting learning behavior, no success close to that of TD-Gammon has been repeated. Other games that have been studied include Go (Schraudolph, Dayan & Sejnowski 1994) and Chess (Thrun 1994). It is still an open question as to if and how the success of TD-Gammon can be repeated in other domains.

7.2 Robotics

In recent years there have been many robotics applications that have used reinforcement learning. Here we will concentrate on the following four examples, although many other interesting ongoing robotics investigations are underway.

1. In (Schaal & Atkeson 1994) a two armed robot, shown in Figure 9, learns
   to juggle a device known as a devil-stick. This is a complex non-linear
   control task involving a six-dimensional state space and less than 200
   msecs per control decision. After about 10 initial attempts the robot
   learns to keep juggling for hundreds of hits. A typical human learning the
   task requires an order of magnitude more practise to achieve proficiency
   at mere tens of hits.

   The juggling robot learned a model of the world from experience, which
   was generalized to unvisited states by a function approximation scheme
   known as locally weighted regression (Cleveland & Delvin 1988, Moore
   & Atkeson 1992). Between each trial, a form of dynamic programming
   specific to linear control policies and locally linear world transitions was
   used to improve the policy. The form of dynamic programming is known
   as linear quadratic regulator design (Sage & White 1977).

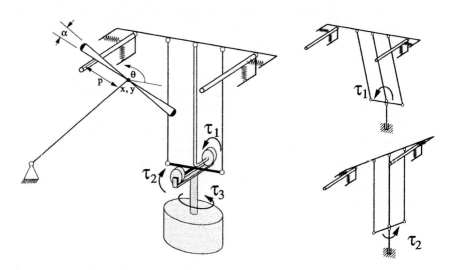

Fig. 9. Schaal and Atkeson's Devil Sticking robot. The tapered stick is hit alternately by each of the two hand sticks. The task is to keep the devil stick from
falling for as many hits as possible. The robot has three motors indicated by torque
vectors τ_1, τ_2, τ_3.

2. In (Mahadevan & Connell 1991 a) the task is for a mobile robot to push
   large boxes for extended periods of time. Box-pushing is a well-known
   difficult robotics problem, characterized by immense uncertainty in the
   results of actions. Q-learning was used in conjunction with some novel
   clustering techniques designed to enable a higher-dimensional input than
   a tabular approach would have permitted. The robot learned to perform very competitively with the performance of a human-programmed

solution. Another aspect of this work, described in Section 5.5, was a pre-programmed breakdown of the monolithic task description into a set of lower-level tasks to be learned.

3. In (Mataric 1994) a robotics experiment was performed in what, from the viewpoint of theoretical reinforcement learning, is an unthinkably high dimensional state space, containing many dozens of degrees of freedom. Four mobile robots traveled within a small enclosure collecting small disks and transporting them to a destination region. There were three enhancements to the basic Q-learning algorithm. Firstly, pre-programmed signals called *progress estimators* were used to break the monolithic task into subtasks. This was achieved in a robust manner in which the robots were not forced to use the estimators, but had the freedom to profit from the inductive bias they provided. Secondly, control was decentralized. Each robot learned its own policy independently without explicit communication with the others. Thirdly, state space was brutally quantized into a fairly small number of discrete states according to to values of a small number of of pre-programmed boolean features of the underlying sensors.

4. The final example concerns an application of reinforcement learning by one of the authors of this survey to a packaging task from a manufacturing industry. The problem involves filling containers with variable numbers of non-identical products. The product characteristics also vary with time, but can be sensed. Depending on the task, various constraints are placed on the container-filling procedure. Here are three examples:

   - The mean weight of all containers produced by a shift must not be below the manufacturer's declared weight W.
   - The number of containers below the declared weight must be less than $P\%$.
   - No containers may be produced below weight W'.

Such tasks are controlled by machinery which operates according to various *setpoints*. Conventional practice is that setpoints are chosen by human operators, but this choice is not easy as it is dependent on the current product characteristics and the current task constraints. The dependency is often difficult to model and highly non-linear. The task was posed as a finite-horizon Markov decision task in which the state of the system is a function of the product characteristics, the amount of time remaining in the production shift and the mean wastage and percent below declared in the shift so far. The system was discretized into 200,000 discrete states and local weighted regression was used to learn and generalize a transition model. Prioritized sweeping was used to maintain an optimal value function as each new piece of transition information was obtained.

In simulated experiments the saving was considerable, typically with wastage reduced by a factor of ten. The algorithm is now being used in regular production, with an extremely conservative exploration strategy to allay fears of wildly overweight or underweight containers and now reduces wastage by between a factor of two and three. Since then, the same learning algorithm has been applied to other packaging tasks in the same company without needing human tweaking between tasks.

Some interesting aspects of practical reinforcement learning come to light from these examples. The most striking is that in all cases, to make a real system work it proved necessary to supplement the the fundamental algorithm with extra pre-programmed knowledge. Supplying extra knowledge comes at a price: more human effort and insight is required and the system is subseqently less autonomous. But it is also clear that for tasks such as these a knowledge-free approach would not have achieved worthwhile performance within the finite lifetime of the robots.

What forms did this pre-programmed knowledge take? It included an assumption of linearity for the juggling robot's policy, a manual breaking up of the task into subtasks for the two mobile robot examples, while the box-pusher also used a clustering technique for the Q-values which assumed locally consistent Q values. The four disk collecting robots additionally used a manually discretized state space. The packaging example was far lower dimensional and so required correspondingly weaker assumptions, but there too the assumption of local piecewise continuity in the transition model enabled massive reductions in the amount of learning data required.

The exploration strategies are interesting too. The juggler used careful statistical analysis to judge where to profitably experiment. However, both mobile robot applications were able to learn well with greedy exploration— always exploiting without deliberate exploration. The packaging task used optimism in the face of uncertainty. None of these strategies mirrors theoretically optimal (but computationally intractable) exploration, and yet all proved adequate.

Finally, it is also worth considering the computational regimes of these experiments. They were all very different, which indicates that the differing computational demands of various reinforcement learning algorithms do indeed have an array of differing applications. The juggler needed to make very fast decisions with low latency between each hit, but had long periods (30 seconds and more) between each trial to consolidate the experiences collected on the previous trial and to perform the more aggressive computation necessary to produce a new reactive controller on the next trial. The box pushing robot was meant to operate autonomously for hours and so had to make decisions with a uniform length control cycle. The cycle was sufficiently long for quite substantial computations beyond simple Q-learning backups. The four disk collecting robots were particularly interesting. Each robot had a short life of less than 20 minutes (due to battery constraints) meaning that substantial number crunching was impractical, and any significant combinatorial search would have used a significant fraction of the robot's learning lifetime. The packaging task had easy constraints. One decision was needed every few minutes. This provided opportunities for fully computing the optimal value function for the 200,000-state system between every control cycle, in addition to performing massive cross-validation-based optimization of the transition model being learned.

A great deal of further work is currently in progress on practical implementations of reinforcement learning. The insights and task constraints that they produce will have an important effect on shaping the kind of algorithms that are developed in future.

8 Conclusions

There are a variety of reinforcement-learning techniques that work effectively on a variety of small problems. But very few of these techniques scales well to larger problems. This is not because researchers have done a bad job of inventing learning techniques, but because it is very difficult to solve arbitrary problems in the general case. In order to solve highly complex problems, we must give up *tabula rasa* learning techniques and begin to incorporate bias that will give leverage to the learning process.

The necessary bias can come in a variety of forms, including the following:

shaping: The technique of shaping is used in training animals; the agent is given very simple problems to solve first, then gradually exposed to more complex problems. Technically, shaping can be used to train hierarchical systems from the bottom up, and to alleviate problems of delayed reinforcement by decreasing the delay until the problem is well understood.

local reinforcement signals: Whenever possible, agents should be given reinforcement signals that are local. If it is possible to compute a gradient, reward the agent for taking steps up the gradient, rather than just for achieving the final goal.

imitation: An agent can learn by "watching" another agent perform the task. For real robots, this requires perceptual abilities that are not yet available. But another strategy is to have a human supply appropriate motor commands to a robot through a joystick or steering wheel, for example.

problem decomposition: Decomposing a huge learning problem into a collection of smaller ones, and providing useful reinforcement signals for the subproblems is a very powerful technique for biasing learning. Most interesting examples of robotic reinforcement learning employ this technique to some extent.

reflexes: One thing that keeps agents that know nothing from learning anything is that they have a hard time even finding the interesting parts of the space; they wander around at random never getting near the goal, or they are always "killed" immediately. These problems can be ameliorated by programming a set of "reflexes" that cause the agent to act initially in some way that is reasonable. These reflexes can eventually be overridden by more detailed and accurate learned knowledge, but they at least keep the agent alive and pointed in the right direction while it is trying to learn.

With appropriate biases, supplied by human programmers or teachers, complex reinforcement-learning problems will eventually be solvable. There is still a lot of work to be done on learning techniques and especially on methods for approximating, decomposing, and incorporating bias into problems.

Acknowledgements

Leslie Pack Kaelbling's work was supported in part by a National Science Foundation National Young Investigator Award IRI-9257592 and in part by ONR Contract N00014-91-4052, ARPA Order 8225. Michael L. Littman's work was supported in part by Bellcore. Andrew W. Moore's work was supported by an NSF Research Initiation Award and 3M Corporation.

References

Ackley, D. H. & Littman, M. L. (1989), Generalization and scaling in reinforcement learning, in Advances in Neural Information Processing 2, Morgan Kaufmann, San Mateo, CA

Anderson, C. W. (1986), Learning and Problem Solving with Multilayer Connectionist Systems, PhD thesis, University of Massachusetts, Amherst, MA

Barto, A. G., Bradtke, S. J. & Singh, S. P. (1993), Learning to act using real-time dynamic programming, Technical Report 93-02, Department of Computer and Information Science, University of Massachusetts, Amherst, MA

Barto, A. G., Sutton, R. S. & Anderson, C. W. (1983), Neuronlike adaptive elements that can solve difficult learning control problems, IEEE Transactions on Systems, Man, and Cybernetics SMC-13(5), 834–846

Bellman, R. (1957), Dynamic Programming, Princeton University Press, Princeton, NJ

Berry, D. A. & Fristedt, B. (1985), Bandit Problems: Sequential Allocation of Experiments, Chapman and Hall, London, UK

Bertsekas, D. P. (1987), Dynamic Programming: Deterministic and Stochastic Models, Prentice-Hall

Bertsekas, D. P. & Tsitsiklis, J. N. (1989), Parallel and Distributed Computation: Numerical Methods, Prentice-Hall, Englewood Cliffs, NJ

Cassandra, A. R., Kaelbling, L. P. & Littman, M. L. (1994), Acting optimally in partially observable stochastic domains, in Proceedings of the Twelfth National Conference on Artificial Intelligence, Seattle, WA

Chapman, D. & Kaelbling, L. P. (1991), Input generalization in delayed reinforcement learning: An algorithm and performance comparisons, in Proceedings of the International Joint Conference on Artificial Intelligence, Sydney, Australia

Chrisman, L. (1992), Reinforcement learning with perceptual aliasing: The perceptual distinctions approach, in Proceedings of the Tenth National Conference on Artificial Intelligence, AAAI Press, San Jose, CA, pp. 183–188

Cleveland, W. S. & Delvin, S. J. (1988), Locally Weighted Regression: An Approach to Regression Analysis by Local Fitting, Journal of the American Statistical Association 83(403), 596–610

Dayan, P. (1992), The convergence of TD(λ) for general λ, Machine Learning 8(3), 341–362

Dayan, P. & Hinton, G. E. (1993), Feudal reinforcement learning, in Advances in Neural Information Processing Systems 5, Morgan Kaufmann, San Mateo, CA

Dayan, P. & Sejnowski, T. J. (1994), TD(λ) converges with probability 1, Machine Learning

Dean, T., Kaelbling, L. P., Kirman, J. & Nicholson, A. (1993), Planning with deadlines in stochastic domains, in Proceedings of the Eleventh National Conference on Artificial Intelligence, Washington, DC

Gullapalli, V. (1990), A stochastic reinforcement learning algorithm for learning real-valued functions, Neural Networks 3, 671–692

Gullapalli, V. (1992), Reinforcement Learning and its application to control, PhD thesis, University of Massachusetts, Amherst, MA

Howard, R. A. (1960), Dynamic Programming and Markov Processes, The MIT Press, Cambridge, MA

Kaelbling, L. P. (1993 a), Hierarchical learning in stochastic domains: Preliminary results, in Proceedings of the Tenth International Conference on Machine Learning, Morgan Kaufmann, Amherst, MA

Kaelbling, L. P. (1993 b), Learning in Embedded Systems, The MIT Press, Cambridge, MA

Kaelbling, L. P. (1994 a), Associative reinforcement learning: A generate and test algorithm, Machine Learning

Kaelbling, L. P. (1994 b), Associative reinforcement learning: Functions in k-DNF, Machine Learning

Lin, L.-J. (1993 a), Hierachical learning of robot skills by reinforcement, in Proceedings of the International Conference on Neural Networks

Lin, L.-J. (1993 b), Reinforcement Learning for Robots Using Neural Networks, PhD thesis, Carnegie Mellon University, Pittsburgh, PA

Lin, L.-J. & Mitchell, T. M. (1992), Memory approaches to reinforcement learning in non-markovian domains, Technical Report CMU-CS-92-138, Carnegie Mellon University, School of Computer Science

Littman, M. L. (1994), Memoryless policies: Theoretical limitations and practical results, in From Animals to Animats 3, Brighton, UK

Lovejoy, W. S. (1991), A survey of algorithmic methods for partially observed markov decision processes, Annals of Operations Research 28(1), 47–65

Maes, P. & Brooks, R. A. (1990), Learning to coordinate behaviors, in Proceedings Eighth National Conference on Artificial Intelligence, AAAI, Morgan Kaufmann, pp. 796–802

Mahadevan, S. & Connell, J. (1991 a), Automatic programming of behavior-based robots using reinforcement learning, in Proceedings of the Ninth National Conference on Artificial Intelligence, Anaheim, CA

Mahadevan, S. & Connell, J. (1991 b), Scaling reinforcement learning to robotics by exploiting the subsumption architecture, in Proceedings of the Eighth International Workshop on Machine Learning, pp. 328–332

Mataric, M. J. (1994), Reward Functions for Accelerated Learning, in W. W. Cohen & H. Hirsh (eds.) Proceedings of the Eleventh International Conference on Machine Learning, Morgan Kaufmann

Meeden, L., McGraw, G. & Blank, D. (1993), Emergent control and planning in an autonomous vehicle, in Proceedings of the Fifteenth Annual Conference of the Cognitive Science Society, Erlbaum, Hillsdale, NJ, pp. 735–740

Monahan, G. E. (1982), A survey of partially observable Markov decision processes: Theory, models, and algorithms, Management Science 28(1), 1–16

Moore, A. W. (1991), Variable resolution dynamic programming: Efficiently learning action maps in multivariate real-valued spaces, in Proc. Eighth International Machine Learning Workshop

Moore, A. W. (1994), The parti-game algorithm for variable resolution reinforcement learning in multidimensional state spaces, in S. J. Hanson, J. D. Cowan & C. L. Giles (eds.) Advances in Neural Information Processing Systems 6, Morgan Kaufmann, San Mateo, CA

Moore, A. W. & Atkeson, C. G. (1992), An Investigation of Memory-based Function Approximators for Learning Control, Technical Report, MIT Artifical Intelligence Laboratory

Moore, A. W. & Atkeson, C. G. (1993), Prioritized Sweeping: Reinforcement Learning with Less Data and Less Real Time, Machine Learning

Narendra, K. & Thathachar, M. A. L. (1989), Learning Automata: An Introduction, Prentice-Hall, Englewood Cliffs, NJ

Peng, J. & Williams, R. J. (1993), Efficient learning and planning within the dyna framework, Adaptive Behavior 1(4), 437–454

Peng, J. & Williams, R. J. (1994), Incremental multi-step Q-learning, in Proceedings of the Twelfth International Conference on Machine Learning, Morgan Kaufmann, New Brunswick, New Jersey

Sage, A. P. & White, C. C. (1977), Optimum Systems Control, Prentice Hall

Samuel, A. L. (1959), Some studies in machine learning using the game of checkers, IBM Journal of Research and Development 3, 211–229

Schaal, S. & Atkeson, C. (1994), Robot Juggling: An Implementation of Memory-based Learning, Control Systems Magazine

Schmidhuber, J. H. (1991), Reinforcement learning in markovian and non-markovian environments, in D. S. Lippman, J. E. Moody & D. S. Touretzky (eds.) Advances in Neural Information Processing Systems 3, Morgan Kaufmann, San Mateo, CA, pp. 500–506

Schraudolph, N. N., Dayan, P. & Sejnowski, T. J. (1994), Using the td(lambda) algorithm to learn an evaluation function for the game of go, in Advances in Neural Information Processing Systems 6, Morgan Kaufmann, San Mateo, CA

Singh, S. P. (1992 a), Reinforcement learning with a hierarchy of abstract models, in Proceedings of the Tenth National Conference on Artificial Intelligence, AAAI Press, San Jose, CA, pp. 202–207

Singh, S. P. (1992 b), Transfer of learning by composing solutions of elemental sequential tasks, Machine Learning 8(3), 323–340

Singh, S. P., Jaakkola, T. & Jordan, M. I. (1994), Model-free reinforcement learning for non-Markovian decision problems, in Proceedings of the Machine Learning Conference

Sutton, R. S. (1984), Temporal Credit Assignment in Reinforcement Learning, PhD thesis, University of Massachusetts, Amherst, MA

Sutton, R. S. (1988), Learning to predict by the method of temporal differences, Machine Learning 3(1), 9–44

Sutton, R. S. (1990), Integrated architectures for learning, planning, and reacting based on approximating dynamic programming, in Proceedings of the Seventh International Conference on Machine Learning, Morgan Kaufmann, Austin, TX

Sutton, R. S. (1991), Reinforcement learning architectures for animats, in Proceedings of the International Workshop on the Simulation of Adaptive Behavior: From Animals to Animats, The MIT Press, Cambridge, MA, pp. 288–296

Tesauro, G. (1992), Practical issues in temporal difference learning, Machine Learning 8, 257–277

Tesauro, G. (To appear), TD-Gammon, a self-teaching backgammon program, achieves master-level play, Neural Computation

Thrun, S. (1994), Personal Communication

Thrun, S. B. (1992), The role of exploration in learning control, in D. A. White & D. A. Sofge (eds.) Handbook of Intelligent Control: Neural, Fuzzy, and Adaptive Approaches, Van Nostrand Reinhold, New York, NY

Watkins, C. J. C. H. (1989), Learning from Delayed Rewards, PhD thesis, King's College, Cambridge, UK

Watkins, C. J. C. H. & Dayan, P. (1992), Q-learning, Machine Learning 8(3), 279–292

Williams, R. J. (1992), Simple statistical gradient-following algorithms for connectionist reinforcement learning, Machine Learning 8(3), 229–256

Cognition - Perspectives from Autonomous Agents

Rolf Pfeifer

AI Lab, Computer Science Department, University of Zurich
Winterthurerstrasse 190, CH-8057 Zurich, Switzerland
Phone: +41 1 257 43 20/31, Fax: +41 1 363 00 35, e-mail: pfeifer@ifi.unizh.ch

Abstract. The predominant paradigm in cognitive science has been the cognitivistic one, exemplified by the "Physical Symbol Systems Hypothesis". The cognitivistic approach generated hopes, that one would soon understand human thinking — hopes that up till now have still not been fulfilled. It is well-known that the cognitivistic approach, in spite of some early successes, has turned out to be fraught with problems. Examples are the frame problem, the symbol grounding problem, and the problems of interacting with a real physical world.

In order to come to grips with the problems of cognitivism the study of embodied autonomous systems has been proposed, for example by Rodney Brooks. Brooks' robots can get away with no or very little representation. However, this approach has often been criticized because of the limited abilities of the agents. If they are to perform more intelligent tasks they will need to be equipped with representations or *cognition* — is an often heard argument. We will illustrate that we are well-advised not to introduce representational or cognitive concepts too quickly. As long as we do not understand the basic relationships between simple architectures and behavior, i.e. as long as we do not understand the dynamics of the system-environment interaction, it is premature to spend our time with speculations about potentially useful architectures for so-called high-level processes.

This paper has a tutorial and a review aspect. In addition to presenting our own research we will review some of the pertinent literature in order to make it usable as an introduction to "New AI".

Keywords: Cognition, autonomous agents, cheap designs, "New AI"

1. Introduction

The interest in cognition in the context of autonomous agents is twofold. On the one hand we want to *design* robots, on the other we want to *understand* cognition in humans, animals, and robots. The former goal is one of engineering, the latter one of cognitive science. Both aspects have also been present throughout the history of artificial intelligence.

The two perspectives are intimately related but different. In order to design a robot to achieve a particular task (like collecting objects while avoiding obstacles and maintaining a certain energy level) we can get inspiration from the study of biological systems. But in addition we can use all the technology available whether anything resembling it exists in nature or not. This holds for the physical set-up (sensors, body, effectors) but also for the control architecture. Even if there is no evidence whatsoever that animals ever run Lisp programs in their brains we are free to endow our creatures with programs of any sort. However, there are normally good reasons why things are as they are in the animal world which is why we are interested in nature: it is helpful in our design efforts.

The cognitive science perspective is oriented towards understanding existing systems. The practice of the research community over the recent decades has shown that the synthetic approach, i.e. understanding by building, has been very productive and is now widely used and generally accepted. The general idea of synthetic modelling is independent of a particular paradigm of cognitive science.

Although cognition is the main topic of this paper, it should be pointed out that we are not interested in cognition per se but in understanding behavior: we are trying to understand a certain behavior and in order to explain it we resort to cognition. In robot design we implement processes relating to cognition whenever we believe that we can make our robots perform better in this way.

The underlying motivation for this focus on cognition is as follows. We have now experienced more than half a decade of "New AI" research. If we look at what our robots can do we are not very impressed. Some robots can move towards a light source, most can avoid obstacles, others can perhaps sort red and blue objects, yet others can follow walls. If a robot can even do something useful like collecting soda cans we are surprised. Although we realize the difficulties in getting the robots to do these relatively simple tasks we would like to get them to do more sophisticated sorts of things. Traditional artificial intelligence systems have been able to perform tasks that we find more interesting like problem solving, medical and technical diagnosis, chess, scheduling, natural language understanding, etc. The utility of most of these tasks is obvious. If our autonomous agents are to enter the market place they need to be able to do something useful, too.

It is undisputed that if we had a better understanding of cognition — or more precisely of how cognition relates to behavior — we would be able to design better robots. In this paper we will argue that our current understanding of behavior in general is insufficient. We will also argue that the synthetic approach can help us improve this understanding. In particular we need to investigate the dynamics of the system-environment interaction which has been incredibly neglected in the past. When looking for explanations of behavior we have to be careful not to attribute too much to the systems we observe. We should only attribute representations or cognition if there is no other way. We have a strong tendency to project our own introspection onto other systems and we like to anthropomorphize (e.g. Turkle, 1984; Pfeifer,

1993). "How else could it be?" is a remark frequently made in the field. We hope to demonstrate that in many cases it can very well be different, even if we intuitively think this to be impossible.

First the cognitivistic paradigm will be outlined. Then some of its major problems will be discussed. The hypothesis put forward is that these problems occur, in essence, because of the neglect of the dynamics of the system-environment interaction and an inappropriate conceptualization of the relation between observed agent, observer/designer, to-be-developed artifact and environment (also called the "frame-of-reference"-problem). Then an approach will be shown which takes care of at least some of the fundamental problems. In particular it will be demonstrated that the notion of representation can only be appropriately understood if we know the physical set-up of the agent (the hardware), which is quite in contrast to most traditional views. Moreover, we will demonstrate that we often can achieve designs which are much simpler and cheaper in terms of the control structures they require if we take the physics of the agent into account. In particular if we are interested in "self-organization" we must consider mechanisms which are beyond the immediate computational model. If we are to understand cognition we have to take "non-computational" processes into account which have to do with the hardware of the agent and the system-environment interaction.

Because of the unfamiliarity of this aspect of intelligent agents we will present a series of case studies. In the last section we will draw some preliminary conclusions about cognition, cheap designs, and the "Zen of robot programming."

2. Cognitive Science

Defining cognition and cognitive science: Typically cognitive science is defined as the interdisciplinary study of cognition. This corresponds roughly to Gardner's view who defines cognitive science as the (interdisciplinary) study of epistemological issues, i.e. the study of knowledge (Gardner, 1987). Cognition has been defined in various ways. Table 2.1 lists some of them.

Within the AI community cognition is sometimes considered as the manipulation of symbols. The first and the forth definition mention as the main ingredient of cognition the manipulation of knowledge. The trouble with these definitions is that they resort to an equally ill-defined concept, namely knowledge. The second and third are broader. In fact, Neisser's definition is so broad that it includes much of what is being studied in psychology today. Knowledge is typically associated with high-level processes. Neisser's definition also includes more low-level processes.

The study of these definitions, although it provides us with some idea what people mean by the term, does not get us very far, and we will not add another one to the list. One problem with all of them is that they implicitly or explicitly endorse a cognitivistic view (see below): knowledge must be represented in the system in some form if it is to be manipulated, an assumption which can be challenged on good grounds (e.g. Clancey, 1993). Rather than defining cognition we will try to understand behavior and the respective control architectures, using various examples.

Table 2.1. Defining cognition

	Definition	*Source*
Def 1	"In my view ..., it is better to reserve the term cognition for the manipulation of declarative knowledge." (p. 160).	McFarland, 1991
Def 2	"... cognition refers to all the processes by which the sensory input is transformed, reduced, elaborated, stored, recovered, and used [including] terms as sensation, perception, imagery, retention, recall, problem solving, and thinking" (p. 4).	Neisser, 1967
Def 3	"Cognition is the collection of mental processes and activities used in perceiving, remembering, thinking, and understanding, as well as the act of using those processes" (p. 12).	Ashcraft, 1994
Def 4	"The use and handling of knowledge. Those who stress the role of cognition in perception underline the importance of knowledge-based processes in making sense of the 'neurally coded' signals from the eye and other sensory organs. It seems that man is different from other animals very largely because of the far greater richness of his cognitive processes. Associated with memory of individual events and sophisticated generalizations, they allow subtle analogies and explanations — and the ability to draw pictures and speak and write ..." (p. 149).	The Oxford Companion to the Mind; Gregory, 1987

We will demonstrate that it is difficult for cognitive science to maintain a pure knowledge view which is confined to high-level processes and which, at least implicitly, subscribes to cognitivism. Cognitive science must include the study of physical processes concerned with the system-environment interaction. If cognitive science is to serve the goal of understanding behavior its definition must be extended to include research on autonomous agents irrespective of whether these agents explicitly represent knowledge or not.

3. The Cognitivistic Paradigm

The cognitivistic paradigm can be very briefly characterized by the terms computation, representation, and functionalism. It is exemplified by the so-called "Physical Symbol Systems Hypothesis" which has been the foundation of the research program of artificial intelligence (Newell & Simon, 1976).

Let us briefly review these ideas. The idea of computation has been formalized by Alan Turing (Turing, 1939). In fact, other mathematicians have developed similar ideas. We refer to Turing since he is well-known in artificial intelligence for inventing the Turing test. The Turing test has been suggested as an empirical means to determine whether computers can be called intelligent (Turing, 1950). It is assumed that the notion of a Turing machine is known. Roughly a Turing machine has internal states and a read-write head which can read a tape, move it to the left or right and write on it. Its actions are determined by the internal state and by what it reads from the tape. The tape is potentially infinite. Computation is anything that can be carried out on a Turing machine.

Turing machines were not invented because they can do particular things. What is exciting about them is that they can model or simulate any machine whatsoever: Turing machines are *universal*. In other words, they are the only thing that needs to be studied.

The Turing machine is an abstract machine — the physical realization is irrelevant. What counts are only the steps the machine executes. It is of no concern how much time one step takes and how it is physically performed. It is interesting to note that if we are thinking of a Turing machine being physically realized and in fact having an infinitely long tape it may turn out to be a formidable engineering problem to deal with this infinitely long tape. This problem might even be harder than the computational ones which can be solved by an abstract Turing machine.

The idea of a Turing machine is illustrated in Fig. 3.1. Normally Turing machines are discussed because of their *abstract* properties and their universality. To set the stage for the discussions to come we chose a cartoon of a Turing machine which nicely illustrates that what in the abstract sounds unproblematic, such as a potentially infinitely long tape, turns out to be a significant problem if we are thinking about physical realization. This should be kept in mind in the following discussions of the cognitivistic paradigm. Throughout the paper we will stress the point of physical realization because it has been neglected for a long time.

The cognitivistic paradigm is *functionalistic*. Functionalism as proposed by Putnam (1975) means that thinking and other intelligent functions need not be carried out by means of the same specified machinery in order to reflect the same kinds of processes. In other words, the idea is that intelligence or cognition can be studied at the level of algorithms or computational processes without having to consider the underlying structure of the device on which the algorithm is performed. Briefly we can view cognition as computation. Or, to use Gardner's terms:

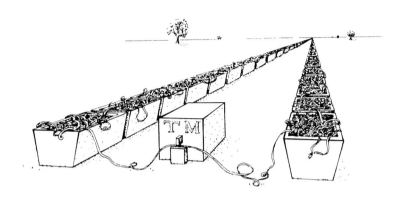

Fig. 3.1. Illustration of a Turing machine. A strict Turing machine requires a potentially infinite tape (Penrose, 1989; reprinted with permission of Oxford University Press).

"First of all, there is the belief that, in talking about human cognitive activities, it is necessary to speak about mental representations and to posit a level of analysis wholly separate from the biological or neurological ..." (Gardner, 1987, p. 6).
This idea is illustrated in Fig. 3.2.

Cognitivism is nicely exemplified by the so-called "Physical Symbol Systems Hypothesis" or PSSH (Newell and Simon, 1976). Computational processes operate on representations, the symbol structures. Cognition is viewed as symbol manipulation. The PSSH states that in order for a system to be capable of general intelligent action a necessary and sufficient condition is that it be a physical symbol system, i.e. a system capable of manipulating symbol structures. Necessary means that a system cannot be intelligent unless it is a physical symbol system, sufficient means that if something is a physical symbol systems it has the potential for general intelligent action. Typical examples of physical symbol systems are "Newell machines", i.e. production systems, general purpose programming languages, or — according to Newell and Simon — humans.

Before turning to the criticisms of the cognitivistic view it should be mentioned that there are many good reasons why it has been so widely adopted. If we consider perception we find that there are never two identical stimulus situations: a cup never looks exactly the same, although it is the same object. If we introduce a symbol to designate the cup we can abstract from the enormous variety in which a particular cup will stimulate the sensory system.

Moreover, if we want to reason about the real world it is sensible to represent it in terms of symbol structures. Representations enable us to manipulate these structures without having to act directly in the world, and so on. Logic and language obviously do employ symbols which has been another important reason to postulate symbol systems as the basis of intelligence. The branch of artificial intelligence which is based on the PSSH has also been called GOFAI by John Haugeland (1985), for "good old-fashioned artificial intelligence".

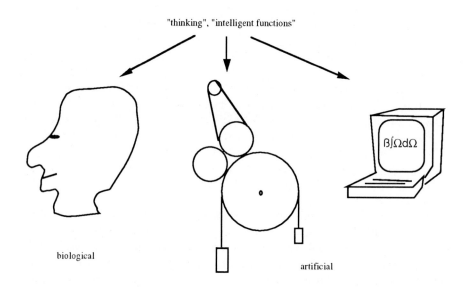

Fig 3.2. Functionalism expresses the conviction that cognition (thinking) can be studied at the level of algorithms or computational processes and that there is no need to study the underlying structure.

Many other researchers adhere, at least in essence, to this paradigm. In psychology similar views have been expressed by Pylyshyn (1984). In linguistics and philosophy Fodor can be seen as the ultimate functionalist (e.g. Fodor, 1975) (see Gardner, 1987).

4. Problems with the cognitivistic paradigm

Over the last few years the cognitivistic paradigm has been increasingly criticised (e.g. Brooks, 1991; Clancey, 1991; Dreyfus, 1992; Dreyfus and Dreyfus, 1986; Harnad, 1990; McClelland and Rumelhart, 1986; Suchman, 1987; Varela et al., 1991; Winograd and Flores, 1986 to mention but a few). Systems which are constructed within this paradigm, i.e. systems which are based on the PSSH, suffer from lack of robustness, and a failure to perform in real time (e.g. Brooks, 1991; Pfeifer and Verschure, 1994). Lack of robustness means lack of fault and noise tolerance, as well as the inability to perform adequately in novel situations.

Other criticisms contend that in traditional systems there is a lack of integrated learning, and they are not sufficiently "brain-like" (e.g. McClelland and Rumelhart, 1986) and suggest connectionism as an alternative. We have dealt with this issue elsewhere and will not pursue it further in this paper (Pfeifer and Verschure, 1992a).

There are also a number of related fundamental issues which are entailed by a classical symbol processing approach, namely the symbol grounding problem (e.g. Harnad, 1990), the frame problem (e.g. Pylyshyn, 1987), and the problem of lack of situatedness (e.g. Clancey, 1993). Roughly the symbol grounding problem refers to how symbols acquire meaning. The frame problem has to do with modeling change, i.e. how models can be kept consistent with the real world. Finally, a situated system, or a situated agent is one which can bring to bear its own experience onto a particular situation, and the interaction of its experience with the current situation will determine the agent's actions. The agent is tightly coupled to the situation in which it has to act. Traditional artificial intelligence systems lack this property of situatedness. Note that a situated agent is different from a reactive one. A reactive agent does not incorporate experience over time — it will always react the same in the same situation.

The problems mentioned are due to an insufficient understanding of the nature of intelligent systems. The hypothesis put forward in this paper is that this lack of understanding is due to (a) a fundamental neglect of the physical set-up of the agents which constitutes the intelligent system, (b) an almost complete disregard of the dynamics of the system-environment interaction, and (c) a failure to appropriately conceptualize the relation between the observed agent, the observer/designer, the constructed or to-be-constructed artifact, and the environment (also called the "frame-of-reference" or FOR problem).

We will capitalize on the fact that we currently have only a very poor understanding of the relation between behavior and internal mechanisms. It is precisely this problem which cannot be appropriately dealt with within the cognitivistic paradigm, and therefore not within the physical symbol systems assumption.

Before going into the case studies we will just briefly explain the hypotheses. Cognitivism, by its very nature of being functionalist, deliberately focuses on algorithms. It is obvious that it ignores the physical level since the physical instantiation — in this paradigm — is irrelevant. Consequently, within the cognitivistic paradigm, the system-environment interaction has also been neglected. The conviction always has been: "let's focus on the higher level processes (thinking, reasoning, etc.) and if we understand those we can add components to deal with perception and action." Our hypothesis is that this argument is flawed and that the interaction with the environment is fundamental and not merely an "addition later on."

Because the system-environment interaction has not been seriously taken into account, the "frame-of-reference"-problem has also been neglected. In other words, in artificial intelligence and cognitive science the relation between the various "participants" in the research process and the environment has not been appropriately conceptualized. For a very detailed treatment, see Clancey (1991). For the purposes of the present argument let us only mention one point. It concerns the relation between behavior — which has to do with the system-environment interaction — and the internal mechanisms which are responsible for the behavior.

Behavior of an agent can be explained in terms of goals and knowledge. For example, we can say that John goes out the door and up the stairs because he has the goal to get coffee and knows that a coffee machine is upstairs. Such a description is called a knowledge level description (Clancey, 1989) or an intentional description (Dennett, 1971). But this description is produced by the observer, rather than being "in the head of John", the agent. Goals and knowledge are *attributed* to the agent. What mechanisms are indeed responsible for John's behavior we don't know. We will show an example of how knowledge level descriptions can differ entirely from the internal mechanisms which are responsible for the respective behaviors[1].

5. The dynamics of the system-environment interaction

The remainder of this paper is devoted to the study of the relation between behavior and internal mechanisms. First we will illustrate the issue using an architecture which we have developed in our own lab, "distributed adaptive control". It has been described in detail elsewhere and we only give a very coarse outline here (Pfeifer and Verschure, 1992b; Verschure et al., 1992). The physical set-up of our robot is fairly standard (Fig. 5.1). The robot is roughly circular. It has a ring of collision sensors distributed over the front semi-circle. There is another ring of sensors which might be called proximity sensors. They are like range finders (yielding distance) but rather than providing distance their output is $exp(-d)$ where d is the distance of the robot to an object. Thus, nearby objects have high proximity whereas distant ones have low proximity. It is essential that the sensors return proximity rather than distance. What matters is not distance but risk of hitting (the latter is expressed by proximity, given that the agent moves at a constant speed). $exp(-d)$ is sometimes called the transduction function: it translates physical events in the environment into neural pulses for the control system. On both sides, left and right, there are two target sensors.

The transduction function depends on the type of sensor. If the sensor signal correlates with distance we have to take some kind of inverse to get proximity

---

[1] On this issue various positions are possible. They have been discussed in detail by Fodor (1984). For our argument the important distinction is the one between attribution and mechanism.

("nearness"), e.g. $exp(-d)$ or $1/d$. If the sensor signal correlates with proximity a scaling operation is sufficient. Very roughly the former (i.e. distance) is achieved by sonar (time of flight), the latter (i.e. proximity) by IR (intensity of reflection). In this paper we have chosen the $exp(-d)$ version to make the point about transduction explicit.

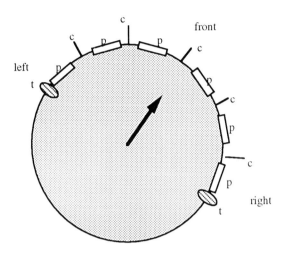

Fig. 5.1. Illustration of the basic setup of the robot used in the simulations. t: target sensors, p: proximity sensors, c: collision sensors. Only a few sensors are shown — in the simulations more are used. The arrow indicates in which direction the robot is facing.

The agent has the following basic reflexes: "if collision on left, reverse and turn right" (and similar for collision on right), "if target is sensed on right turn right" (and similar for left). There is a basic motivation to keep moving. The control architecture is shown in Fig. 5.2. There are four layers of units, collision (C), target (T), motor (M), and proximity (P). They are associated with the corresponding sensors. The connections to the motor layer are hard-wired — they implement the reflexes. The connections from proximity layer to the collision and target layers are modifiable by Hebbian learning with active forgetting (i.e. whenever something is learned, there is also forgetting). The inhibitory element (I) accumulates activation from the collision layer which eventually inhibits the output of the target layer. In other words, avoidance dominates approach. There is a decay process operating on the activation of I.

This architecture implicitly encodes part of the *value system* of the agent. One of the values encoded states "not hitting obstacles is better than approaching targets". Note that this is the way an observer would state it. For the agent this value is implicit — it does not "know" about it. Other values are encoded in the reflexes. They (implicitly) tell the organism what is good for it: if it hits something it should reverse and turn to the other side and analogously for the other reflexes. This value system is essential for the self-organization. It constrains the self-organizing process in the sense that it mediates between the (computational) processes inherent in the control architecture and the environment. This can be illustrated by an example.

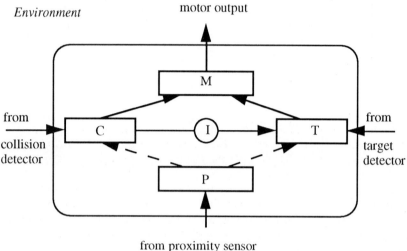

Fig. 5.2. Control architecture of "distributed adaptive control". The solid arrows indicate hard-wired connections, the broken arrows modifiable ones. There is full connectivity between the proximity layer (P) and the collision and target layers (C and T). M: motor layer, I: inhibitory element. Details: see text.

In most connectionist models a set of patterns is presented to the model. Implicitly this means that all these patterns are relevant. Not so for autonomous agents. There is simply a continuous stimulation of the sensors as the agent is moving through the environment. Determining which patterns are relevant and which are not, is a fundamental problem. It can be resolved by introducing a value system. In "distributed adaptive control", whenever an action is triggered which is associated with one of the reflexes, Hebbian learning will take place. In other words, learning takes place whenever something happens which is relevant to the organism. Note that what is important to the organism is not only determined by the computational mechanisms themselves but by the value system[2].

Let us now look at the behavior of the agent. One type of behavior, illustrated in Fig. 5.3, can be described as "anticipation." Initially the agent collides with obstacles (Fig. 5.3a). Then it starts turning before it collides. And after a while it turns when it is still a certain distance from an obstacle.

---

[2]For more detail on this issue, see Pfeifer and Verschure (1992b)

In other words, it somehow "anticipates" the obstacles (Fig. 5.3b). How does this anticipatory behavior come about?

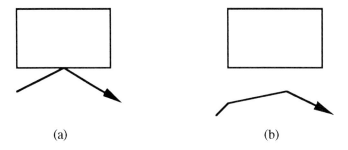

(a)   (b)

Fig. 5.3. Idealized trajectories: initially (a), and after the robot has been moving around in the environment for some time (b). How quickly the behavior in the right panel comes about is a matter of the parameter settings.

Most people, when confronted with this problem, would suggest an architecture like the one depicted in Fig. 5.4. There are a number of buffers. Inputs are copied into successive buffers so that information from the agent's earlier position can be used to control its behavior which then leads to the expected anticipatory behavior. When starting on the project this is what we had in mind. It turned out not to be necessary: the architecture shown in Fig. 5.2 does the job. This is how it works.

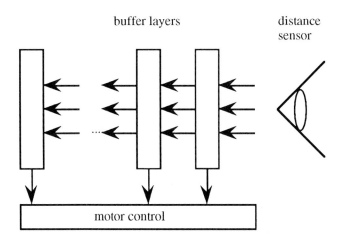

Fig. 5.4. Architecture for "anticipation". It turns out that in order to achieve the behavior depicted in Fig. 5.3 one layer is sufficient if the system-environment interaction is appropriately taken into account.

The initial situation is shown in Fig. 5.5. The agent is approaching an obstacle. It does not react since none of its motor units are triggered by the current levels of activation. Thus the agent moves straight ahead and will eventually hit the obstacle. Fig. 5.6 shows the situation when the agent is actually hitting it. Through the collision the nodes associated with the corresponding collision sensors are activated (their activation is set to 1). On the one hand this triggers the action "reverse-and-turn-right" on the other this enables Hebbian learning to take place since now there is activation in both the collision and the proximity layer. Since collisions are correlated in systematic ways with the activation patterns in the proximity layer this association is picked up by the associative network mechanism.

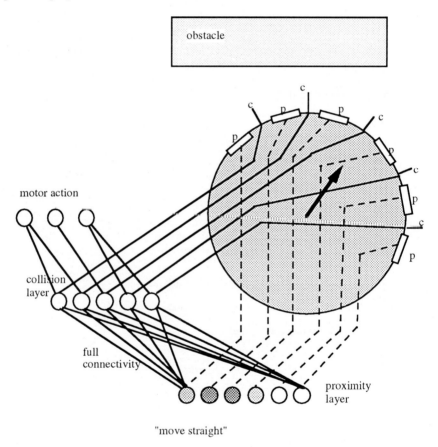

Fig. 5.5. Explanation of how "anticipation" comes about. The levels of activation at the beginning of the experiment are shown. The darkness of the nodes indicates their level of activation. The robot is not hitting, and the activation provided by the proximity sensors is insufficient to trigger a reflex.

Fig. 5.7 shows the situation after the agent has hit obstacles a number of times. The activation provided by the proximity sensors to a node in the collision layer is sufficient to bring its activation above threshold (which then initiates the motor action). Learning takes place whenever an action which is associated with a reflex is triggered since then there is activation in both the collision and the proximity layer. Learning is permanent; there is no distinction between a learning and a performance phase.

The conditions for this mechanism to work are as follows:

1. The two sensors, i.e. the collision and the proximity sensor, are activated in the interaction of the agent with its environment in different but similar ways. This overlap which represents the redundancy of the sensor readings is a prerequisite for learning to take place.

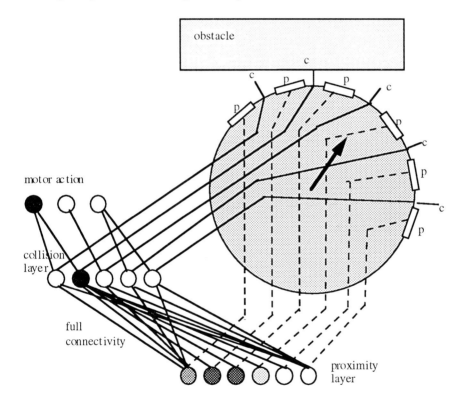

"reverse-and-turn-right"

Fig. 5.6. The situation in which the robot is hitting an obstacle is shown Because a collision sensors has been triggered a node in the collision layer is now active and Hebbian learning can take place.

2. The agent should not indefinitely continue to anticipate otherwise it will at some point always reverse and turn, no matter how far it is away from an obstacle. This is taken care of by the fact that the proximity sensor has a limited range. In our simulation experiments, this limited range is achieved by taking a transduction function of the sort $exp(-d)$ which diminishes quickly with increasing d.

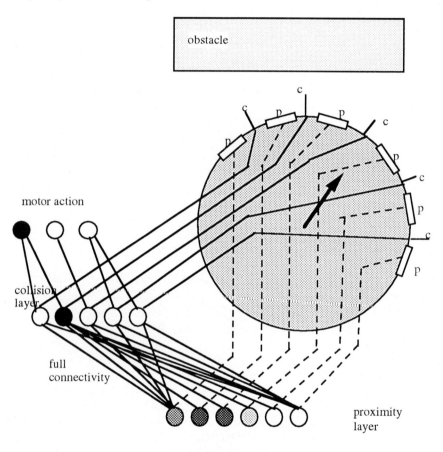

"reverse-and-turn-right"

Fig. 5.7. This shows the situation in which enough activation is provided by the nodes of the proximity sensor to the nodes of the collision sensor to trigger the reflex. Thus in this situation no collision is needed any longer for a reflex to take effect.

3. There must be a process constraining generalization in the associative network. Various solutions have been suggested for this problem (e.g. Oja, 1982). We have been working with *active forgetting* (as used in Verschure and Coolen, 1991), and with an algorithm which works on the basis of

differences between incoming and "stored" patterns, rather than on the input pattern itself (Verschure and Pfeifer, 1993).

4. The environment must match the sensory capabilities. This concerns the resolution of the sensors: it is obvious that smaller objects than the resolution of the sensors cannot be detected. It also pertains to the distribution of the sensors on the robot: if the space between two of them is too large it may not detect relevant obstacles, even though the resolution of the sensors would be sufficient. A similar point holds for motor capabilities.

In summary the "anticipatory behavior" of the agent is the result of the interplay of computational mechanisms, environmental characteristics, and physical properties of the robot: a purely computational account would not be sufficient. This implies that the classical idea of viewing cognition as computation, i.e. as manipulation of representations, neglects important factors necessary for an in-depth understanding of behavior. Studying Hebbian learning in the abstract using patterns to train the neural network as a computational process is not sufficient.

In order to illustrate the complex interplay between computational mechanisms, environmental characteristics, and physical properties of the agent a number of case studies will be presented in the next section.

6. "Non-computational" aspects: case studies

First, a remark concerning the term "non-computational" is in place here. In cognitive science computation is used in two different ways. On the one hand it is a kind of descriptive tool, as it is used in many other sciences: certain processes are reproduced and "run" on a computer, or, as Beer (1992) put it "... computation is a formalism ... for expressing mechanical or automatic processes" (p.2). In this sense all of our descriptions of the robots given so far can be seen as computational models: the properties of the sensors and how they are activated by physical events in the environment can be simulated in principle on a computer (although it may not be practically feasible).

On the other hand computation is seen as in fact *being* cognition, not merely a simulation of it. This position corresponds to the cognitivistic paradigm as introduced earlier. When we use the term "non-computational" in this paper we mean it in contrast to the cognitivistic position. We leave open the difficult question of whether there are phenomena which cannot in principle be simulated on a computer (but see Penrose, 1989, who considers processes related to consciousness as non-computational in this latter sense, i.e. not simulable on a computer).

Another way of looking at the problem of understanding behavior (or cognition) is in terms of information processing. What we argue here is that in order to understand behavior it is necessary to take phenomena into account which are not subsumed under the term "information processing". Examples are the position of the sensors on the robot, the properties of the sensors, etc.

Information processing psychology which has emerged from traditional artificial intelligence — by its very nature as an information processing theory — does not capture such phenomena.

Table 6.1: Case studies to be discussed in this section

Nb.	topic	author	goal
1	representation	Verschure and Pfeifer, 1993	What are the assumptions underlying representation?
2	topology preservation	-"-	How to get topology preservation by exploitation of physical characteristics.
3	non-goal-directed navigation	Pfeifer and Verschure, 1993	How to exploit regularities in the environment for navigation purposes.
4	visually guided navigation	Franceschini et al., 1992	Exploiting insights from the visual navigation system of the housefly.
5	collecting soda cans	Connell, 1992	Minimalist approach to robot design.
6	cricket phonotaxis	Webb, 1993a, 1993b	Postulating mechanisms by means of a biomimetic approach.

The case studies that will be discussed are summarized in Table 6.1. They are meant to elucidate various "non-computational" aspects of agent models.

6.1. Representation

The goal of this case study is to demonstrate that representation can only be understood in relation to the physical properties of the agent.

Fig. 6.1 shows what is meant by a representation in the traditional sense. According to Newell a representation follows the "representation law":

$decode[encode(T)(encode(X))] = T(X),$

where X is the original external situation and T is the external transformation (putting the block on the table) (Newell, 1991, p. 59). In other words, a situation is assumed to exist in the environment. The environment is mapped onto an internal representation (the encode function). On this representation operators are applied which correspond to the external transformation. The resulting representation is then mapped onto the real world situation (the decode function).

This way of looking at representation has a lot of intuitive appeal. But it turns out that if we are thinking of a system which has to interact with the environment we need to define the encode and decode functions. If we start from designer-defined categories (like block A, block B, table Ta) we are faced with the symbol grounding problem, i.e. we have to map configurations of the sensory space onto the internal categories. We will describe another way of viewing representation which does not presuppose the existence of an environmental state nor of objects present in the situation, but which starts from the agent-environment interaction.

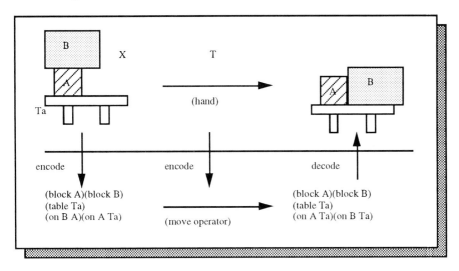

Fig. 6.1. Representation according to Newell (1991).

In order to understand what the agent in the previous example has learned about its environment we analyzed its weight matrix after it had moved around for a while. Analyzing weight matrices directly is often less illuminating than analyzing activation patterns. Learning in our agent has been contingent upon triggering of an action, i.e. the node associated with a particular motor neuron had to be active (see e.g. Fig. 5.6). So, one might want to call what the agent has acquired an *action prototype*. The following exercise illustrates the nature of these prototypes (for more detail, see Verschure and Pfeifer, 1993). We only give a synopsis here which suffices for our argument.

The action prototypes can be appreciated by calculating distance profiles. We assume that a particular collision node, X, is activated just above threshold by uniform contributions from all proximity sensor nodes. In other words we assume that the agent is in a position in which all proximity sensor nodes (continuous valued nodes associated with each proximity sensor) contribute the same amount of activation. Since the weights on all the connections are different, the physical stimulation required in the sensors is also different and therefore the distances of the objects producing this stimulation will be different. In Fig. 6.2 on the right higher levels of activation in the range finder nodes are required to contribute the same activation to node X as for those proximity sensor nodes on the left. In other words, the objects on the right have to be closer which is reflected in the distance profile. In order to calculate the distance profile the external situation corresponding to this uniform contribution, the physical properties of the sensors (i.e. the $exp(-d)$ law, which turns the range finder output d into a proximity measure) have to be taken into account.

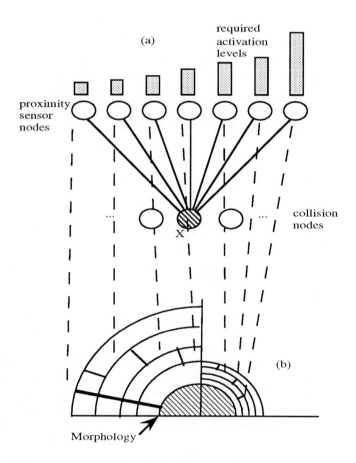

Fig. 6.2. Calculation of the distance profiles: the basic idea. Fig. 6.2a shows one collision node, X, which receives inputs from all proximity layer nodes. In order for the inputs to be equal the activation of the proximity layer nodes has to vary with the inverse (or rather the $exp(-d)$) of the weight. Fig. 6.2b shows the corresponding distance profile. i.e. what the environment of the agent would have to look like to just trigger a response, as seen from an observer's point of view, i.e. in world coordinates. A line is drawn from the collision node to the maximum distance delivering the uniform input. This maximum distance is closely correlated with the collision detector adjacent to the proximity sensor.

A distance profile taken from one of our simulations after 5000 steps is shown in Fig. 6.3. The simulations were performed using an agent with 37 collision and 37 range finder sensors (1 in the front, 18 on each side). For each of the collision nodes on the left, i.e. those nodes which trigger the motor action "reverse-and-turn-right" a profile is calculated (i.e. there are 37 profiles shown, but some at least partially overlap) and a line is drawn to the maximum distance. There is a clear asymmetry in these profiles, as is to be expected.

The weights giving rise to these profiles have been acquired by the agent over time. They can be viewed as prototypes of various situations. We can make a distinction between the *meaning* and the *extension* of a prototype. The meaning is, for example, "a situation in which it is a good idea to turn right", i.e. the meaning is associated with a particular action. The extension then is the particular set of weights which represents — via the calculations presented — the distance profile.

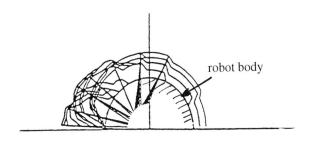

robot body

Fig. 6.3. The distance profiles calculated from the simulations after 5000 steps. The surface (morphology) of the robot is shown by the shaded area. Details, see text.

From this study we can draw the following conclusion: If we want to understand representation, i.e. if we want to understand the weight configurations in the neural network, we must take the physical set-up of the agent into account. In our calculation we use the fact that the (physical) sensor yields distance, that the transduction function of the sensors is $exp(-d)$, and we have to know the positions of the sensors on the robot. If we know nothing about the physics of the robot it is not possible to interpret what the network has acquired. The prototypes have been learned by the agent through self-organization mediated by the value system which includes physical aspects, as pointed out earlier.

It should be mentioned that often we mean by representation something which does not directly depend on sensory inputs. The point made here is independent of whether what the agent has acquired (i.e. the weight matrix) relates directly to sensory inputs or not. Anything the agent will be able to represent, ultimately has to build on basic representations acquired directly through the system-environment interaction.

6.2. Topology preservation

The previous study illustrates another interesting point. What the agent evolves is a kind of topology-preservation. Topology-preservation in this case means that if a node in the neural network which is connected to a particular proximity sensor is activated it will spread its activation to nodes in the neural network which are associated with collision sensors which — on the robot — are adjacent to the proximity sensor.

This implies that we can get topology preservation via the morphology of the agent: there is no need to provide for topology preservation in terms of mechanisms within the neural network (as one would have to do in the case of the well-known topology preserving feature maps of Kohonen — Kohonen, 1982). In other words, there is no *information process* responsible for the topology preservation.

An important conclusion from this study is that if we take the physical set-up into account we can get certain properties for free. In our example we get topology-preservation for free because of the way in which the collision and the proximity sensors are physically positioned on the robot and because the two types of sensors are activated by the environment in similar but different ways.

6.3. Non-goal-directed navigation

The example can be used to illustrate an additional point. The environment for our simulation experiments is shown in Fig. 6.4. The circle indicates the range of the target source: if the agent is outside of it, as in the figure, it cannot sense the target. If we look at the behavior of the agent in "distributed adaptive control" after it has been moving about in its environment for a while (say, 5000 steps) we see that it starts following walls. Note that the agent follows walls even if it is outside the range of the target source. Why?

As the agent moves about in the environment it will at times encounter a target source. If the target is normally behind a hole in the wall it will be pulled towards the wall by one of the reflexes. But there is also an avoidance reflex. The inhibitory element gives preference to avoidance — the agent turns away. Then the approach reflex gets active again etc. This is what accounts for the wiggly trail. Through the Hebbian learning target detection and proximity sensor stimulation on the side are associated. Subsequently the agent will follow walls even if there is no target source.

There is no component for "wall-following" in the agent: it evolves as the agent interacts with its environment. In an environment in which there is the regularity that targets are normally behind holes in walls, following walls is a perfectly suitable strategy for finding targets: irrespective of where the agent is released it will find its target, as long as it is somewhere along the wall. In environments where this regularity does not hold the agent will not start following walls. The disadvantage of this behavior is that it will start circling around large objects. Notice that the agent is *not goal-directed*, i.e. it does not work towards a goal[3]. It only detects when it is at the target source. Its behavior could be described in terms of goals and knowledge about location of targets, but clearly there is no such knowledge represented within the agent. Traditional systems in artificial intelligence are mostly goal-directed: there is an explicit goal representation and through some mechanism (means-ends analysis, plans) the agent is trying to achieve the goal. Theorizing in terms of goal-directed systems is also widespread in cognitive science (for more detail, see Pfeifer and Verschure, 1993).

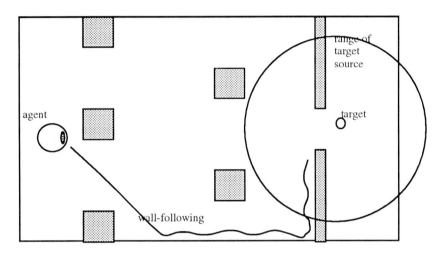

Fig. 6.4. Robot environment. The trajectory shows the robot's behavior after about 5000 steps. The circle on the right indicates the range of the target source. If the robot is on the left, as shown in the figure, it cannot sense the target. The robot follows walls even if it is outside the range of the target source. Explanation: see text.

What can we learn from this example? First, contrary to our intuition, we can get agents to achieve tasks without them being goal-directed. Second, it is beneficial to exploit regularities in the environment. Third, there is no need to program the agent with all competences at the very beginning, it is sufficient that there be the potential for the agent to evolve them. This is cheaper since certain behaviors (such as wall following) are useful in some environments but not in others. Fourth, it is not necessary to represent everything in the system itself. In our example, the agent does not know that targets are found behind holes in walls. But because it behaves in particular ways it will find

---

[3]This is the meaning of the term "goal-directed" as used by McFarland, 1989.

them. The agent needs no map of the environment nor does it need the explicit knowledge of location of food. This point is central to the argument and requires some elaboration.

Often our reaction is that the behavior of an agent could be significantly improved by introducing explicit representations or cognition. By improved in this case we might mean that the number of target sources visited per unit time is higher. If the agent had the possibility to evaluate its steps before actually taking them it might perform better. In other words, an agent equipped with the capability of *planning* would have an advantage (McFarland, 1991). This capability of planning is a cognitive one in its traditional meaning. Precisely because of this advantage of planning over non-planning systems, many traditional artificial intelligence approaches to robotics have employed substantial planning components. The implications have been insufficient real-time behavior and lack of robustness. The former is due to the computationally intractable nature of planning (e.g. Chapman, 1987), the latter is caused by the problems of maintaining a model of the real world (the frame problem) which includes the problem of relating the elements of the representation to the sensory information (the grounding problem). There are always trade-offs. Simply using an artificial intelligence planning system as normally employed (see McDermott, 1992 for a review) is likely to run into difficulties. The fact that some animals behave as if they were planning does by no means imply that their mechanisms reflect in any sense the ones we are using in artificial intelligence.

This case study concludes our discussion of "distributed adaptive control". We now summarize how an approach which starts from the system-environment interaction — rather than postulating high-level symbolic processes — takes care of the fundamental problems mentioned initially, namely grounding, situatedness, and the frame problem. The agent's categorizations of the environment, i.e. its prototypes, are *grounded* since they are acquired through its interaction with the environment and therefore built up from its own point of view, not from the one of the observer. The agent is *situated*: it's behavior is determined by the current system-environment interaction and its experience (as expressed in the weights of the connection matrix) influences its actions. The frame problem is irrelevant since no complex models of the world are maintained.

6.4. Visually guided navigation

The goal of this case study is to show that implicit assumptions which are often made can substantially influence the complexity of the control architecture. For example, if pixel arrays are used for navigation purposes certain operations may be orders of magnitude more complex than if motion detectors are used. It is shown how the "right" physics of the sensors automatically resolves certain difficult problems.

Our presentation is inspired by work of Franceschini et al. (1992) on the development of a robot which navigates based on principles derived from the housefly. It has been demonstrated that the impressive navigational skills of the housefly are largely due to the fact that its visual system can detect optical flow. In what follows we will not investigate the housefly but discuss a particular question which is relevant for robot navigation but which is often neglected.

If we are dealing with a fast robot with inertia we would want the following behavior: if it is moving fast it should turn earlier to avoid obstacles than if moving slowly, in order to minimize risk of hitting. Intuitively one would think that this requires (a) an assessment of the distance and the speed of the robot, and (b) a mechanism which adjusts the distance (and perhaps the angle) at which the agent should divert. This turns out to be unnecessary if motion detection is used.

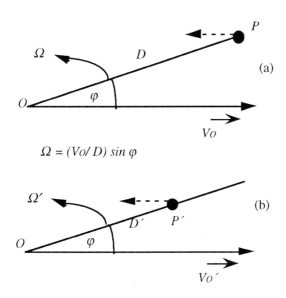

Fig. 6.5. Navigation based on optical flow. (a) high speed, far away object, (b) low speed, nearby object. Allowing for certain idealizations, the angular speed Ω is the same in both situations if $(Vo/D) = (Vo'/D')$. So, if $D' < D$, then, if Vo' is proportionally smaller than Vo, the situation is identical to the agent (adapted from Franceschini et al., 1992).

Our simplified argument is illustrated in Fig. 6.5. Fig. 6.5a shows the optical flow, as indicated by the angular speed Ω, induced by an object P at some distance D from the agent when moving at speed Vo. Fig. 6.5b is the analogue for a lower speed Vo' and a proportionally less distant object P'. It is obvious from the figure that if $(Vo/D) = (Vo'/D')$, the angular speeds Ω and Ω' are the same. Let us assume that there is a hard-wired mechanism for obstacle avoidance which will cause the agent to turn, given a certain Ω. It follows that whatever the speed V of the agent there is no need for any adaptive mechanism *within* the agent to adjust for its speed — it is taken care of by the mechanism determining motion relative to the environment. As an aside we see here another instance of the frame-of-reference problem: adaptivity is not something which is purely a property of an agent, but rather of its interaction with the environment. What we perceive as two different situations, namely Figs. 6.5a and 6.5b, is one and the same situation to the agent. This is, of course, only strictly true in our idealization, i.e. if the environment only consists of the objects shown in the figures.

Again we see that by taking the system-environment interaction into account we can arrive at an effective and cheap mechanism for navigation which adjusts for speed without there being an information process or algorithm to perform the adaptation.

What is shown in Fig. 6.5 is called the principle of motion parallax. There is a sine law for calculating angular speed. In the eye of the fly there is a non-uniform layout of the visual axes such that sampling of the visual space is finer towards the front than laterally. This gradient can be said to compensate for the sine law inherent in the optic flow field. The introduction of the sine gradient allows the underlying motion detection system to be built uniformly by elements each displaying the same temporal properties as its neighbours. Once again we see that if we want to understand the behavior of the agent it is not sufficient to look at the control architecture (which suggests something linear and homogeneous, whereas the motion detection problem is highly non-linear). The advantages of having a sine gradient are so obvious that they may in retrospect contribute to explain the gradient in resolution that is observed in the peripheral retina of so many creatures, including humans and flies (Franceschini et al., 1992).

This example illustrates another important point. What seems most natural to at least some of us may often not be the most efficient. While it seems natural to use pixel arrays in visual sensors it turns out that using flow sensors as in the housefly is much simpler for certain navigational tasks such as obstacle avoidance and the related control architecture is much cheaper and more robust. What appears to the observer as a second-order effect (a phenomenon which requires a sequence of static images) is a first-order effect from the point of view of an agent equipped with flow sensors. The correlation apparatus which is normally part of the algorithm is delivered directly from the retinal output or from the "elementary motion detectors (EMDs)" (Franceschini et al., 1992). This example again illustrates that control algorithms only make sense with respect to the system-environment interaction and cannot be studied in isolation, as in the symbolic approach to cognition. In fact this point is so obvious that it is surprising how often it is not given sufficient consideration.

6.5. Collecting soda cans

Herbert is a robot whose task it is to collect soda cans in the MIT robotics lab. It was developed by Jon Connell (Connell, 1992) with the explicit goal to employ a minimalist design. Minimalist means that no element is introduced in the control architecture unless it is absolutely needed to achieve a certain behavior.

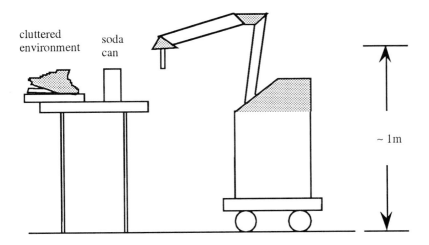

Fig. 6.6. Basic set-up of the robot Herbert (after Connell, 1992).

Roughly, Herbert works as follows. Its basic set-up is shown in Fig. 6.6. It is equipped with a number of sensors: two rings of IR sensors, a compass, and a laser light striper on its body, as well as IR, touch and simple torque sensors on its arm. It has only one task, namely to collect soda cans. It wanders through the laboratory looking for cans. For navigation purposes it uses the two rings of IR sensors and the onboard compass. There is a tactical and a strategic navigation system. The tactical system keeps the robot within the range of the IR sensors from some part of the environment. In other words the robot "wants" to stay near an object which implies that it will end up tracking along the "coast" of its world. This form is — or was — also used by human navigators in order not to get lost in the open sea. The strategic navigation system ensures that the robot will explore the environment in sensible ways and that it will find its way back home (see below).

Detecting soda cans is very demanding since we are dealing with a cluttered environment. Herbert identifies soda cans by means of the laser light stripper. In trying to detect cans it exploits various constraints. For example, in the laser image cans appear relatively isolated, a fact which greatly helps identify them. This is due to the fact that table surfaces (where cans normally stand) are largely parallel to the incoming laser beam. Moreover, their surfaces are typically shiny which leads to only little diffuse scattering. The "representation" needed to identify the cans exploits these constraints. It is related directly to the ways in which the laser information is processed. Consequently, it is much simpler than those required, say, in a general purpose vision system.

The grasping is done by exploiting the environment as "its own representation", without relying on any sort of internal representation. It also exploits the physical interaction with the environment to position its inaccurate motor system. For example, before grasping a can it touches the table on which the can stands to get information about the approximate height at which the can is located.

Once Herbert has grabbed a can it heads back. Herbert does not "know" where it is or where it has to go: there is no map in the robot. But how does it find its way back then? As Herbert explores the environment its movements are already constrained. During the exploration phase it has to pass through every door it can. But in order to ensure that it can always find its way home, before turning into the door it must first test if it has been traveling opposite to the home vector. It can do this using the onboard compass. If it finds that indeed it has not been traveling opposite to the home vector it immediately turns around, terminates its wandering, and heads back home. This guarantees that the robot never executes a path segment that it does not know how to invert later on. However, this implies that certain areas will never be explored. An example is given in Fig. 6.7. While this is obviously a disadvantage this constraint enables Herbert to find its way back by very simple means, namely by simply turning south. Once it can no longer turn south at a door it has arrived home. It then drops the can and starts exploring again.

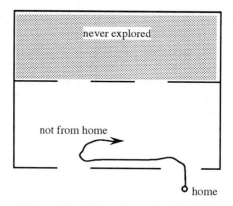

Fig. 6.7. Because of constraints in the exploration algorithm certain areas of the office environment are never explored (after Connell, 1992, p. 131).

How does Herbert perform? One criterion is that is should find all the cans. This point concerns the exploratory behavior. As illustrated in Fig. 6.7 there are environments in which Herbert will not be able to do thorough exploration — some cans will be missed. Once it has found a can it must be able to find its way back. From informal discussions it can be inferred that in this respect Herbert's behavior is not very robust. It seems that the lack of virtually any internal state is a factor contributing to its brittleness. Its almost complete dependence on the environment puts its task-achieving behavior in jeopardy if, for example, the environment slightly changes during task execution, such as a door being opened or closed. Moreover, it might turn out to be beneficial to have at least some very crude representation of the starting point where the cans need to be taken back to. But, as pointed out above, we have to be careful to remain minimalist, i.e. to introduce only as much internal state as is really necessary (which is often hard to find out). Another source of potential problems of robustness is due to the assumptions contained in the can-detection system since these assumptions do not always hold.

Herbert illustrates the following points. First, tasks which at first glance seem to require considerable internal representations or cognition, can be achieved without — or almost without — the latter. This fact is all the more surprising, given the fundamental importance attributed to representation in classical artificial intelligence. For example, Herbert does not need a map of the environment. Second, good task performance can be achieved with relatively cheap sensors and with imprecise motor control. This requires the agent to take advantage of the system-environment interaction. Third, there are a number of trade-offs. Herbert is quite brittle in some respects. For example, certain areas are never explored, and slight changes in the environment may prevent it from finding its way back to drop the cans. This is, of course, precisely the reason why classical artificial intelligence introduces the concept of representation, and therefore cognition: to give the agent independence of environmental situations. Another way of putting this is to say that cognition gives the agent additional *autonomy*. Having little internal state is cheap in terms of control architecture, but leads to reduced robustness in changing environments. However, introducing cognition entails problems relating to updating models of the real world, which can in turn induce behavioral instabilities. We will come back to this point in the last section.

6.6. Cricket phonotaxis

Cricket phonotaxis provides an excellent example of a design which incorporates and exploits physical processes rather than relying entirely on computational ones. By phonotaxis we mean those processes by which animals, in our case the crickets, move towards the calling song of a potential mate. Our description will be short, just sufficient to make our point. For more detail, see Webb (1993a, 1993b).

The male crickets produce a particular sound by rubbing one wing against the other. Females can find a male by this cue over distances of twenty meters through rough vegetation. Clearly — one would think — the cricket will need mechanisms for recognizing the sound to distinguish it from the song of other species, and for analyzing the direction of the sound. Recognizing a sound would also imply some sort of internal representation of the sound. To detect the direction of the sound a solution which immediately comes to mind is exploiting intensity differences or phase differences in the sound waves.

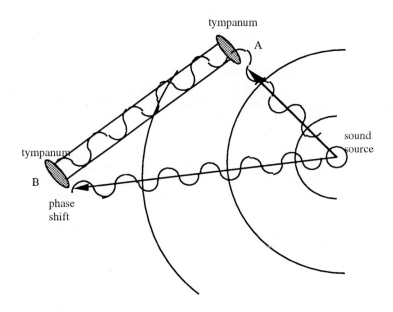

Fig. 6.8. Schematic representation of how phonotaxis can be achieved by exploiting a physical mechanism rather than by means of sophisticated neural circuitry. At the tympanum at point B the sound waves traveling directly to B and those travelling via A interfere depending on the phase shift.

The way phonotaxis works is roughly as follows (Webb, 1993b). The size of a cricket is about 1-2cm, the wavelength of the song about 5-6cm. Thus, there will certainly not be enough intensity difference between the ears (which are located on their front legs). Phase differences are on the order of microseconds which would require large scale neural processing to detect. Given the limited neural system of the cricket it seems hard to imagine how this might be accomplished. The "trick" of the cricket is the following (see Fig. 6.8): instead of using a neural mechanism for this purpose, or an *information process* it uses a *physical* process. The ears are connected to one another by a tracheal tube, so that vibration outside one tympanum is transferred through to the inside of the opposite tympanum. The resulting vibration of the tympanum is a superposition of the directly arriving sound, and the sound arriving indirectly via the trachea, the relative phase of which depends on the orientation of the cricket to the sound source. Thus the intensity of the response is altered substantially by the direction of the sound. Note that such a system fails if the sound is of the wrong wavelength, as the trachea length is constant. Thus, without "analyzing" the sound the cricket reacts only to the appropriate songs, namely those having the right wavelengths. This is an instance of what biologists call "matched filters."

What can we learn from this example? First, we have to be careful not to postulate mechanisms which are more complex than required to achieve the task. There is no need to perform a general analysis of the sound, there is no need for an internal representation of the sound pattern, and the detection of the direction is not performed by an information process only, but involves a physical process (phase cancellation of sound waves). Again, exploiting the physics, as in the case of topology preservation, sometimes leads to much cheaper control mechanisms. Second, considering the task to be achieved rather than construing mechanisms in isolation will often lead to simpler designs. In the case of the cricket the only task of the sound system is to move towards the source of the right sound. For this purpose the system of the cricket in which there is no separation of sound detection and action is highly efficient. Only if we separate — as we often do when studying cognition — the task of perception from the action we are (mis-) lead to think there would have to be a separate perceptual system.

Concerning the methodology, we should perhaps mention the following point. Webb's approach is an example of "biomimetics", i.e. a biological system is modeled as closely as possible. In other words, a robot which mimics the phonotactic behavior of the crickets as closely as possible is built. The goal is on the one hand to get inspiration about robot design and on the other to come up with testable hypotheses about potential mechanisms for phonotaxis in crickets. If we read Webb (1993a) it is surprising how little is known about the actual biological mechanisms in spite of many years of intensive research. Thus this synthetic methodology might turn out to be highly fruitful for biologists. And it has proved fruitful for robotics or artificial intelligence already because it provides us with an interesting perspective on behavior vs. mechanism.

7. Cognition, cheap designs and the "Zen of robot programming"

It seems that we have come quite far afield from our main topic, cognition. In this section we will try to summarize what we have said so far and draw some conclusions. As mentioned initially we have two goals, namely understanding cognition (or rather, behavior) and designing agents. Let us first look at design.

Here are the main messages. They can be seen as a set of general design principles for autonomous agents:

Cheap designs: Although one often gets the impression that task-performance and robustness can be increased by introducing more sophisticated designs, there is always a trade-off, a price to pay: the designs cost more in terms of machinery needed. But there are more important implications. When introducing cognition, e.g. if we introduce a certain representation in the form of a map, this enables the robot to navigate in ways it could not without. For example, it enables the agent to eventually infer shortcuts once it has acquired sufficient knowledge about the environment which it has added to the map. The trade-off is that a map requires the robot to always know where it currently is located. Moreover, all the information about the environment needs to be stored (represented) on the map, otherwise the map is not going to be very useful. This in turn implies that the map has to be updated accordingly. In addition to computational issues updating a model of the environment faces the following problem. The agent only has limited information about the environment. Because of its limited sensory-motor capabilities it can only acquire local information. If the environment changes this implies that inconsistencies will be accumulated over time. This is an instance of the frame problem mentioned earlier. Notice that the frame problem is a fundamental one which cannot be simply compensated for by more sophisticated computation.

The issue of cheap designs has been illustrated in many examples: anticipation, topology preservation, non-goal-directed navigation, collecting soda cans, etc. Cheap designs typically work well in the niches they are designed for, but they are not very robust if the environment changes. More expensive designs which include symbolic representations (and therefore cognition in the cognitivistic paradigm) give the agent obvious advantages but they entail the problem of symbol grounding. Maps, as just mentioned, lead to the frame problem, etc. Thus, we cannot simply say that introducing cognition will lead to better task performance: we have to be aware of the consequences.

Exploiting the dynamics of system-environment interaction: Cheap designs typically take advantage of the dynamics of the system-environment interaction in interesting ways. We have shown several examples of this principle. It will be an essential part of any robot design strategy. Focusing on the computational mechanisms is not enough. There need be no component for "anticipation" in the "distributed adaptive control" paradigm, no "recognition component" for particular sounds in phonotaxis, and no map in Herbert, in order for it to find its way back. The real world "is there" which implies that certain things do not have to be represented *because* they "are there". Herbert relies on heavily on this principle. The agents perform well because they are "fitted" to the environment in which they have to function, i.e. their niches.

Constraints provided by the real world: The real world provides constraints which can be exploited. Because we are dealing with a physical world things normally do not change too abruptly. Thus, successive sensor readings tend to be highly similar. If at some point a sensor reading is lost this is not tragic since simply another one can be taken. If it is taken shortly after the first one chances are that it will be highly similar to the first. Note that this philosophy would be a nightmare in computer science where one of the basic axioms is that no information ever be lost.

Constraints of this sort can also be exploited in the learning mechanisms. While in computer science — in the virtual world — it is always possible to manipulate independently each and every bit of a representation in memory, when dealing with the real world, the individual parts of a representation tend to be highly correlated. In other words, though Hebbian learning has been shown to be limited in its learning capacity, in the real world, for many tasks, more powerful mechanisms may not be required. This fact is exploited by the example of Hebbian learning in our "distributed adaptive control" model.

Ecological validity: One point that we have so far not discussed explicitly, but which has been contained in what we have said, is the choice of task. We have always assumed the task as given. As pointed out at the beginning in traditional artificial intelligence tasks like problem solving, medical diagnosis, chess, scheduling, or natural language are common. In the area of autonomous agents typical tasks are navigation, obstacle avoidance, wall following and collecting objects. But the latter do not seem very interesting to us. We are therefore quick to suggest more complex ones. Why not endow our robots with the capability of communicating in human natural language? The reason is that if we want to avoid the problems we have had with traditional systems (discussed at length earlier) the agents have to be "fitted" to their environments. This implies that the tasks have to match their sensory-motor complexity. As Rod Brooks put it: "Elephants don't play chess", or we might add: "ants don't have natural language". The great danger is that if we tackle "un-ecological" tasks we are likely to fall into the same traps as the traditional cognitivists. The cartoon of Fig. 7.1 illustrates this development. Initially there is great hope for the cognitivistic paradigm. Then some severe problems having to do with the interaction with the real world are recognized. Fortunately the field of autonomous agents appears (or re-appears) with great intensity. In this paradigm cognition is a relatively marginal topic. However, soon problems with this paradigm become evident. The researcher on the right tries to combine the two, i.e. he develops autonomous agents and endows them with cognition. He should be careful.

The "Zen of robot programming": Robot programming is very different from traditional programming, the main reason being that we are dealing with the real world rather than a virtual one. Designing autonomous agents is very different from building traditional artificial intelligence models. The real world is there to be exploited. There are regularities which can be taken advantage of, matters don't change abruptly, certain things "are there". This implies that we can be more relaxed about many aspects we had to worry about earlier. But if we want to exploit the real world the agents have to fit harmoniously into it which also implies choosing the right tasks for them, i.e. tasks which match their sensory-motor complexity. And we have to start thinking differently about control: in terms of system-environment interaction rather than information processes only, in terms of self-organization and value systems, rather than by "giving orders" as in traditional programming. This is my interpretation of what Rod Brooks has called the "Zen of robot programming".

Fig. 7.1. Cartoon of the development of cognitive science over recent years. Initially there is great hope for the cognitivistic paradigm. Then some severe problems having to do with the interaction with the real world are recognized. Furtunately the field of autonomous agents re-appears with great intensity. However, soon problems with this paradigm become evident. The researcher in frame 5 tries to combine the two ...

Minimalist explanations: Let us now look at the second goal, namely understanding behavior (or cognition) in existing agents, i.e. humans, animals, and, of course, robots. We have seen that through building robots we can develop a new level of understanding behavior which was not possible previously. We can now build the robots ourselves, i.e. we know their control structures and their physical set-up, we can conduct experiments, and observe their behavior. We saw that we can sometimes get robots to do things using mechanisms which are surprisingly simple and cheap. The same principle is to be applied to explaining the behavior of humans and animals: only introduce the bare minimum. By the way, this is reminiscent of Occam's Razor, a principle generally accepted in the empirical sciences. The main message is that a lot of behavior can be explained without resorting to cognition. Something like cognition in the traditional sense should only be introduced as an explanatory concept if there is no other way.

Conclusion: We are worried that our robots can only do relatively simple things. If we want to make progress we must move towards more difficult tasks. But we have to be careful in the choice of the tasks. For example, we should not be blinded by the apparent complexity of tasks tackled in traditional artificial intelligence. Human medical diagnosis is an extremely complex task. But the models of medical expertise as exemplified by expert systems hardly reflect human behavior at all (e.g. Lamberts and Pfeifer, 1993): they typically abstract from the truly difficult problems, namely those having to do with the system-environment interaction. They are cognitive models and therefore suffer from the problems discussed. Thus, it should not be the goal of our research to try and do the same sorts of tasks as in traditional artificial intelligence, just this time using autonomous agents instead of workstations. That would not meet our criterion of ecological validity and is likely to fail. Rather we should start with the difficult problems of system-environment interaction using very simple systems, increasing task complexity in small steps, while always keeping the ecological and minimalist perspective in mind. Only by remaining minimalist we will really learn to understand the dynamics of the system-environment interaction. At some point we may have to introduce something we might want to call cognition but then it will be well-founded rather than being based on vague intuitions from our introspection. If we pursue this line of research we will eventually be able to derive design principles and move from the stage of a purely exploratory discipline to a mature science. The fact that our agents currently do not look very impressive and that we do not have a general theory yet does not imply that there are no achievements.

Acknowledgement

This research was supported in part by grant nb. 21-34119.92 of the Swiss National Science Foundation. I would like to thank Luc Steels for suggesting to me to think about the issue of cognition in the context of autonomous agents, and to David McFarland and Christian Scheier for many stimulating discussions on this topic and their comments on the paper.

References

Ashcroft, M.H. (1994). Human memory and cognition. New York, NY: Harper Collins College Publishers (second edition)

Beer, R. (1992). A dynamical systems perspective on autonomous agents. Techreport, CES-92-11, Case Western Reserve University, Cleveland, OH

Brooks, R.A. (1991). Intelligence without reason. IJCAI-91, Proc. of the 12th International Joint Conference on Artificial Intelligence, vol 1, 569-595

Chapman, D. (1987). Planning for conjunctive goals. Artificial Intelligence, **32**, 333-337

Clancey, W.J. (1989). The knowledge level reconsidered: Modeling how systems interact. Machine Learning, **4**, 285-292

Clancey, W.J. (1991). The frame of reference problem in the design of intelligent machines. In: K. van Lehn (ed.). Architectures for intelligence. The 22nd Carnegie Mellon Symposium on Cognition. 357-423

Clancey, W.J. (1993). Situated action: a neuropsychological interpretation. Response to Vera and Simon. Cognitive Science, **17**, 87-116

Connell, J. (1992). Minimalist mobile robotics. Boston: Academic Press.

Dennett, D. (1971). Intentional systems, Journal of Philosophy, **68**, 87-106. Reprinted in J. Haugeland (ed.), Mind Design. Montgomery, Vermont: Bradford Books, 1981, 220-242

Dreyfus, H.L. (1992). What computers still can't do. Cambridge, MA: MIT Press

Dreyfus, H.L., and Dreyfus, S. (1986). Mind over machine: the power of human intuitive expertise in the era of the computer. New York: Free Press

Fodor, J.A. (1975). The language of thought. Cambridge, MA: Harvard University Press

Fodor, J.A. (1984). Fodor's guide to mental representation: the intelligent Auntie's vade-mecum. Mind, **93**, 76-100

Franceschini, N., Pichon, J.M., and Blanes, C. (1992). From insect vision to robot vision. Phil. Trans. R. Soc. Lond. B, **337**, 283-294.

Gardner, H. (1987). The mind's new science. New York: Basic Books (first published 1985)

Gregory, R.L. (1987). The Oxford companion to the mind. Oxford, UK: Oxford University Press

Harnad, S. (1990). The symbol grounding problem. Physica D, **42**(1-3), 335-346.

Haugeland, J. (1985). Artificial intelligence: the very idea. Cambridge, MA: The MIT Press

Kohonen, T. (1982). Self-organized formation of topologically correct feature maps. Biological Cybernetics, **43**, 59-69

Lamberts, K., and Pfeifer, R. (1993). Computational models of expertise: accounting for routine and adaptivity in skilled performance. In: K. Gilhooly, and M. Keane (eds.). Advances in the psychology of thinking. New York: Simon and Schuster.

McClelland, J. and Rumelhart, D. (1986). Parallel distributed processing. Cambridge, MA: MIT Press

McDermott, D. (1992). Robot planning. AI Magazine, Summer 92, 55-79

McFarland, D. (1989). Goals, no-goals and own goals. In A. Montefiore and D. Noble (eds.). Goals, no-goals and own goals. A debate on goal-directed and intentional behavior. London: Unwin Hyman, 39-57

McFarland, D. (1991). Defining motivation and cognition in animals. International Studies in the Philosophy of Science, **5**, 153-170

Neisser, U. (1967). Cognitive psychology. New York: Appleton-Century-Crofts
Newell, A. (1991). Unified theories of cognition. Cambridge, MA: Harvard University Press
Newell, A., and Simon, H.A. (1976). Computer science as empirical inquiry: symbols and search. Comm. of the ACM, **19**, 113-126
Oja, E. (1982). A simplified neuron model as a principal component analyser. Journal of Mathematical Biology, **15**, 267-273
Penrose, R. (1989). The emperor's new mind. Oxford: Oxford University Press
Pfeifer, R. (1993). Emotions in robot design. In: Proc. of the 2nd IEEE International Workshop on Robot and Human Communication. November, 1993, Tokyo, Japan, 408-413
Pfeifer, R., and Verschure, P.F.M.J. (1992a). Beyond rationalism: symbols, patterns and behavior. Connection Science, **4**, 313-325
Pfeifer, R., and Verschure, P.F.M.J. (1992b). Distributed adaptive control: a paradigm for designing autonomous agents, in Toward A Practice of Autonomous Systems: Proceedings of the First European Conference on Artificial Life. Cambridge, MA: MIT Press, 21-30.
Pfeifer, R., and Verschure P.F.M.J. (1993). Designing efficiently navigating non-goal-directed robots. Proceedings SAB-92. From animals to animats, 31-39
Pfeifer, R., and Verschure, P. (1994). The challenge of autonomous systems: pitfalls and how to avoid them. In L. Steels, and R. Brooks (eds.). The artificial life route to artificial intelligence. Cambridge, MA: MIT Press
Putnam, H. (1975). Philosophy and our mental life. In H. Putnam (ed.) Mind, language and reality: philosophical papers, vol. 2. Cambridge: Cambridge University Press
Pylyshyn, Z.W. (1984). Computation and cognition. Toward a foundation for cognitive science. Cambridge, MA: MIT Press, Bradford Books (paperback edition 1986)
Pylyshyn, Z.W. (ed.) (1987). The robot's dilemma. The frame problem in artificial intelligence. Norwood, NJ: Ablex (2nd printing 1988)
Suchman, L. (1987). Plans and situated action. Cambridge University Press
Turing, A. M. (1939). On computable numbers, with an application to the Entscheidungsproblem. Proc. of the London Mathematics Society, (ser. 2) **42**, 230-265, **43**, 544 (1937)
Turing, A.M. (1950). Computing machinery and intelligence. Mind, **59**, 433-460. Reprinted in: E.A. Feigenbaum, and J. Feldman (eds.) (1963). Computers and thought. New York: McGraw-Hill, 11-35
Turkle, S. (1984). The second self. Computers and the human spirit. New York: Simon & Schuster
Varela, F.J., Thompson, E., and Rosch, E. (1991). The embodied mind. Cognitive science and human experience. Cambridge, MA: MIT Press
Verschure, P., and Coolen, A. (1991). Adaptive fields: distributed representations of classically conditioned associations. Network, **2**, 189-206

Verschure, P.F.M.J., Kröse, B.J.A., and Pfeifer, R. (1992). Distributed adaptive control: the self-organization of structured behavior. Robotics and Autonomous Systems, **9**, 181-196

Verschure, P.F.M.J, and Pfeifer, R. (1993). Categorization, representation, and the dynamics of system-environment interaction. Proceedings SAB-92. From animals to animats, 210-217

Webb, B. (1993a). Perception in real and artificial insects: A robotic investigation of cricket phonotaxis. PhD Thesis, University of Edinburgh

Webb, B. (1993b). Modeling biological behavior or 'Dumb animals and stupid robots'. In Proc. of the 2nd European Conference on Artificial Life, 1090-1002

Winograd, T. and Flores, F. (1986). Understanding computers and cognition. Reading, MA: Addison-Wesley

Lifelong Robot Learning[1]

Sebastian Thrun[2] and Tom M. Mitchell[3]

[2] University of Bonn, Institut für Informatik III, Römerstr. 164, 53117 Bonn, Germany
[3] School of Computer Science, Carnegie Mellon University, Pittsburgh, PA 15213, USA

Abstract. Learning provides a useful tool for the automatic design of autonomous robots. Recent research on learning robot control has predominantly focussed on learning single tasks that were studied in isolation. If robots encounter a multitude of control learning tasks over their entire lifetime there is an opportunity to transfer knowledge between them. In order to do so, robots may learn the invariants and the regularities of their individual tasks and environments. This task-independent knowledge can be employed to bias generalization when learning control, which reduces the need for real-world experimentation. We argue that knowledge transfer is essential if robots are to learn control with moderate learning times in complex scenarios. Two approaches to lifelong robot learning which both capture invariant knowledge about the robot and its environments are presented. Both approaches have been evaluated using a HERO-2000 mobile robot. Learning tasks included navigation in unknown indoor environments and a simple find-and-fetch task.

1 Why Learning?

Traditionally, it has often been assumed in robotics research that accurate a priori knowledge about the robot, its sensors, and most important its environment is available. By assuming the availability of accurate models of both the world and the robot itself, the kind of environments such robots can operate in, and consequently the kind of tasks such robots can solve, is limited. Limitations especially arise from four factors:

- **Knowledge bottleneck.** A human designer has to provide accurate models of the world and the robot.
- **Engineering bottleneck.** Even if sufficiently detailed knowledge is available, making it computer-accessible,i.e., hand-coding explicit models of robot hardware, sensors and environments, has often been found to require unreasonable amounts of programming time.
- **Tractability bottleneck.** It was early recognized that many realistic robot domains are too complex to be handled efficiently (Schwartz et al., 1987), (Canny, 1987), (Bylander, 1991). Computational tractability has turned out to be a severe obstacle for determining optimal control for complex robots in complex domains.

---

[1] This paper is also available as Technical Report IAI-TR-93-7, University of Bonn, Dept. of Computer Science III, March 1993.

- **Precision bottleneck.** The robot device must be precise enough to accurately execute plans that were generated using the internal models of the world.

Recent research on autonomous robots has changed the design of autonomous agents and pointed out a promising direction for future research in robotics (see for example (Brooks, 1989) and several papers in (Maes, 1991)). Reactivity and real-time operation have received considerably more attention than, for example, optimality and completeness. Many approaches have dropped the assumption that perfect world knowledge is available–some systems even operate in the extreme where no domain-specific initial knowledge is available at all. Consequently, today's robots are facing gradually unknown, dynamic environments, they have to orient themselves, explore their environments autonomously, recover from failures, and they have to solve whole families of tasks.

If robots lack initial knowledge about themselves and their environments, learning becomes inevitable. The term *learning* refers to a variety of algorithms that are characterized by their ability to replace missing or incorrect world knowledge by experimentation, observation, and generalization. Learning robots thus collect parts of their domain knowledge themselves and improve over time. They are less dependent on a human instructor to provide this knowledge beforehand. Most approaches to robot learning are flexible enough to deal with a whole class of environments, robots and/or tasks (goals). Consequently the internal prior knowledge, if available at all, is often too weak to solve a concrete problem off-line. In order to reach a goal, learning robots rely on the interaction with their environment to extract information. Different learning strategies differ mainly in three aspects: their way to do experimentation (exploration), their way to generalize from the observed experience, and the type and amount of prior knowledge that constrains the space of their internal hypothesizes about the world. There will be no optimal, general learning technique for autonomous robots, since learning techniques are characterized by a trade-off between the degree of flexibility, given by the size of the "gaps" in the domain knowledge, and the amount of observations required for filling these gaps. Generally speaking, the more universal a robot learning approach, the more experimentation we expect the robot to take to learn successfully, and vice versa. However, the advantage of applying learning strategies to autonomous robot agents is obvious: Learning robots can operate in a whole class of initially unknown environments, and they can compensate for changes, since they can re-adjust their internal belief about the environment and themselves. Moreover, all empirically learned knowledge is grounded in the real world. It has long been recognized in AI that learning is a key feature for making autonomous agents capable of solving more complex tasks in more realistic environments. Recent research has produced a variety of rigorous learning techniques that allow a robot to acquire huge chunks of knowledge by itself. See for example (Mel, 1989), (Moore, 1990), (Pomerleau, 1989), (Tan, 1991), (Mahadevan and Connell, 1991), (Lin, 1992b), and (Kuipers and Byun, 1990).

What exactly is the concrete problem addressed by robot learning? Let us outline a general definition of learning robot control. Assume the robot acts in an environment

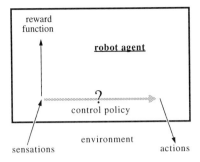

Figure 1 The robot control learning problem. A robot agent is able to perceive the state of its environment by its sensors, and change it using its effectors (actions). A reward function that maps sensations to rewards measures the performance of the robot. The control learning problem is the problem of finding a control policy that generates actions such that the reward is maximized over time.

W (the *world*). Each time step, the environment is in a certain state $z \in Z$. By state we mean the universe of all quantities in the environment that may change during the lifetime of the robot. The robot is able to (partially) perceive the state of its environment through its sensors. It is also able to act using its effectors, as shown in Figure 1. Let S defines the set of all possible sensations, and A the set of all actions that the robot can execute. Actions (including zero-actions, if the robot does nothing) change the state of the world. Hence, the environment can be understood as a mapping $W : Z \times A \longrightarrow Z$ from states and actions to states. For example, imagine an autonomous mobile robot whose task it is to keep the floor clean. The worlds such a robot will be acting in are buildings, including the obstacles surrounding the robot, the floors, the dirt on the floors, humans that walk by, and so on. Appropriate sensors might include a camera mounted on the robot's arm, and the set of all sensations might be the space of all camera images. Actions, which change the robot's location and hence the state of the world, may include "go forward," "turn," "switch on/off vacuum," and "lift arm."

In order to define the goals of the robot, we assume that the robot is given a *reward function* $R : S \longrightarrow I\!R$, that maps sensations to scalar reward values. The reward evaluates the success of the robot to solve its tasks: In the most simple form the reward is positive (say 100), if the robot reaches its goals, it is negative (-100) if the robot fails, and zero otherwise. Positive reward corresponds to pleasure, and negative reward represents pain. In the mobile robot domain, for example, the reward may be positive if there is no dirt on the floor, and negative reward may be received if the robot collides with the furniture or runs out of battery power. The control learning problem is the problem of finding a control function F that generates actions such that the reward is maximized over time. More formally, a *control learning problem* can be described in the following way:

Control learning problem:
$$\langle S, A, W, R \rangle \longrightarrow F : S^* \to A$$
such that F maximizes R over time

This formulation of the robot control learning problem has been extensively studied in the field of "Reinforcement Learning" (Sutton, 1984), (Barto *et al.*, 1991). Thus far, most approaches that emerged from this field have studied robot learning with a minimal set of assumptions: The robot is able to sense, it is able to execute actions, actions have an effect on future sensations, and there is a pre-given reward function that defines the goals of the robot. The goal of reinforcement learning algorithms is to maximize the reward over time. Note that this general definition lacks any specifications about the robot at hand, its environment, or the kind of reward functions the robot is expected to face. Hence, approaches that are able to adaptively solve such a robot control learning problem are *general* learning techniques. Perhaps the most restricting assumption found in most approaches to reinforcement learning is that the robot is able to sense the state of the world reliably. If this is the case, it suffices to learn the policy as a function of the most recent sensation to action: $F : S \to A$, i.e., the control policy is purely reactive. As Barto et al. pointed out (Barto *et al.*, 1991), the problem of learning a control policy can then be attacked by Dynamic Programming techniques (Bellman, 1957).

As it turns out, even if robots have access to complete state descriptions of their environments, learning control in complex robot worlds with large state spaces is practically not feasible. This is because it often takes too much experimental time to acquire the knowledge required for maximizing reward, i.e., to fill the huge knowledge gaps. One might argue that better learning techniques have to be invented that decrease the learning complexity, while still being general. Although there is certainly space for better learning and generalization techniques that reduce the amount of experimentation required, it seems unlikely that such techniques will ever be applicable to complex, real-world robot environments with sparse reward and difficult tasks. The complexity of knowledge acquisition is inherent in the formulation of the problem. Real-world experimentation will be the central bottleneck of any general learning technique that does not utilize prior knowledge about the robot and its environment.

Why do natural agents such as humans learn so much better than artificial agents? Does the learning problem faced by natural agents differ from that of artificial ones? And, if so, might such differences provide possible explanations for the superior learning capabilities of natural agents? We will not attempt to give general answers to these general questions. We will, however, point out the importance of knowledge transfer and lifelong learning problems in order to make robots to learn more complex control.

2 The Necessity of Lifelong Agent Learning

Humans typically encounter a multitude of control learning problems over their entire lifetime, and so may robots. Henceforth, in this article we will be interested in the lifelong (control) learning problem faced by a robot agent, in which it must learn a collection of control policies for a variety of related control tasks. Each of these control problems, $\langle S, A, W_i, R_i \rangle$ involves the same robot with the same set of sensors, effectors, and may vary only in the particular environment W_i, and in the reward function R_i that defines the goal states for this problem. For example, an industrial mobile robot might face multiple learning tasks such as shipping packages, delivering mail, supervising critical processes, guarding its work-place at night, and so on. In this scenario the environment will be the same for all tasks, but the reward function varies. Alternatively, a window-cleaning robot that is able to climb fronts of buildings and move arbitrarily on walls and windows might have the very single task of cleaning windows. It will, however, face multiple fronts and windows (environments) over its lifetime.

In the lifelong learning problem, for each different environment and each reward function the robot agent must find a different control policy, F_i. The lifelong learning problem of the agent therefore corresponds to a set of control learning problems:

Lifelong learning problem:
$$\{\langle S, A, W_i, R_i \rangle \longrightarrow F_i | F_i : S \rightarrow A\}$$
such that F_i maximizes R_i over time

Of course the agent could approach the lifelong learning problem by handling each control learning problem independently. However, there is an opportunity for the agent to do considerably better. Because these control learning problems are defined in terms of the same S, A, and potentially the same W, the agent should be able to reduce the difficulty of solving the i-th control learning problem by using knowledge it acquired from solving earlier control learning problems. For example, a robot that must learn to deliver laser printer output and to collect trash in the same environment should be able to use much of what it has learned from the first task to simplify learning the second. The problem of lifelong learning offers the opportunity for synergy among the different control learning problems, which can speed-up learning over the lifetime of the agent.

Viewing robot learning as a lifelong learning problem motivates research on "bootstrapping" learning algorithms that transfer learned knowledge from one learning task to another. Bootstrapping learning algorithms might start with low-complexity learning problems, and gradually increase the problem solving power to harder problems. Similar to humans, future robots might first have to learn simple tasks (such as low-level navigation, hand-eye coordination), and, once successful, draw their attention to increasingly more difficult and complex learning tasks (such as picking up and delivering laser printer output). As, for example, Singh (Singh, 1992b) and Lin (Lin, 1992b) have demonstrated, learning related control tasks with increasing complexity can result in a remarkable synergy between these control learning tasks,

Figure 2 The robot we used in all experiments described in this article is a wheeled HERO-2000 robot with a manipulator and a gripper. It is equipped with two sonar sensors, one on the top of the robot that can be directed by a rotating mirror to give a full 360° sweep (24 values), and one on the wrist that can be rotated by the manipulator to give a 90° sweep. Sonar sensors return approximate echo distances. Such sensors are inexpensive but very noisy.

and an improved problem solving power. In their experiments, simulated mobile robots were able to solve more complex navigation tasks when the robots were given simpler, related learning tasks beforehand. They also report that their systems were unable to learn the same complex tasks in isolation, pointing out the importance of knowledge transfer for robot learning.

The remainder of the article is organized as follows. In the next two sections, we will briefly present two approaches to the lifelong learning problem. Both approaches have been implemented and evaluated using a HERO-2000 mobile robot with a manipulator shown in Figure 2. In the first approach, called explanation-based neural network learning (EBNN), we will assume that the environment of the agent stays the same in all control learning tasks. This allows the robot to learn task-independent action models. Once learned, these action models provide a means of transferring knowledge across control learning tasks. In EBNN, such action models are used to *explain and analyze* observations. In Section 4, we drop the assumption of static environments and describe a mobile autonomous robot that has to solve the same task in different, related environments. We demonstrate how this robot might learn and transfer environment-independent knowledge that captures the characteristics of its sensors, as well as invariant characteristics of the environments. In Section 5, we will review some related approaches to robot learning which also utilize previously

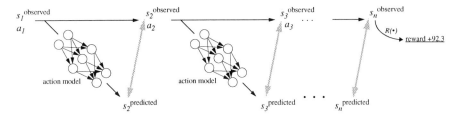

Figure 3 Episode: Starting with the initial state s_1, the action sequence $a_1, a_2, \ldots, a_{n-1}$ was observed to produce the final reward R_n. The robot agent uses its action models, which capture previous experiences in the same environment, to *explain* the observed episode and thus bias learning. See text for an explanation.

learned knowledge. As we will see, there are several types of techniques that can be grouped into categories. The article is concluded by Section 6

3 Explaining Observations in Terms of Prior Experiences

As defined in the previous section, the lifelong (control) learning problem faced by a learning robot agent is the problem of learning collections of tasks in families of environments over the entire lifetime. In this section we will draw our attention to a particular sub-type of lifelong learning problems, namely to the lifelong learning problem of robots that spend their whole life exclusively in the same environment. This restricted class of lifelong learning scenarios plays an important role in autonomous agent research. Many prospective robot agents, such as housekeeping robots, industrial robot arms or artificial insects, face a variety of learning problems in the very same environment. In such scenarios, knowledge about the environment can be reused, since the environment stays the same for each task. The explanation-based neural network learning algorithm (EBNN), which is presented in this section, transfers task-independent knowledge via learned action models which are learned empirically during problem solving.

3.1 Learning Action Models

Robots observe their environments (and themselves) by the effects of their actions. In other words, each time an action is performed, the robot may use its sensors to sense the way the world has changed. Let us assume for simplicity that the robot is able to accurately sense the state of the world. As pointed out in the previous section, learning control reduces to learning a reactive controller[4]. If the environment is sufficiently predictable, neural network learning techniques such

---

[4] We intentionally avoid the complex problem of incomplete and noisy perception perception, since the algorithm presented in this section is kind of orthogonal to research on these issues.

as the Back-propagation training procedure (Rumelhart *et al.*, 1986) can be used to model the environment. More specifically, each time an action is performed, the previous state denoted by s, the action a and the next state s' form a training example for the action model network, denoted by M:

$$\langle\, s, a\, \rangle \longrightarrow s'$$

Action models[5] are thus functions of the type $M : S \times A \longrightarrow S$. Since we assume that the environment is the same for all tasks, action models capture invariants that allow for transferring knowledge from one task to another.

Each individual control learning problem requires a different policy, i.e., learning a control function $F_i : S \longrightarrow A$ that, when employed by the agent, maximizes the corresponding reward R_i. In general, F_i is difficult to learn directly. Following the ideas of reinforcement learning ((Samuel, 1959), (Sutton, 1988), (Barto *et al.*, 1990), (Watkins, 1989)), we approach the problem of learning F_i by learning an *evaluation function*, Q_i, defined over states and actions.

$Q_i : S \times A \longrightarrow I\!R$, where $Q_i(s, a)$ is the expected future cumulative reward achieved after executing action a in state s.

Suppose the agent has already learned an accurate evaluation function Q_i. Then, it can easily use this function to select optimal actions that maximize reward over time. Given some state, s, which the agent finds itself in, it computes its control action a^* by considering its available actions to determine which of them produces the highest Q_i value:

$$a^* = \mathrm{argmax}_{a \in A}\ Q_i(s, a)$$

Since $Q_i(s, a)$ measures the future cumulative reward, a^* is the optimal action. The problem of learning a policy is thus reduced to the problem of learning an evaluation function.

How can an agent *learn* the evaluation function Q_i? To see how training data for learning Q_i might be obtained, consider the scenario depicted in Figure 3. Starting at the initial state s_1, the robot agent performs the action sequence $a_1, a_2, \ldots, a_n$. After action a_n it receives the reward $r_n = 92.3$ which, in this example, indicates that the action sequence was considerably successful. At first glance, this episode can be used by the agent to derive training examples for the evaluation function Q_i by associating the final reward[6] with each state-action pair:

$$\langle\, s_1, a_1\, \rangle \longrightarrow r_n = 92.3$$

See (Bachrach and Mozer, 1991), (Chrisman, 1992), (Lin and Mitchell, 1992), (Mozer and Bachrach, 1989), (Rivest and Schapire, 1987), (Tan, 1991), (Whitehead and Ballard, 1991) for approaches to learning with incomplete perception.

[5] See for example (Barto *et al.*, 1989), (Jordan, 1989), (Munro, 1987), (Thrun, 1992) for more approaches to learning action models with neural networks.

[6] For simplification of the notation, we assume that reward will only be received at the end of an episode. EBNN can be applied to arbitrary reward functions. See (Mitchell and Thrun, 1993b) for more details.

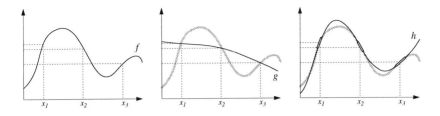

Figure 4 Fitting slopes: Let f be a target function for which three examples $\langle x_1, f(x_1) \rangle$, $\langle x_2, f(x_2) \rangle$, and $\langle x_3, f(x_3) \rangle$ are known. Based on these points the learner might generate the hypothesis g. If the output-input derivatives are also known, the learner can do much better: h.

$$\langle s_2, a_2 \rangle \longrightarrow r_n = 92.3$$

$$\vdots \tag{1}$$

$$\langle s_{n-1}, a_{n-1} \rangle \longrightarrow r_n = 92.3$$

An inductive learning method, such as the Back-propagation training procedure, can use such training examples to learn Q_i. As the number of training examples grows, the agent's internal version of Q_i will improve, resulting in it choosing increasingly more effective actions.[7] Notice that for each control learning problem $\langle S, A, W_i, R_i \rangle$, the agent must learn a distinct Q_i, since the reward may differ for different tasks.

3.2 The Explanation-Based Neural Network Learning Algorithm

How can the agent use its previously learned knowledge, namely the neural network action models, to guide learning of the evaluation function Q_i? Since we assume that the environment is the same for all individual control learning problems, neural network action models capture important domain knowledge that is independent

---

[7] Note that more sophisticated learning schemes for learning evaluation functions have been developed. In his dissertation, Watkins (Watkins, 1989) describes Q-Learning, a scheme for learning evaluation function $Q_i(s_k, a_k)$ *recursively*. In Q-Learning training patterns are derived based on the maximum possible Q value at the next state: $\langle s_k, a_k \rangle \longrightarrow \max_a Q(s_{k+1}, a)$. Indeed, in all our experiments we applied a linear combination of Watkins' recursive scheme and the non-recursive scheme described in the paper. This combination is strongly related to Sutton's $TD(\lambda)$ algorithms (Sutton, 1988). Since the exact procedure is not essential for the ideas presented in this article, we will omit any details. A second extension, also widely used, is to discount reward over time. If actions are to be chosen such that the number of actions is minimal, reward is typically discounted with a *discount factor* $\gamma \leq 1$. The resulting control policy consequently prefers sooner reward to more distant reward. See (Mitchell and Thrun, 1993b) or (Thrun and Mitchell, 1993) for a more detailed description of these issues in the context of EBNN.

of the particular control learning problem at hand. In the explanation-based neural network learning algorithm (EBNN), the agent uses these action models to bias learning of the control functions.

EBNN works as follows. Suppose the robot faces the control learning problem number i, i.e., it has to learn the evaluation function Q_i. The learning scheme described above provides training values for the desired evaluation function. Repetitive experimentation allows the robot to collect enough data to learn the desired Q_i. However, this process does not utilize the knowledge represented in the action models. Assume the agent has already learned accurate action models that model the effect of actions on the state of the environment. Of course, these action models will only be approximately correct, since they are learned inductively from a finite amount of training data. In EBNN, the agent employs these action models to explain, analyze and generalize better from the observed episodes. This is done in the following three steps:

1. **Explain.** An *explanation* is a post-facto prediction of the observed state-action sequence (DeJong and Mooney, 1986), (Mitchell *et al.*, 1986), (Mitchell and Thrun, 1993a). Starting with the initial state-action pair (s_1, a_1), the agent post-facto predicts subsequent states up to the final state s_n using its neural network action models. Since the action models are only approximately correct, predictions will deviate from the observed states.

2. **Analyze.** Having explained the whole episode, the explanation is *analyzed* to extract further information that is useful for learning the evaluation function Q_i. In particular, the agent analyzes how small changes of the state features will affect the final reward, and thus the value of the evaluation function. This is done by extracting *partial derivatives (slopes)* of the target function Q_i with respect to the observed states in the episode: First, the agent computes the partial derivative of the final reward with respect to the final state s_n. Notice that the reward function $R(s)$, including its derivative, is assumed to be given to the agent. These slopes are now propagated backwards through the chain of action model inferences. Neural network action models represent differentiable functions. Using the chain rule of differentiation, the agent computes the partial derivative of the final reward with respect to the preceding state s_{n-1} by multiplying the partial derivative of the reward with the derivative of the neural network action model. This process is iterated, yielding all partial derivatives of the final reward along the whole episode:

$$\frac{\partial Q_i}{\partial s_n}(s_{n-1}, a_{n-1}) \approx \frac{\partial R}{\partial s}(s_n) \cdot \frac{\partial M}{\partial s}(s_{n-1}, a_{n-1})$$

$$\vdots$$

$$\frac{\partial Q_i}{\partial s_1}(s_1, a_1) \approx \frac{\partial R}{\partial s}(s_n) \cdot \frac{\partial M}{\partial s}(s_{n-1}, a_{n-1}) \cdot$$
$$\cdots \cdot \frac{\partial M}{\partial s}(s_2, a_2) \cdot \frac{\partial M}{\partial s}(s_1, a_1)$$

Here $M : S \times A \longrightarrow S$ denotes the neural network action model. The reward-state slopes analyze the importance of the state features for the final reward.

State features believed (by the action models) to be irrelevant for achieving the final reward will have partial derivatives of zero, whereas large derivative values indicate the presence of strongly relevant features.

3. **Learn.** The analytically extracted slopes approximate the slopes of the target evaluation function Q_i. Figure 4 illustrates the importance of the slope information of the target function. Suppose the unknown target function is the function f depicted in Figure 4a, and suppose that three training examples are given: x_1, x_2 and x_3. An arbitrary continuous function approximator, for example a neural network, might hypothesize the function g shown in Figure 4b. If the slopes at these points are known as well, then the resulting function might be much better, as illustrated in Figure 4c.

EBNN uses a combined learning scheme utilizing both types of training information. The target values (c.f. Equation 2) for learning Q_i are generated from observation, whereas the target slopes, given by

$$\nabla \langle s_{n-1}, a_{n-1} \rangle \longrightarrow \frac{\partial R}{\partial s}(s_n) \cdot \frac{\partial M}{\partial s}(s_{n-1}, a_{n-1})$$

$$\vdots$$

$$\nabla \langle s_1, a_1 \rangle \longrightarrow \frac{\partial R}{\partial s}(s_n) \cdot \prod_{k=1}^{n-1} \frac{\partial M}{\partial s}(s_k, a_k),$$

are extracted from analyzing the observations using domain knowledge acquired in previous control learning tasks. Both sources of training information, the target values and the target slopes, are used to update the weights and biases of the target network.[8] Consequently, the domain knowledge is used to bias the generalization. Since this bias is knowledgeable, it will partially replace the need for real-world experimentation, hence accelerate learning.

3.3   Accommodating Imperfect Action Models

Initial experiments with EBNN on a simulated robot navigation task showed a significant speedup in learning when the robot agent had access to highly accurate action models (Mitchell and Thrun, 1993b). If the action models are not sufficiently accurate, however, the robot performance can seriously suffer from the analysis. This is because the extracted slopes might very well be wrong and mislead generalization. This observations raises an essential question for research on lifelong agent learning and knowledge transfer: How can a robot agent deal with incorrect prior knowledge? Clearly, if the agent lacks training experience, any inductively learned bias might be poor and misleading. But even in the worst case, a learning mechanism that employs previously learned domain knowledge should not take more time for learning control

---

[8] As Simard and colleagues pointed out, the Back-propagation algorithm can be extended to fit target slopes as well as target values (Simard *et al.*, 1992). Their algorithm *tangent prop* incrementally updates weights and biases of a neural network such that both the value and the slope error are simultaneously minimized.

than a learning mechanism that does not utilize prior knowledge at all. How can the learner avoid the damaging effects arising from poor prior knowledge?

In EBNN, malicious slopes are identified and their influence is gradually reduced. More specifically, the accuracy of the extracted slopes is estimated based upon the observed prediction error of the action models. For example, if the action models perfectly post-facto predict the observed episode, the estimated accuracy of all slopes will be 1. Likewise, if for some of the action models in the chain of model derivatives have produced inaccurate state predictions, the corresponding estimated accuracy will be close to zero. The accuracies of the slopes are used as a trade-off factor between value and slope fitting when training the target network. Since tangent prop allows to weight each training pattern individually, the estimated accuracies can be used to determine the ratio with which value learning and slope learning are weighted when learning the target concept. More specifically, in EBNN the step-size for weight updates is multiplied by the estimated slope accuracy when learning slopes.

As illustrated elsewhere (Mitchell and Thrun, 1993b), weighting slope training by their accuracies was found to successfully reduce the impact of malicious slopes resulting from inaccurate action models. We evaluated EBNN using nine different sets of action models that were trained with different amounts of training data. With well-trained action models in the simulated robot domain, the same speedup was observed as before. With increasing inaccurate action models, the performance of EBNN approached that of standard reinforcement learning without knowledge transfer. In these experiments, EBNN degraded gracefully with increasing errors in the action models. These results are intriguing, because they indicate the feasibility of lifelong learning algorithms that are able to benefit from previously learned bias, even if this bias is poor.

3.4   A Concrete Example: Learning to Pick up a Cup

In what follows, we will present some initial results obtained with our HERO-2000 robot. Thus far, we predominantly investigated the effect of previously learned knowledge on the learning speed for new control learning tasks. We therefore trained the action models beforehand with manually collected training data. The robot agent had to learn a policy for approaching and grasping a cup. The robot at hand, shown again in Figure 5, used a hand-mounted sonar sensor to observe its environment. In our experiments, sonar data was pre-processed to estimate the direction and distance to the closest object, which was used as the world state description. The robot's actions were `forward(inches)`, `turn(degrees)`, and `grab`. Positive reward was received for successful grasps, and a penalty was imposed for unsuccessful grasps as well as for losing sight of the cup. Actions were modeled by three separate networks, one for each action. The networks for the parameterized actions `forward(inches)` and `turn(degrees)` predicted the distance and the orientation to the cup (one hidden layer with 6 units), whereas the model for `grab` predicted the probability that a pre-given, open-loop grasping routing would man-

Figure 5 The robot uses its manipulator and a sonar sensor to sense the distance to a cup and to pick it up.

age to pick up the cup (four hidden units). All action models were pre-learned from approximately two hundred training episodes containing an average of five steps each, which were manually collected beforehand. Figure 6 shows as an example the action model for the `grab` action. Since this particular action model modeled the probability of success of the grasping routine, the reward function was simply the identity mapping $R(s) = s$ (with the constant derivative 1). The evaluation function Q was also modeled by three distinct networks, one for each action (8 hidden units each). After learning the action models, the six training episodes for the evaluation networks shown in Figure 7 were provided by a human teacher who controlled the robot. We applied a version of $TD(\lambda)$ (Sutton, 1988) with $\lambda = 0.7$ and Watkins' Q-Learning (Watkins, 1989) with experience replay (Lin, 1992a) for learning control.

Figure 8 illustrates the learned Q function for the `grab` action with (right row) and without (left row) employing the action models and EBNN. In this initial stage of learning, when little data is yet available, the generalization bias imposed by the pre-learned action models is apparent. Although none of the Q functions has yet converged, the Q functions learned using EBNN have a shape that is more correct, and which is unlikely to be guessed based solely on the few observed training points with no initial knowledge. For example, even after presenting six episodes the plain learning procedure predicts positive reward solely based upon the angle of the cup, whereas the combined EBNN method has already learned that grasping will fail if the cup is too far away. This information, however, is not represented in the training episodes (Figure 7), since there is no single example of an attempt to grasp a cup

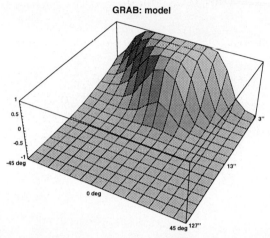

Figure 6 Action model for the action model `grab`. The x and y axis measures again the angle and distance to the cup. The z axis plots the expected success of the `grasp` action, i.e., the probability that the grasping succeeds.

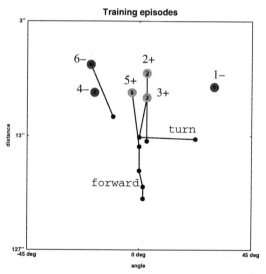

Figure 7 Six training episodes for learning control, labeled by 1 to 6. The horizontal axis measures the angle of the cup, relative to the robot's body, and the vertical axis measures the distance to the cup in a logarithmic scale. Successful grasps are labeled with "+", unsuccessful with "−". Notice that some of the episodes included forwarding and turning.

far away. It rather seems that the slopes of the model were "copied" into the target

GRAB: Q-function

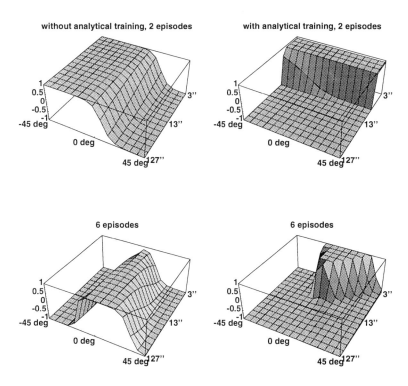

Figure 8 Evaluation functions for the action `grab` after presenting 2 (upper row) and 6 (lower row) training episodes, and without (left column) and with (right column) using the action model for biasing the generalization.

evaluation function. This illustrates the effect of the slopes in EBNN: The evaluation functions learned with EBNN discovered the correlation of the distance of the cup and the success of the `grab` action from the neural network action model. If these action models are learned in earlier control learning tasks, there will be a significant synergy effect between them.

The EBNN results in this section are initial. They are presented because they indicate that task-independent knowledge, once learned, can be successfully transfered by EBNN when learning the grasping task. They are also presented since they evaluated EBNN in a real robot domain, unlike the results presented in (Mitchell and Thrun, 1993b), (Thrun and Mitchell, 1993). However, thus far we did not collect enough training data to learn a complete policy for approaching and grasping the cup with either method. Future research will characterize the synergy during the full course of learning until convergence. It will also experiment with families of related tasks, such as approaching and grasping different objects, including cups that lie on the side.

3.5 EBNN and Lifelong Robot Learning

What lesson does EBNN teach us in the context of lifelong robot learning? In this section we made the restricting assumption that all control learning problems of the robot agent play in the very same environment. If this is the case, any type of models of the robot and its environment are promising candidates for transferring task-invariant knowledge between the individual control learning problems. In EBNN, task-independent information is represented by neural network action models. These models bias generalization when learning control via the process of explaining and analyzing observed episodes. EBNN is a method for lifelong agent learning, since it learns and re-uses learned knowledge that is independent of the particular control learning problem at hand. Although the initial experiments described in this article do not fully demonstrate this point, EBNN is able to efficiently replace real-world experimentation by previously learned bias. In related experiments with a simulated robot presented elsewhere (Mitchell and Thrun, 1993b), we observed a significant speed-up by a factor of 3 to 4 when pre-learned action models were employed.

It is important to mention that we expect EBNN to be even more efficient if the dimension of the state space is higher. This is because for each single observation EBNN extracts a d-dimensional slope vector from the environment, if d if the dimension of the input space. Our conjecture is that with accurate domain theories the generalization improvement scales linearly with the number of instance features (e.g., we would expect a three order of magnitude improvement for a network with 1000 input features), since each derivative extracted by EBNN can be viewed roughly as summarizing information equivalent to one new training example for each of the d input dimensions. This conjecture is roughly consistent with our experiments in the simulated robot domain. The relative performance improvement might increase even more if higher-order derivatives are extracted, which is not yet the case in EBNN. It is important to notice, however, that the quality of the action models is crucial for the performance of EBNN. With inaccurate action models, EBNN will not perform better than a purely inductive learning procedure. This does not surprise. Approaches to the lifelong learning problem will always be at most as efficient as non-transferring approaches the robot solves its first control learning problem in its life. The desired synergy effect occurs later, when the agent has learned an appropriate domain-specific bias, hence can reason more rationally about its observations.

4 Lifelong Learning in Multiple Environments

In the previous section we focussed on a certain type of lifelong robot learning problems, namely problems which deal with a single environment. We will now focus on a more general type of lifelong robot learning scenarios, in which the environments differ for the individual control learning tasks. As pointed out in Section 2, there are quite a few robot learning scenarios where a robot has to learn control in whole families of environments. For example, a vacuum cleaning robot might have to learn to clean different buildings. Alternatively, an autonomous vehicle might have to learn navigation in several types of terrain.

Multi-environment scenarios provide less possibilities for transferring knowledge. At a first glance, transferring knowledge seems hard, if the environment is not the same for all control learning tasks. But even in this type problems there are invariants that may be learned and used as a bias. The key observation is that all lifelong learning scenarios involve the same robot, the same effectors, the same sensors, although they might have to deal with a variety of environments and control learning tasks therein. Approaches to this type of lifelong learning problems thus aim at learning the characteristics of the sensors and effectors of the robot, as well as invariants in the environments, if there are any. The principle of learning and transferring task-independent knowledge is the same as in the previous section–just the type of knowledge that is transfered differs.

In the remainder of this section we will not describe a general learning mechanism, but a particular approach to learning the characteristics of the robot's sensors and the environments. Using the HERO-2000 robot as a testbed, we will demonstrate how inverse sensor models can represent a knowledgeable bias which is somewhat independent of the particular environment at hand.

4.1 Learning to Interpret Sensations

What kind of invariants can be learned and transfered across multiple environments? Two observations are crucial for the approach described here. First, the robot and its sensors are the same for each environment. Second, there might be regularities in the environments that can be learned as well. In this section we will describe a neural network approach to learning the characteristics of the robot's sensors, as well as those of typical indoor environments.

The task of the mobile HERO-2000 robot is to explore unknown buildings (Thrun, 1993). Facing a new indoor environment such as the laboratory environment depicted in Figure 9, the robot has to wander around and to use its sensors to avoid collisions. In the exploration task the robot uses two of its sensors: A rotating sonar sensor is mounted on the head of the robot, as shown in Figure 2. The robot also monitors its wheels encoders to detect stuck or slipping wheels. Negative reward is received for collision which can be detected using the wheel encoders. Positive reward is received for entering regions where the robot has not been before. Initially, the robot does not possess any knowledge about its sensors and the environments it will face throughout its life. Sensations are uninterpreted 24-dimensional vectors of floats (sonar scans), along with a single bit that encodes the state of the wheels. In order to simplify learning, we assume that the robot has access to its x-y-θ coordinates

Figure 9 The robot explores an unknown environment. Note the obstacle in the middle of the laboratory. Our lab causes many malicious sonar values, and is a hard testbed for sonar-based navigation. For example, some of the chairs absorb sound almost completely, and are thus hard to detect by sonar.

in a global reference frame (θ measures the orientation of the robot).[9] Navigation in unknown environments is clearly a lifelong robot learning problem. Initially, the robot has to experience collisions, since the initial knowledge does not suffice to prevent from them. Collisions will be penalized by negative reward. In order to transfer knowledge, the robot then has to learn how to interpret its sensors in order to prevent collisions. This knowledge is re-used for each environment the robot will face over its lifetime. After some initial experimentation, the robot should be able to maneuver in new, unknown world while successfully avoiding collisions with obstacles.

We will now describe a pair of networks which learn sensor-specific knowledge that can be re-used across multiple environment. The *sensor interpretation function*, denoted by $\mathcal{R}$, maps sensor information—in our case a sonar scan—to reward information. More specifically, $\mathcal{R}$ evaluates for arbitrary locations close to the robot the probability for a collision, based on a single sonar scan. Figure 10a shows $\mathcal{R}$. Input to the network is a vector of sonar values, together with the coordinates of the query point (Δx, Δy) relative to the robot's local coordinate frame. The output of the network is 1, if the interpretation predicts a collision for this point, and 0, if the network

---

[9] If the robot wheels are perfect, this x-y-θ position can be calculated internally by dead reckoning. The robot at hand, however, is not precise enough, and after 10 to 20 minutes of operation the real coordinated usually deviate significantly from the internal estimates. An approach to compensate such control errors is described in (Thrun, 1993).

(a)

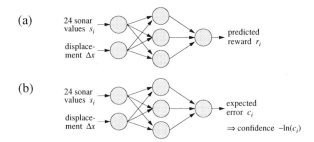

(b)

Figure 10 Task-independent knowledge: (a) sensor interpretation network $\mathcal{R}$ and (b) confidence network $\mathcal{C}$.

predicts free-space. This function can be learned by standard supervised learning algorithms, if the robot keeps track of all sensor readings and of all locations where it collided (and where it did not collide). As the robot operates, it constructs maps that label occupied regions and free-space, which is used to form training examples for the sensor interpretation network $\mathcal{R}$. As usual, the robot uses Back-propagation to learn $\mathcal{R}$.

In order to prevent the robot hardware from real-world collisions, we designed a robot simulator and used it for generating the training patterns for the interpretation network. In the simulator the whole robot environment is known, and training examples that map sensations to occupancy information can be generated easily. In a few minutes the simulated robot explored its simulated world, collecting a total of 8 192 training examples. Six examples for sensor interpretation using $\mathcal{R}$ are shown in Figure 11a. The circle in the center represents the robot, and the lines orthogonal to the robot represent distances sensed by the sonars. The probability of negative reward, as predicted by the $\mathcal{R}$, is displayed in the circular regions around the robot: The darker the region, the higher the probability of collision. As can be seen from this figure, the network $\mathcal{R}$ has successfully learned how to interpret sonar signals. If sonar values are small (meaning that the sonar signal bounced back early), the network predicts an obstacle nearby. Likewise, large value readings are interpreted as free-space. The network has also learned invariants in the training environments. For example, "typical" sizes of walls are known. In Figure 11a-1, for example, the robot predicts a long obstacle of a certain width, but behind this obstacle it predicts a considerably low probability for collision. At first glance, this prediction surprises, given that sonar sensors cannot "see through obstacles." In the training environments, however, regions behind walls happened to be free fairly often, which explains the X-ray predictions by the network. This provides clear evidence that the sensor interpretation network does not only represent knowledge about the robot's sensors, but also knowledge about certain invariants of the environments at hand.

(a) sensor interpretation

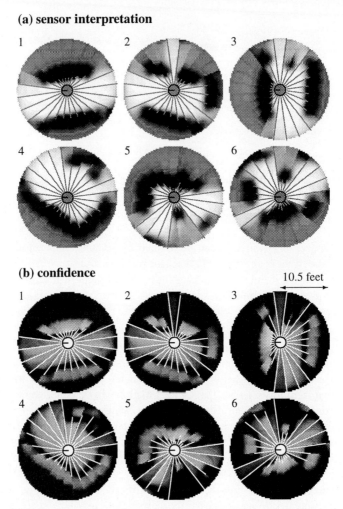

(b) confidence

10.5 feet

Figure 11 Capturing domain knowledge in the networks $\mathcal{R}$ and $\mathcal{C}$. (a) Sensor interpretations and (b) confidence values are shown for the following examples: 1. hallway, 2. hallway with open door. 3. hallway with human walking by, 4. corner of a room, 5. corner with obstacle, 6. several obstacles. Lines indicate sonar measurements (distances), and the region darkness represents in (a) the expected collision reward for surrounding areas (dark values indicate negative reward), and in (b) the confidence level (dark values indicate low confidence).

4.2 Building Maps

We will now motivate the need for a second network, the so-called *confidence network $\mathcal{C}$*, which is related to $\mathcal{R}$. As can bee seen from Figure 11a, a single sonar scan can be used to build a small local map around the robot. If the robot moves

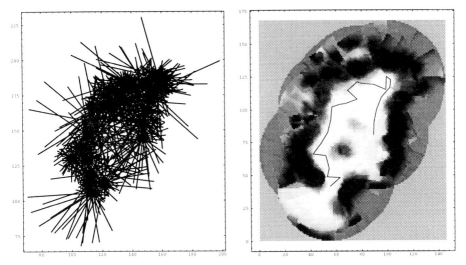

Figure 12 Map building: (a) Raw, noisy sensor input on an exploration path with 27 measurements. (b) Resulting model, corresponding to the lab shown in Figure 1b (lab doorway is toward bottom left). The path of the robot is also plotted (from right to left), demonstrating the exploration of the initially unknown lab. In the next steps, the robot will pass the door of the lab and explore the hallway.

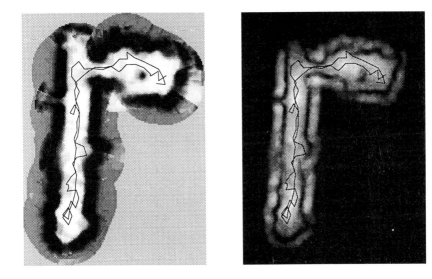

Figure 13 Lab and floor: (a) model, and (b) confidence map.

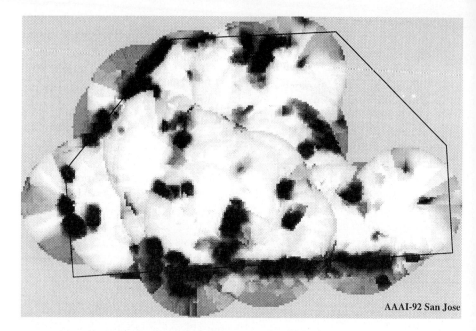

AAAI-92 San Jose

Figure 14 Map compiled during the first AAAI autonomous robot competition in San Jose, July 1992. The map reflects the knowledge of the robot after Stage 3 of the competition. In this and the previous stage, the robot had to find and approach ten visually marked poles, while avoiding collision with obstacles. Lines mark the boundary of the competition area. The map is somehow inaccurate since it was built on two separate days, and the locations of the robot and the obstacles were not quite identical. We compensated for such inaccuracies by decaying the confidence over time.

around and takes several readings, the resulting interpretations can be combined to form a *map*[10] of the world. However, there will be points in the world where the interpretations from multiple sensations will disagree. There are two main reasons for conflicting interpretations: First, sonar sensors, like any sensor, are noisy devices. They often fail to detect certain objects such as smooth surfaces or objects with absorbing surfaces like some of our office chairs. Second, sonars are blind for areas behind obstacles. Hence interpretations for regions behind an obstacle will be inaccurate–they truly reflect the average probability of occupancy (the *prior*). This observation makes it necessary to design a mechanism that resolves conflicts between different predictions.

We will approach this problem by explicitly modeling the reliability of the interpretations, and using the reliability as weighting factor when combining multiple

---

[10] See (Moravec, 1988) and (Elfes, 1987) for related approaches to map building and robot navigation.

interpretations. More specifically, consider a testing phase for the interpretation network $\mathcal{R}$. For some of the testing examples, $\mathcal{R}$ will manage to predict the target value closely. For others, however, there will be a significant residual error for the reasons given above. This error is used to train a second network $\mathcal{C}$, which is called *confidence network*. The input to $\mathcal{C}$ is the same as the input to $\mathcal{R}$. The target output is the (normalized) prediction error of $\mathcal{R}$. After training, $\mathcal{C}$ thus predicts the expected deviation of the sensor interpretations, denoted by $\mathcal{C}(s, \Delta x, \Delta y)$. The confidence into these interpretation $\mathcal{R}(s, \Delta x, \Delta y)$ is thus given by $-\ln \mathcal{C}(s, \Delta x, \Delta y)$. When multiple sensor interpretations are combined, the individual interpretation values are averaged, weighted by their confidence value. Figure 11b shows confidence values for the interpretations showed in Figure 11a. Here dark regions correspond to low confidence. Likewise, light regions indicate high confidence. As can be seen from these examples, the confidence in regions behind obstacles is generally low. Low confidence is also predicted for boundary regions between free-space and obstacles, This does not surprise, since sonar values detect objects in a 15° cone. They do not tell *where* in the cone an object is found. Consequently, the fine-structure of objects is hard to predict. Similar to the sensor interpretation network, the confidence network represents knowledge about the sensors and the environments of the robot that can be transfered across environments. It is important to notice that both networks, $\mathcal{R}$ and $\mathcal{C}$, represent learned knowledge that is independent of the particular environment at hand. These networks act as a bias in the modeling process, when the robot constructs an internal model of a new environment.

The Figures 12 to 14 show some example maps that were obtained for different environments. We used the networks $\mathcal{R}$ and $\mathcal{C}$ to find models of our lab (9). A simple non-adaptive algorithm was used for exploration that basically planned minimal-cost plans to the closest unexplored region. Models of the lab and the hallway are shown in Figures 12 and 13. Although both networks were trained in simulation, they successfully prevented the robot from colliding with obstacles. The same networks $\mathcal{R}$ and $\mathcal{C}$ were used on a autonomous robot competition, that was held during the AAAI conference in July 1992 in San Jose. Here the task and environment differed from the environments the robot had seen previously: The environment was a large arena filled with paper boxes, and the task was to find and to navigate to 10 visually marked poles. The robot had no information about the location of the obstacles and the poles. It had to explore and model the environment by itself. The final implementation (we named the robot "Odysseus") was far more complex than what is described here. Odysseus' navigation was map-based, and the adaptive map building procedure was a component of Odysseus' control. The robot employed the same networks that were generated by the simulator and tested in our experiments in the lab. After a total of 40 minutes operation during two separate stages of the competition the robot produced the map shown in Figure 14. The networks produced maps that were accurate enough to protect the robot from any collision with an obstacle. They also allowed the robot to find close-to-optimal paths to the goal locations.

4.3 Sensor Interpretation and Lifelong Robot Learning

What is the contribution of this approach to the problem of lifelong agent learning? Of course, learning sensor interpretation does not necessarily have to be viewed as a method for learning bias, and the presented method is by no means a general learning scheme for lifelong learning problems. However, for the purpose of this article we will characterize map building in terms of lifelong robot learning. Obviously in each new environment the robot clearly has to learn control. This is because initially, when the robot faces a new, unknown environment, its knowledge does not suffice to generate actions that maximize reward. The robot then gradually learns a policy for action generation step-by-step, and the internal maps provide the freedom for learning, the knowledge "gaps" filled during the course of interaction with the world. Building internal two-dimensional occupancy maps, together with the static planning routine that generates action using these maps, is learning control. Thus, learning neural network sensor interpretations offers a promising perspective for research in lifelong robot learning in multiple environments. Although both the environments and the goals differed for the individual robot control tasks in our experiments, we have demonstrated that knowledge transfer could drastically reduce real-world experimentation. The robot did know about collisions in both real-world environments without ever having experienced one, solely based on previously learned knowledge. We believe that methods which acquire and utilize models of the sensors, effectors (not demonstrated here), and invariants in the environments are promising candidates for efficient robot learning in more complex lifelong learning scenarios.

5 Related Work

Approaches to knowledge transfer in the Lifelong Learning Problems can be viewed as techniques for acquiring function approximation bias. Various researchers have noted the importance of learning bias and transferring knowledge across multiple robot control learning tasks. They can roughly be grouped into the following categories:

1. **Learning models.** Action models are perhaps the most straightforward way to learn and transfer task-independent knowledge, if all individual control learning tasks deal with a single environment. Approaches that utilize action models differ in the type of action models they employ, and the way the action models are used to bias learning control. Sutton (Sutton, 1990) presents a system that learns action models, like EBNN. He uses these models for synthesizing hypothetical experiences that refine the control policy. In his experiments he found a tremendous speedup in learning control when using these action models. EBNN differs from this approach in that it uses its action models for explaining real-world observations, and in that provides a mechanism to recover from errors in the action

models. Lin (Lin, 1992a) describes a mechanism where past experience is memorized and repeatedly replayed when learning control. The collection of past experience forms a non-generalizing action model. Lin also reports significant speedups when using his replay mechanism. As mentioned above, experience replay was also used for neural network training in EBNN. Thus far, there has been little research on learning models that can act as a bias in lifelong learning problems with multiple robot environments.

2. **Learning behaviors and abstractions.** A second way of learning and transferring knowledge across tasks are behaviors. Behaviors are controllers (policies) with low complexity–Often the term behavior refers to reactive controllers. Reinforcement learning, for example, can be viewed as a technique for learning behaviors. Behaviors form abstract action spaces, since the basic actions of the robot might be replaced by the action of invoking a behavior. Thus, with appropriate behaviors abstract action spaces can be formed, and hierarchies of actions can be identified. Learning behaviors accelerates learning control by restricting the search space of all possible policies, mainly for two reasons. First, the number of behaviors is often smaller than the number of actions. Second, behaviors typically are selected for longer periods of time. The latter argument is usually more important and provides a stronger bias for learning control.

Singh (Singh, 1992b), (Singh, 1992a) reports a technique to learn complex tasks by first learning controllers for simple tasks based on reinforcement learning. These controllers represent reactive behaviors. The high-level policy is learned in the abstract action space formed by the low level behaviors. In order to represent even the high-level controller as a purely reactive function, Singh makes several restricting assumptions on the type of tasks and sensor information. In his doctoral thesis, Lin (Lin, 1992b) describes a related scheme for learning behaviors, action hierarchies and abstraction. He assumes that a human instructor teaches a robot a set of elemental behaviors which suffice for all tasks which the robot will face over its lifetime. Unlike Singh, his approach does not guarantee that optimal controller can be learned in the limit. Recently, Dayan and Hinton (Dayan and Hinton, 1993) proposed a system that uses a pre-given hierarchical decomposition to learn control on different levels of abstraction. Each level of abstraction differs in the grain-size of the sensory information, resulting in differently specialized controllers. Since their system learns reactive controllers on each level using reinforcement learning algorithms, it is unclear for what type of problems this procedure will learn successfully, since essential information may be missed when providing incomplete sensor information to purely reactive controllers.

3. **Learning inductive function approximation bias.** Another, more straightforward approach to learning and transferring knowledge is to learn the inductive bias of the function approximators used for learning control directly. Atkeson (Atkeson, 1991) presents a scheme for learning distance measures for instance-based, local approximation schemes. In his algorithm, scaling factors are learned that allows to weight different input features differently. Sutton (Sutton, 1992)

reports a family of learning schemes that allow to learn inductive bias similar to Kalman filters (Kalman, 1960). Although he did not describe his methods in the context of learning control, his research has been motivated by transferring knowledge across multiple control learning tasks.

4. **Learning representations.** Representations, together with inductive bias, determine the way a function approximator generalizes from examples. Many researchers have focussed on learning appropriate representations in order to learn bias. For example, Pratt (Pratt, 1993) describes several approaches that allow to re-use learned representations in hidden units of neural networks. Although she could empirically demonstrate that this transfer could significantly reduce the number of training epochs required for the convergence of the Back-propagation algorithm, she only found occasional improvements in the generalization. A similar technique is reported by Sharkey and Sharkey (Sharkey and Sharkey, 1992). Some researchers have studied knowledge transfer if several tasks are learned simultaneously. For example, Suddarth and Kergosien (Suddarth and Kergosien, 1990) demonstrated that multiple learning tasks of certain types can successfully guide and improve generalization. In his approach, he gives *hints* to neural networks in form of additional output units that learn a closely related task. These hints constrain the internal representation developed by the network. In a more general way, Caruana (Caruana, 1993) recently proposed to learn whole collections of tasks in parallel, using a shared internal representation. He conjectures that multi-task learning will make neural network learning algorithms scale to more complex learning tasks.

Both approaches to the lifelong learning problem described in this article fall into the first category. In EBNN, the robot learns action models which bias generalization in the evaluation functions. Sensor interpretation functions are inverse models of the sensors and the environments of the robot. In the robot exploration tasks, the robot thus learns models of itself and typical aspects of its environments, which bias the construction of the individual maps. As pointed out earlier, action models are well-suited if the environment is the same for all learning tasks, whereas sensor models are appropriate for lifelong agent learning in multiple environments.

6   Conclusion

This article presents a lifelong learning perspective for autonomous robots. We propose not to study robot learning problems in isolation, but in the context of the multitude of learning problems that a robot will face over its lifetime. Lifelong robot learning opens the opportunity for transfer of learned knowledge. This knowledge may be used as a bias when learning control. Although control learning methods that allow the transfer of knowledge are more complex as most algorithms that solve isolated control learning problems, robot learning itself becomes easier. Robots that memorize and transfer knowledge rely less on real-world experimentation and thus

Figure 15 (a) The University of Bonn robot "Rhino" (manufactured by Real World Interface, Inc.), (b) Map (approximately 20 × 30 meters) constructed by Rhino at the AAAI-94 robot competition, using the technique described in this article.

learn faster. This is because previously learned knowledge may act as a knowledgeable bias that may partially replace the pure syntactic bias of inductive learning algorithms.

We have demonstrated with two concrete approaches the potential synergy effect of knowledge transfer. These approaches addressed two main types of lifelong agent learning scenarios, namely those that are defined in a single environment, and those that are not. We strongly believe that knowledge transfer is essential for scaling robot learning algorithms to more realistic and complex domains. Exploiting previously learned knowledge simplifies learning control. These results support our fundamental claim that learning becomes easier, if it is embedded into a lifelong learning context.

Acknowledgment

We thank the CMU Robot Learning Group and the Odysseus team at CMU for invaluable discussion that contributed to this research. We also thank Ryusuke Masuoka for his invaluable help in refining EBNN.

This research was sponsored in part by the Avionics Lab, Wright Research and Development Center, Aeronautical Systems Division (AFSC), U. S. Air Force, Wright-Patterson AFB, OH 45433-6543 under Contract F33615-90-C-1465, Arpa Order No. 7597 and by a grant from Siemens Corporation. The views and conclusions contained in this document are those of the authors and should not be interpreted as representing the official policies, either expressed or implied, of the U.S. Government or Siemens Corp.

Addendum

Since this article was submitted, EBNN was successfully applied to a variety of real-world learning tasks. In (Mitchell and Thrun, 1995, Thrun, 1994, Thrun, 1995a), result of applying EBNN to mobile robot navigation using the CMU Xavier robot are reported. EBNN has also been applied to robot perception (Mitchell *et al.*, 1994), (O'Sullivan *et al.*, 1995), object recognition (Thrun and Mitchell, 1994) and the game of chess (Thrun, 1995b). In (Thrun and Mitchell, 1994), a definition of the lifelong learning problem in the context of supervised learning can be found.

The approach to interpreting sonar sensors for building occupancy maps reported in Sect. 4 has been, with slight modifications, successfully employed in the University of Bonn's entry "Rhino" at the 1994 AAAI mobile robot competition (Buhmann *et al.*, 1995). Currently, maps are routinely built for large indoor areas.

References

Christopher A. Atkeson, 1991. Using locally weighted regression for robot learning. In Proceedings of the 1991 IEEE International Conference on Robotics and Automation, pp. 958–962, Sacramento, CA

Jonathan R. Bachrach and Michael C. Mozer, 1991. Connectionist modeling and control of finite state systems given partial state information

Andrew G. Barto, Richard S. Sutton, and Chris J. C. H. Watkins, 1989. Learning and sequential decision making. Technical Report COINS 89-95, Department of Computer Science, University of Massachusetts, MA

Andrew G. Barto, Richard S. Sutton, and Chris J. C. H. Watkins, 1990. Learning and sequential decision making. In M. Gabriel and J.W. Moore, (eds.), Learning and Computational Neuroscience, pp. 539–602, Cambridge, MA. MIT Press

Andrew G. Barto, Steven J. Bradtke, and Satinder P. Singh, 1991. Real-time learning and control using asynchronous dynamic programming. Technical Report COINS 91-57, Department of Computer Science, University of Massachusetts, MA

R. E. Bellman, 1957. Dynamic Programming. Princeton University Press, Princeton, NJ

Rodney A. Brooks, 1989. A robot that walks; emergent behaviors from a carefully evolved network. Neural Computation, 1(2): 253

Joachim Buhmann, Wolfram Burgard, Armin B. Cremers, Dieter Fox, Thomas Hofmann, Frank Schneider, Jiannis Strikos, and Sebastian Thrun, 1995. The mobile robot Rhino. AI Magazine, 16(1)

Tom Bylander, 1991. Complexity results for planning. In Proceedings of IJCAI-91, pp. 274–279, Darling Habour, Sydney, Australia. IJCAI, Inc

John Canny, 1987. The Complexity of Robot Motion Planning. MIT Press, Cambridge, MA

Richard Caruana, 1993. Multitask learning: A knowledge-based of source of inductive bias. In Paul E. Utgoff, (ed.), Proceedings of the Tenth International Conference on Machine Learning, pp. 41–48, San Mateo, CA. Morgan Kaufmann

Lonnie Chrisman, 1992. Reinforcement learning with perceptual aliasing: The perceptual distinction approach. In Proceedings of 1992 AAAI Conference, Menlo Park, CA. AAAI Press/MIT Press

Peter Dayan and Geoffrey E. Hinton, 1993. Feudal reinforcement learning. In J. E. Moody, S. J. Hanson, and R. P. Lippmann, (eds.), Advances in Neural Information Processing Systems 5, San Mateo, CA. Morgan Kaufmann

Gerald DeJong and Raymond Mooney, 1986. Explanation-based learning: An alternative view. Machine Learning, 1(2): 145–176

Alberto Elfes, 1987. Sonar-based real-world mapping and navigation. IEEE Journal of Robotics and Automation, RA-3(3): 249–265

Michael I. Jordan, 1989. Generic constraints on underspecified target trajectories. In Proceedings of the First International Joint Conference on Neural Networks, Washington, DC, San Diego. IEEE TAB Neural Network Committee

R.E. Kalman, 1960. A new approach to linear filtering and prediction problems. Trans. ASME, Journal of Basic Engineering, 82: 35–45

Benjamin Kuipers and Yung-Tai Byun, 1990. A robot exploration and mapping strategy based on a semantic hierarchy of spatial representations. Technical report, Department of Computer Science, University of Texas at Austin, TX 78712

Long-Ji Lin and Tom M. Mitchell, 1992. Memory approaches to reinforcement learning in non-markovian domains. Technical Report CMU-CS-92-138, Carnegie Mellon University, Pittsburgh, PA

Long-Ji Lin, 1992. Self-improving reactive agents based on reinforcement learning, planning and teaching. Machine Learning, 8

Long-Ji Lin, 1992. Self-supervised Learning by Reinforcement and Artificial Neural Networks. PhD thesis, Carnegie Mellon University, School of Computer Science, Pittsburgh, PA

Pattie Maes, (ed.), 1991. Designing Autonomous Agents. MIT Press (and Elsevier), Cambridge, MA

Sridhar Mahadevan and Jonathan Connell, 1991. Scaling reinforcement learning to robotics by exploiting the subsumption architecture. In Proceedings of the Eighth International Workshop on Machine Learning, pp. 328–332

Bartlett W. Mel, 1989. Murphy: A neurally-inspired connectionist approach to learning and performance in vision-based robot motion planning. Technical Report CCSR-89-17A, Center for Complex Systems Research Beckman Institute, University of Illinois

Tom M. Mitchell and Sebastian Thrun, 1993. Explanation based learning: A comparison of symbolic and neural network approaches. In Paul E. Utgoff, (ed.), Proceedings of the Tenth International Conference on Machine Learning, pp. 197–204, San Mateo, CA. Morgan Kaufmann

Tom M. Mitchell and Sebastian Thrun, 1993. Explanation-based neural network learning for robot control. In S. J. Hanson, J. Cowan, and C. L. Giles, (eds.), Advances in Neural Information Processing Systems 5, pp. 287–294, San Mateo, CA. Morgan Kaufmann

Tom M. Mitchell and Sebastian Thrun, 1995. Learning analytically and inductively. In Steier and Mitchell, (eds.), Mind Matters: A Tribute to Allen Newell. Lawrence Erlbaum Associates

Tom M. Mitchell, Rich Keller, and Smadar Kedar-Cabelli, 1986. Explanation-based generalization: A unifying view. Machine Learning, 1(1): 47–80

Tom M. Mitchell, Joseph O'Sullivan, and Sebastian Thrun, 1994. Explanation-based learning for mobile robot perception. In Workshop on Robot Learning, Eleventh Conference on Machine Learning

Andrew W. Moore, 1990. Efficient Memory-based Learning for Robot Control. PhD thesis, Trinity Hall, University of Cambridge, UK

Hans P. Moravec, 1988. Sensor fusion in certainty grids for mobile robots. AI Magazine, pp. 61–74

Michael C. Mozer and Jonathan R. Bachrach, 1989. Discovering the structure of a reactive environment by exploration. Technical Report CU-CS-451-89, Dept. of Computer Science, University of Colorado, Boulder

Paul Munro, 1987. A dual backpropagation scheme for scalar-reward learning. In Ninth Annual Conference of the Cognitive Science Society, pp. 165–176, Hillsdale, NJ. Cognitive Science Society, Lawrence Erlbaum

Joseph O'Sullivan, Tom M. Mitchell, and Sebastian Thrun, 1995. Explanation-based neural network learning from mobile robot perception. In Katsushi Ikeuchi and Manuela Veloso, (eds.), Symbolic Visual Learning. Oxford University Press

Dean A. Pomerleau, 1989. ALVINN: an autonomous land vehicle in a neural network. Technical Report CMU-CS-89-107, Computer Science Dept. Carnegie Mellon University, Pittsburgh PA

Lorien Y. Pratt, 1993. Discriminability-based transfer between neural networks. In J. E. Moody, S. J. Hanson, and R. P. Lippmann, (eds.), Advances in Neural Information Processing Systems 5, San Mateo, CA. Morgan Kaufmann

Ronald L. Rivest and Robert E. Schapire, 1987. Diversity-based inference of finite automata. In Proceedings of Foundations of Computer Science

David E. Rumelhart, Geoffrey E. Hinton, and Ronald J. Williams, 1986. Learning internal representations by error propagation. In D. E. Rumelhart and J. L. McClelland, (eds.), Parallel Distributed Processing. Vol. I + II. MIT Press

A. L. Samuel, 1959. Some studies in machine learning using the game of checkers. IBM Journal on research and development, 3: 210–229

Jacob T. Schwartz, Micha Scharir, and John Hopcroft, 1987. Planning, Geometry and Complexity of Robot Motion. Ablex Publishing Corporation, Norwood, NJ

Noel E. Sharkey and Amanda J.C. Sharkey, 1992. Adaptive generalization and the transfer of knowledge. In Proceedings of the Second Irish Neural Networks Conference, Belfast

Patrice Simard, Bernard Victorri, Yann LeCun, and John Denker, 1992. Tangent prop – a formalism for specifying selected invariances in an adaptive network. In J. E. Moody, S. J. Hanson, and R. P. Lippmann, (eds.), Advances in Neural Information Processing Systems 4, pp. 895–903, San Mateo, CA. Morgan Kaufmann

Satinder P. Singh, 1992. The efficient learning of multiple task sequences. In J. E. Moody, S. J. Hanson, and R. P. Lippmann, (eds.), Advances in Neural Information Processing Systems 4, pp. 251–258, San Mateo, CA. Morgan Kaufmann

Satinder P. Singh, 1992. Transfer of learning by composing solutions for elemental sequential tasks. Machine Learning, 8

Steven C. Suddarth and Y. L. Kergosien, 1990. Rule-injection hints as a means of improving network performance and learning time. In Proceedings of the EURASIP Workshop on Neural Networks, Sesimbra, Portugal. EURASIP

Richard S. Sutton, 1984. Temporal Credit Assignment in Reinforcement Learning. PhD thesis, Department of Computer and Information Science, University of Massachusetts

Richard S. Sutton, 1988. Learning to predict by the methods of temporal differences. Machine Learning, 3

Richard S. Sutton, 1990. Integrated architectures for learning, planning, and reacting based on approximating dynamic programming. In Proceedings of the Seventh International Conference on Machine Learning, June 1990, pp. 216–224, San Mateo, CA. Morgan Kaufmann

Richard S. Sutton, 1992. Adapting bias by gradient descent: An incremental version of delta-bar-delta. In Proceeding of Tenth National Conference on Artificial Intelligence AAAI-92, pp. 171–176, Menlo Park, CA. AAAI, AAAI Press/MIT Press

Ming Tan, 1991. Learning a cost-sensitive internal representation for reinforcement learning. In Proceedings of the Eighth International Workshop on Machine Learning, pp. 358–362

Sebastian Thrun and Tom M. Mitchell, 1993. Integrating inductive neural network learning and explanation-based learning. In Proceedings of IJCAI-93, Chamberry, France. IJCAI, Inc

Sebastian Thrun and Tom M. Mitchell, 1994. Learning one more thing. Technical Report CMU-CS-94-184, Carnegie Mellon University, Pittsburgh, PA 15213

Sebastian Thrun, 1992. The role of exploration in learning control. In David A. White and Donald A. Sofge, (eds.), Handbook of intelligent control: neural, fuzzy and adaptive approaches. Van Nostrand Reinhold, Florence, Kentucky 41022

Sebastian Thrun, 1993. Exploration and model building in mobile robot domains. In Proceedings of the ICNN-93, pp. 175–180, San Francisco, CA. IEEE Neural Network Council

Sebastian Thrun, 1994. A lifelong learning perspective for mobile robot control. In Proceedings of the IEEE/RSJ/GI International Conference on Intelligent Robots and Systems

Sebastian Thrun, 1995. An approach to learning mobile robot navigation. Robotics and Autonomous Systems. (in press)

Sebastian Thrun, 1995. Learning to play the game of chess. In G. Tesauro, D. Touretzky, and T. Leen, (eds.), Advances in Neural Information Processing Systems 7, San Mateo, CA. MIT Press

Christopher J. C. H. Watkins, 1989. Learning from Delayed Rewards. PhD thesis, King's College, Cambridge, UK

Steven D. Whitehead and Dana H. Ballard, 1991. Learning to perceive and act by trial and error. Machine Learning, 7: 45–83

Cognitive Architectures – From Knowledge Level To Structural Coupling

Walter Van de Velde

Vrije Universiteit Brussel, AI-Lab
Pleinlaan 2, B-1050 Brussels, Belgium

Abstract. This chapter investigates the relation between the two key aspects of intelligent agents: intelligence and agent-hood. Efforts for building general architectures for intelligent agents have typically focussed on one of these aspects. The chapter first reviews some work on cognitive architectures for intelligence and illustrates how these are being applied to physical agents. Secondly, this chapter puts forward a series of hypotheses on the relationship between cognition and agent-hood. These hypotheses are aimed at a unified treatment of behavior and cognition. Central to this discussion in the notion of coordination. In particular the nature and role of representations, internal and external, are explained in relation to their coordinating role.

1 Introduction

This chapter is a tour along a number of architectural options for building *intelligent agents*. The concept of *agent* refers to a system that can be differentiated from its environment and is capable of direct and continued interaction with that environment. Such a system 'exists' in its environment and has with it an observable interaction, which is called its *behavior* (Figure 1). The prototypical examples are autonomous mobile robots, equipped with sensors and effectors that allow for multiple forms of interaction with the real-world (which may include other agents), but also other kinds of environments can be relevant, as for the software agents 'living' in Cyberspace.

The concept of *intelligence* is more tricky. One traditional view holds that intelligence is basically cognition, i.e., the capacity to construct and manipulate symbolic representations that are somehow related to the environment (i.e., represent aspects of that environment) and that are used to determine appropriate actions (e.g. [13]). In another view intelligence relies on self-sufficiency: a behavior is intelligent to the extent that it can be justified as being aimed at sustained behavior in an environment (see Steels, this volume). In a related view the behavior is aimed at sustaining existence of a species, rather than of the individual agent (see, for example McFarland, this volume). In spite of these different options it seems fair to call a behavior intelligent as soon as we, as observers, understand a rationale for it. Whatever the details of this rational, the important point is that intelligence is not in

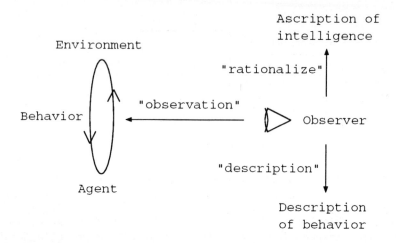

Fig. 1. Behavior is observed interaction between an agent and its environment. Intelligent is ascribed to the agent by the observer.

itself a structural property of a system but a quality ascribed to a system upon observation of its interaction with the world (its behavior) (Figure 1).

The aim of this chapter is to investigate into the connection between the intelligence and the agent-hood of an intelligent agent. Many of the more traditional theories of intelligent agents (see [22] for a number of these) have focussed on the cognitive aspects of intelligence. They are therefore called *cognitive architectures*. Systems like ACT* [1], SOAR [7, 14], THEO [12] and ICARUS [9] – and there are many more –, all provide for symbolic representations and for mechanisms to manipulate and construct these representations. Of those systems that have been used as the cognitive component of autonomous agents, however, the agent-hood is realized in a somewhat ad-hoc fashion: perception and actuation are implemented by front-end modules for input and output to a basic information processing system, and these modules are themselves opaque to the architecture (Figure 2).

On the other hand the behavior oriented architectures, many of which are described in other chapters of this book, aim at reproducing behavioral patterns without any claim about the cognitive level of the agent's activities. In fact, almost all of these architectures leave cognition out of the picture. Whether this is only the case for simplicity's sake is not always clear but little evidence is provided on how behavior-oriented systems are going to scale up to behaviors that can be described as cognitive, for example those that apparently involve complex planning or linguistic interaction.

The dichotomy between the two aspects of intelligent agents – intelligence and agent-hood – and their corresponding architectural proposals leaves one with the impression that behavior and cognition are of a fundamentally dif-

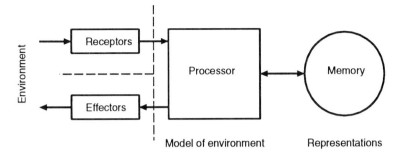

Fig. 2. Cognitive architectures fit the information processing model. A memory contains representations of knowledge about the environment and the effects of actions. Perception and action are treated as separate front-end modules.

ferent nature, and that they should be supported by two different sets of architectural commitments. Are we forced to go for hybrid architectures? Unless one accepts the orthogonality of both aspects this begs the question.

This chapter starts by reflecting upon the concept of a general architecture for intelligence agents (Section 2). In the first main part of this chapter (Section 3) we will have a closer look at some of the cognitive architectures that have been applied for agents. This analysis occurs within a common framework of architectural commitments, and highlighting some of the steps that have been taken to integrate behavior. Then, in Section 4, follows a first conservative re-direction of cognition. It revises the traditional view on the relation between knowledge and behavior in cognitive systems. Afterwards this re-direction is pushed various steps further. The aim is to outline a series of arguments that motivate – although not yet fully justify – the conclusion that cognition and behavior, although phenomenologically very different, can be realized within a single set of architectural commitments. The central notion in this part is *coordination*. An intelligent system consists of coordinated or "coupled" processes, essentially closed and defined in terms of their organization and the dynamics that it implies. All behavior is produced from reaction to perturbations from the environment or from other processes. Within such an architecture, I hypothesize, plasticity is achieved by modifying the coordination or coupling of these processes, rather than by changing these processes themselves.

The key issue, then, is how this coupling between processes can be achieved and maintained. The traditional way, I call it *symbol level coupling*, is to enforce a link between the representations that are involved in each of the processes, for example by a mechanism of consistency maintenance. An alternative, called *structural coupling* [11], occurs when processes are mutually adapted to eachother and to the environment that they both interact with. In this case their behavior is coordinated but there is no direct link between them. The effect of both forms of coupling can be the same, namely an observer's interpretation of coordination of goal-directed behaviors (*knowledge level coupling*). This means that the observer can rationalize the coordination by ascribing bodies of knowledge that are sufficiently "in tune".

Basically, the architecture relies on structural coupling as an alternative to symbol level coupling for realizing knowledge level coupling. If the idea is a viable one then it has a number of implications that may significantly extend its scope: (1) the architecture is independent of whether processes are symbolic in nature or not. The coupling between cognitive and behavioral aspects of intelligence, a difficult issue in existing hybrid architectures, would not require additional technical solutions – although point 3 below is a more fundamental way of integrating cognition and behavior; (2) representations as they are normally used in symbolic coupling can be considered as external elements in the coordination of processes, whether they are embodied in a single or in different agents – this gives a hook on the problem of communication; (3) the role of external representation as coordinating elements can be taken over by internal processes that, by structural coupling, take over the mediating, coordinating role, but are themselves representation less – this is where cognition comes in; (4) consequently the same or at least similar mechanisms can be applied for single-agent as well as for multi-agent learning, coordination and adaptation.

In summary, intelligent behavior involves the coordination between processes. Whereas such coordination may be realized by exchange or sharing of structures between the coordinated processes, we propose that a form of structural coupling originates in the interaction with physical structures in the environment, and is later internalized by mediating processes within the system. Such mediating processes enable a behavior which, under the right circumstances (clues), leads to cognitive effects but is, in fact, itself representation-less.

2   General Architectures

The concept of an architecture of intelligence is illustrated by the abstract of a seminal paper on SOAR [7], which is probably the most well-known such architecture:

> "The ultimate goal of work in cognitive architectures is to provide for a system capable of general intelligent behavior. That is, the goal is to provide the underlying structure that would enable a system to perform the full range of cognitive tasks, employ the full range of problem solving methods and representations appropriate for the task, and learn about all aspects of the tasks and its performance on

them. In this article we present SOAR, an implemented proposal for such an architecture."

The idea is that, whatever aspect of intelligence is under study, whatever problem or whatever application the theory is applied to, this can be realized within the limits of a predefined organization (the 'underlying structure') of architectural elements. For example, in Robo-SOAR the SOAR architecture has been applied to a task that involves robotic interaction with an external environment. The abstract of a paper describing that work [8] reads as follows:

> "This chapter reports on progress in extending the SOAR architecture to tasks that involve interaction with external environments. The tasks are performed using a Puma arm and camera in a system called Robo-SOAR. The tasks require the integration of a variety of capabilities including problem solving with incomplete knowledge, reactivity, planning, guidance from external advice, and learning to improve efficiency and correctness of problem solving. All of these capabilities are achieved without the addition of special purpose modules or subsystems to SOAR."

The first and last sentences are significant: although the SOAR architecture had to be adapted to do Robo-SOAR, the changes are not specific for this application, neither are they locally contained in special purpose modules. They affect the whole of SOAR and should be considered a revision of the theory rather than as an ad-hoc change for a particular application.

The space of manifestations of intelligence is vast and multi-dimensional (Figure 3) and a theory of intelligence is, preferably, not about an individual instance of intelligence. It is a theory about an entire class of such manifestations, expressing basically what is common to all of these. Such a theory must therefore cover a class of intelligent behaviors. It attempts to find what is common to a large collection of intelligent phenomena. It is a common abstraction from a class of intelligent phenomena (Figure 4). This abstraction involves, first, a particular view on the behavior, i.e., a way to describe agent behavior. At the next level structural properties of a system are being described. These refer to the particular entities of construction that realize the behavior described. This level is still specific for that single behavior. Yet more abstract is the level of architecture. It describes the organization of the structures that re-occurs over the variety of systems that realize the wider class of behaviors.[1]

But abstraction is only half of the story: An architecture must also provide *guidance* in creating (or, designing) a particular instance.

An architecture for intelligence is a class of programs that can produce intelligent behavior in a wide variety of situations (environments,

---

[1] The terminology of structure and organization is used as in [11].

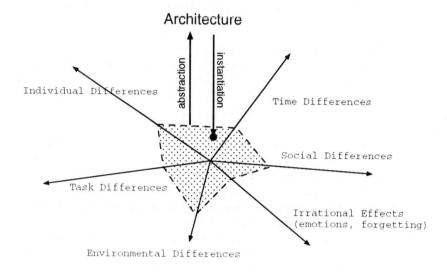

Fig. 3. Variations of intelligence. What is left if we abstract away from all these differences? Architecture of intelligence are more or less ambitious in covering areas of variation. They are common abstractions that can be systematically instantiated to any meaningful variation.

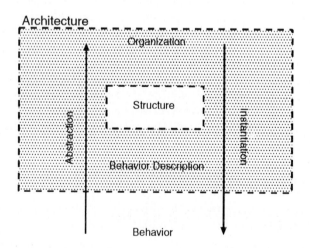

Fig. 4. Abstraction and instantiation in architectures

tasks, over time,...). Such an architecture can be instantiated in a systematic way to account for a particular variety of intelligent behavior.

So, there are three elements to a theory of intelligence: a descriptive framework, the system organization, and guidance on how to instantiate the organization with structures to obtain a system that behaves according to a behavior description. A theory is also an account of how the variation comes about (Figure 4). This idea of instantiation is important. For example, within the class of computational architectures it excludes the Turing model or any general purpose programming language as an architecture of intelligence, since there is no systematic way in which to instantiate the Turing model to achieve a particular instance of intelligent behavior. The descriptive framework for behavior embodies what can be called the *ontological commitment* of a theory. The structure and organization, on the other hand, are the *architectural commitment* of the theory.

The practical value of an architecture is in its use for constructing intelligent agents, i.e., as a theory of design. Ideally it is possible to describe the target behavior using the particular behavior description language associated with the architecture, and to instantiate the architectural organization with structures to realize the described behavior. Both for behavior description entities (conceptual) as for design and construction entities a variety of options exist (Table 1). For example, the knowledge-oriented paradigm uses knowledge as a way to describe behavior: an agent is viewed as acting according to a body of knowledge that is ascribed to it (e.g., [13]). Early rule-based models assumed that the representation of that knowledge in terms of production rules would, upon manipulation by a general inference mechanism, lead to the appropriate behavior. This would fit our notion of systematic instantiation. Although this simplistic view has not stood up, other architectures in this direction propose more complex organizations of symbolic representations. In particular they differentiate between three classes of architectural commitments: for representation, for inference and for learning.[2] We will use these categories for the analysis of three typical cognitive architectures in the next section.

Behavior orientation typically uses descriptions of the interaction dynamics between an agent and its environment. Agents are then constructed in terms of behavior systems, self-contained dynamical systems that can be combined in various ways to yield more complex behaviors.

3 Three Cognitive Architectures

Different architectures can be distinguished according to their *architectural commitments*. An architectural commitment is a decision imposed by the

---

[2] The organization of these three is another architectural commitment, e.g., universal failure driven learning.

	Ontological (analysis and description)	**Architectural** (design and construction)
Cognitive	knowledge and goals	symbols and representations
Behavior oriented	behavior (interaction dynamics)	behavior system (dynamical system)

Table 1. Cognitive and behavior oriented architectures use different conceptual and design entities. The aim is generally the same: to support the instantiation of an architecture with design entities to reproduce behavior describable with the conceptual vocabulary.

architecture. Most cognitive architectures make a distinction between three basic faculties of intelligence, namely representation, inference and learning. The goal in these architectures is to come up with a small set of basic representations, inference and learning mechanisms to account for a wide range of aspects of intelligence in a variety of domains. Such systems stress uniformity of representation, inference and learning but may take very different options for each of these. Also the scope of the systems may differ: problem solving, expert systems, cognitive modeling, autonomous agents.

Table 2 provides an overview of the architectural commitments that are being made by three typical architectures: SOAR, THEO and ICARUS. These examples are interesting because: (1) they all fit into a tradition of representation based intelligence, (2) they make a similar partition of required functionalities for intelligence, and (3) they have all been applied to, or have been designed for physical autonomous agents.

	SOAR	THEO	ICARUS
Representation	Productions Objects	Methods Frames	Probabilistic concept hierarchies
Inference	Search in problem spaces	Slot computation	Heuristic classification
Learning	Chunking	Caching EBG-like Reenforcement	Concept formation

Table 2. Architectural commitments in three cognitive architectures

Of the three architectures, SOAR and THEO have originally been designed for cognitive tasks alone, i.e., tasks not involving ongoing interaction with an environment. Nevertheless, both SOAR and THEO have subsequently been applied in a robotic context. Robo-SOAR [8] and Theo-Agent [2] are the resulting systems. In any case, and this is important, the applications respect the architectural commitments that are made by the basic architectures. ICARUS is less publicized and was never actually finished as a complete system. However, it is an interesting proposal since it makes very different architectural commitments than the other ones. The specificity of ICARUS is that, to some extend, it takes the issue of interaction into account from the start.

3.1 SOAR

The history of SOAR goes back to work on GPS and OPS5. The work culminated into an integrated theory of intelligence in [14]. The main reference for this section is [7].

The architectural commitments in SOAR are summarized in Table 2. With respect to representation SOAR embodies the commitment of representing all long-term memory as productions, all short term memory as objects and any task environment as problem spaces. Especially the latter is significant. The problem space hypothesis postulates that all goal oriented problem solving is done by search in a problem space. Thus, a problem space is the internal representation of the task environment. It consists of states and operators to yield new states (Figure 5).

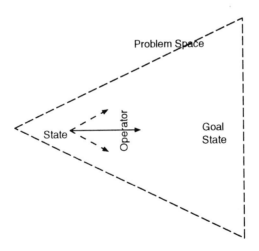

Fig. 5. The SOAR problem space and the four elements of a SOAR context.

The task implementation knowledge, used to create problem spaces, is one kind of long-term memory. The other is control knowledge, knowledge for searching the problem spaces. Control knowledge consists of knowledge to select the goal to work on, the problem space to search, the state from which to search, and the operator to apply to that state. Long term memory contains only these two kinds of knowledge, all represented as production rules. The short term memory is the working memory of a production system in which all long term knowledge is brought to bear on a problem. Production rules fire exhaustively and in parallel.

The activity of the production rules must be understood as search in problem spaces. Globally this search is controlled by the context tree. A context is a quadruple of goal, problem space, state and operator. The SOAR decision cycle manages the context tree by determining which slots in the context tree should have their value changed. The context tree is examined from top to bottom (which avoids insensitivity to changes in the task environment) The decision cycle consists of two phases (Figure 6). In the elaboration phase productions fire exhaustively and in parallel until no more changes occur (quiescence). The production rule, apart from generating new objects (goals, states, operators) or augmenting them, may produce preferences for goals, problem spaces, states and operators. No control is exerted at the level of productions by conflict resolution.

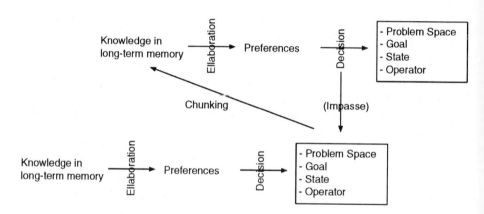

Fig. 6. The SOAR decision cycle and the role of impasses and chunking in it. Note that, despite its duplication in the figure, there is only one long-term memory which is applied for all decisions at all levels in SOAR.

SOAR uses a fixed language of preferences. Preferences express acceptability, rejection or desirability of alternative objects for occupying the slots in the context tree. Preferences are evaluated in the second phase of the decision cycle, called the decision procedure. The result of the decision procedure is a single object that is new, acceptable, not rejected and better than any other choice. With this object the context tree is updated and the next decision cycle begins.

When the decision procedure fails to select a single object based on the preferences an impasse occurs (Figure 6). Each impasse corresponds to one of four ways in which the decision procedure may fail.

1. tie impasse
2. no-change impasse
3. reject impasse
4. conflict impasse

Impasses are the only cases of indecision in the architecture. They generate a subgoal, basically aimed at the resolution of the impasse. This subgoal is solved, just like any goal in SOAR, by a search in a problem space. This search uses exactly the same decision cycle.

Chunking is the only learning mechanism in SOAR (Figure 6). Chunking is aimed at avoiding the impasses the occurred. Whenever an impasse is resolved a chunk is created. A chunk is a production rule that will generate the preference objects that, if they had been present, would have avoided the impasse. The chunk shortcuts the reasoning on the subgoal of resolving the impasse. Everything being the same, the impasse will be avoided subsequently.

Chunking is a general mechanism for learning from experience. It works by analysing the working memory. The working memory elements that were present when the impasse occurred and that have been used in the successful matching during the resolution of the impasse are used as the if-part of the chunk. The resolution of the impasse is interpreted as preferences that are generated in the then-part of the chunk.

The SOAR chunking mechanism is simple but powerful. It converts goal-based problem solving into productions. Similar to caching, except that it caches processes, not data. Moreover chunks are generalized implicitly, since they are applied in other situations than the one from which it is generated. It can be activated in other subgoals of the same problem. The fact that the chunk was ever generated is a guarantee for the practical relevance of the chunk.

SOAR can be viewed as an example theory of intelligence to the extent that the following holds:

- The architecture provides for the representation of a task environment in problem spaces
- Problem solving occurs only in problem spaces
- All long term memory consists of production rules
- The structure of a task environment determines the possible structure of the problem spaces
- The structure of the problem spaces determine the possible methods that can be used for problem solving

Especially the latter two requirements are important. They have to do with SOAR being systematically instantiable for a particular task environment.

3.2 Robo-SOAR

In Robo-SOAR the SOAR architecture has been applied to a task that involves robotic interaction with an external environment. The experiment with Robo-SOAR had two purposes. The first one was to demonstrate the application of SOAR to a robotics problem. The second one is to show how SOAR could be used as an instructable system with human guidance to build up the robot's skills. In [14] Allen Newell has described a way of extending SOAR into a full cognitive system for an integrated agent. He proposes productions that do the coding and decoding for action and perception, i.e., transduction between the cognitive level and the low-level interaction mechanisms (Figure 7). This would include action and perception within the basic architectures, and for example chunking would function for these as well. One could envisage the creation of chunks that make the robot behavior highly reactive, by shortcutting extensive reasoning from perception to action.

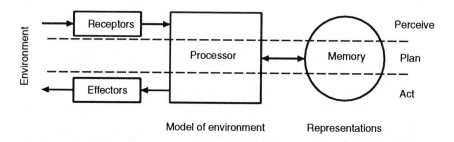

Fig. 7. The information processing model with front-ends for action and perception integrated in the architecture. This is the typical paradigm to integrate action and perception into cognitive architectures, although in practice large blocks of receptor and effector processes remain opaque to the architecture.

However, this is not the way in which this is done in Robo-SOAR. The architecture used is Robo-SOAR uses external modules that call the robot software and the vision software. The interface is at the cognitive level, using terms to refer to the blocks and their positions, and to the specific actions such as approach, close gripper, and so on. Thus the action-perception loop is closed using modules that are opaque to SOAR. Within them architectural commitments are not necessarily respected and, in particular, chunking does not apply.

3.3 THEO

As a general architecture for intelligence THEO makes different architectural commitments than the SOAR architecture (Table 2). We can only describe the architecture in general, as almost any of its behaviors can be changed.[3]

In THEO all representation is frame-based. The information in the system is represented as values of slots of frames. These values can be frames themselves. Frames can also describe information about other slots and frames. This allows for the representation of meta-information. For example, THEO can represent different methods to compute the value of a slot, information about how to select the best method, information that explains the value of a slot or meta-slot, and so on.

Inference in THEO is lazy slot computation. A problem corresponds to a slot without value. Such a problem is described like an access path in a nested frame structure. A slot has itself a slot called method. The value is computed by applying the method. Things start to be more interesting when the method-slot itself has no value. In that case it must itself be computed, for example by selecting a method from the available.methods slot. Typical methods are default.value, which takes the value of slot from the value of its default slot, and from.definition which computes the value from evaluating in context the value of the definitions slot of the slot. Other methods, like inherits and drop.context function as rewrite rules over problems. For example the problem of finding rules to repair the engine of a car would be rewritten by the drop.context method as the problem of find rules to repair an engine, dropping the context within which the engine occurs.

THEO has several forms of learning. The simplest one is caching. Whenever THEO computes the value of a slot, this value is cached in the slot (at least by default). Thus, a problem of computing the same slot will return the cached value. Together with the cached value a slot called explanation is filled. The explanation is the tree of slots that were computed in order to compute the value of slot, together with the method that successfully computed them. The explanation is used to force recomputation of a slot when some of the slots on which its previous computation depended have had their value changed. A second on more sophisticated use of the explanations slot is in a form of explanation-based learning. In some cases THEO can derive a new method to compute a slot, which is a macro composing in an efficient way the different methods that were used to compute the value. This macro is then stored with an object that may be more general than the one it was computed for, thus leading to transfer of experience between different slots. A final form of learning is related to reenforcement learning. THEO will learn to use the most successful methods first in its computations.

3.4 THEO-Agent

THEO was itself designed in a context for knowledge-based systems research. However, it has been applied to a problem of autonomous agents, called THEO-Agent [2]. THEO-Agent adds an agent architecture on top of THEO

---

[3] In fact THEO is a designed as a Representation Language Language.

(Figure 8). The architecture repeatedly assesses the situation, selects a goal, creates a plan for that goal and then performs the first action of that plan. Within THEO this is implemented as a sequence of questions: what do I do next? what plan do I have? what goal do I pursue? what situation am I in? All of these correspond to THEO slots that need to be recursively computed. At all of these levels the learning mechanisms are active. Since the perception slots are constantly updated it is not likely that caching will lead to major benefits (the dependency of cached values on perception slots will regularly uncache values). However the form of explanation-based learning in THEO may lead to interesting generalization which, in effect shortcut part of or all of the reasoning in THEO-Agent. Examples have been reported where THEO-Agent learns to directly connect percepts with motor actions, without going through the goal selection and planning processes [2].

THEO-Agent

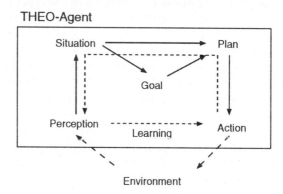

Fig. 8. The THEO-Agent architecture on top of THEO. THEO-Agent asks itself 'what do I do next' and computes the answer by lazy evaluation. Learning is active at all levels but most significantly it shortcuts the 'cognitive' level to become more reactive.

3.5 ICARUS

Although SOAR and THEO make different commitments, they are in a sense quite similar, especially in the way they treat learning (impasse-driven explanation based generalization). The architectural commitments that are embodied in ICARUS are quite different from those in SOAR and THEO. ICARUS is an architecture that has been specifically designed with real-world interaction in mind. It is an interesting proposal because of makes very different architectural commitments than do the other traditional architectures. In ICARUS all knowledge is represented in probabilistic classification hierarchies. ICARUS basically proposes to distinguish between three main components. One component, called Labyrinth, is responsible for the mapping of real world input data to the conceptual level. Here objects in a scene are being recognized. The second important component, called Daedalus, deals with the problem of planning for actions. It takes situations in the world, and derives appropriate actions (reactions) in that situation (we ignore for reasons of simplification the problems of multiple goals, and where the goals come from). This component can be viewed as an adaptive reactive planner that, learns to acquire a more complex situation-action mapping in a learning scenario. The third component, called Maeander, is responsible for the mapping of actions into a real action scheme that drives actuators (motor-schemas).

In ICARUS all three these major components are using a similar representation and reasoning scheme (Figure 9). All knowledge is represented as probabilistic concept hierarchies, and all reasoning is a scheme of heuristic classification in those hierarchies (Table 2). The concepts that are being stored and indexed in the hierarchies are either objects, plans or motor schemata. It is assumed that every instance of any of these concepts is described as a conjunction of attribute-value pairs. Each concept is represented as a set of attributes, their possible values and, for each value, the conditional probability of the attribute having that value, given its membership in the concept. In addition, the probability of an object being in the concept is also represented.

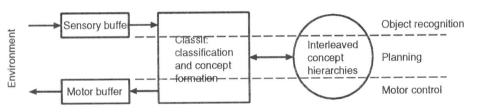

Fig. 9. A single set of architectural commitments underlies all processes in ICARUS.

Inference in ICARUS is probabilistic classification in these hierarchies, and learning occurs at the same time (Figure 10. An instance is sorted down a hierarchy by selecting at each level the class to which to add the instance (and, in that case, updating the probability distribution of the attributes), or possibly by creating a new concept. In some cases ICARUS' learning mechanism may even consider a more drastic reorganization of the concept hierarchy. Classification and concept formation in ICARUS are based on Classit [6].

The concept formation capability in ICARUS is a step toward solving the problem of symbol grounding. Typically an agent designer will prime an agent

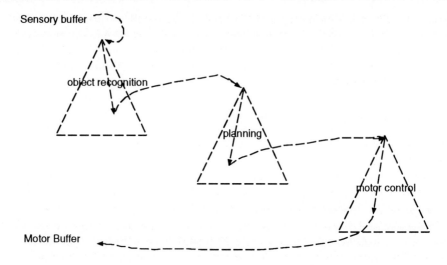

Fig. 10. A simplified view of the heuristic classification processes in three interleaved hierarchies for object recognition, planning and motor control in ICARUS.

with ontological commitments, i.e., a way to view the world (what are the objects, their attributes). In ICARUS these concepts are constructed based on the interaction with the environment.

4   From Knowledge to Structural Coupling, and Back

In previous sections we have described a number of typical cognitive architectures. We have also looked at the way in which they could be applied to physical agents. The typical way is to extend the basic information processing model with front-ends for action and perception integrated in the architecture. In practice, as we have seen, large blocks of receptor and effector processes remain opaque to the architecture. An architecture like ICARUS goes some way to take action and perception seriously, but the information processing model remains central.

In subsequent subsections we reconsider the relation between intelligence and agent-hood by taking the reader through a number of re-interpretations of what's being done in cognitive architectures. Our aim is to move toward a model of coordinated processes. An intelligent system consists of coordinated or "coupled" processes, essentially *closed and defined in terms of their organization and the dynamics that it implies*. This is fundamentally different from the information processing model. All behavior is produced from reaction to perturbations from the environment or from other processes. One consequences is that the difference between action and perception disappears. Both are done (in fact, just happen) for side-effect. The input or output view on both is an artifact of the functional-computational view on intelligent behavior that has been actively explored in Artificial Intelligence. Although it may correspond to a useful epistemological commitment, it has not necessarily also an architectural equivalent.

4.1 Knowledge and behavior

It is amazing how little cognitive architectures have to say about the nature of problem solving. A characterization like 'search in a problem space' is not very helpful. In Newell's view knowledge is a means to select an action. The knowledge level rationalizes behaviour in terms of the reasons that an agent has to believe that certain actions will lead to achieving certain goals. In this sense knowledge is a means to an end, a resource for behaviour [13]. The goal of reasoning is to select one of the possible actions. More recently a different view is being explored, namely the view of problem solving as modelling. The idea is that problem solving is the construction of a situation specific model [5] or case model [15]. My interpretation of this was previously explained in [19] and [21].

From a knowledge level perspective the agent's perception of the world is through knowledge alone. A goal therefore must correspond to a desired state of one's knowledge about the world. Consequently this knowledge must refer to the specific systems that the goal is about. This model - let us call it the case model - at every moment during problem solving summarizes the agent's understanding of the problem, and allows it to eventually conclude that the goal has been reached.

The actions are the means that the agent has for interacting with the world [13]. Again, since at the knowledge level the agent's perception is through knowledge an action must be viewed as a way of obtaining knowledge about the reality. Actions of perception naturally fit in this scheme but also genuine acts of interaction do [19]. For example when a spray-painting robot paints some part then it will assume that the part probably has paint on it afterwards. The action of hitting a nail with a hammer can be assumed to entail that the nail is deeper in the wood, that the nail is deeper in the wood with some probability, or that a loud sound has been produced. The 'effect' of the action simply depends on what assumption the agent makes about it. In the problem solving as modelling view, then, the actions are not the goal of problem solving but are themselves a means to an end. That end is the construction of a model of part of the world that allows the agent to conclude eventually that its goals have been achieved.

The important twist is that, in this view problem solving is no longer an input-output process, neither a means to select actions (as in Newell's knowledge level theory [13]). Rather it is a process of organizing knowledge (obtained through actions) by making assumptions (i.e., constructing a model) that allow one to conclude (in effect, only assume [21]) that a goal is achieved.

Successful problem solving is a matter of making the right assumptions and exploring their consequences. Problem solving is thus viewed as the 'creation' of a suitable case model and the interaction only creates the context for this, by side-effect setting the boundary conditions for the process of maintaining an internal organization and identity [20]. The cognitive agent can be considered a closed and self-contained process. It is subject to influences from the outside that change its state, but all it aims at is to come to terms with its understanding of the world (Figure 11).

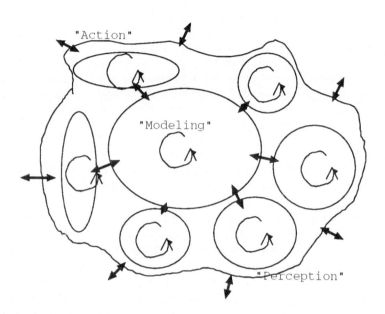

Fig. 11. A system is a collection of processes, essentially closed and self-contained, but subject to pressures from its environment that influence its internal dynamics. Some of these interaction can be called action, others perception. Architecturally there is no difference.

A consequence of the above view is that the difference between action and perception disappears. Both are done for side-effect. The input or output view on both is an artifact of the functional-computational view on intelligent behavior that has been actively explored in Artificial Intelligence. Although it may correspond to a useful epistemological commitment, it has not necessarily also an architectural equivalent.[4]

We are now in a peculiar position by claiming that knowledge is a resource for action and action is a resource for knowledge. This circular view

---

[4] Within THEO an experiment has been done to model action and perception as one and the same category of behavior.

on knowledge and behavior is reminiscent of Maturana's view of autonomous systems [10]. In what follows we will push a little further in this direction.

4.2 Representation for Coordination

In [19] an architecture is described that implements the above model of problem solving. It makes different commitments by distinguishing between 'reasoning' and 'doing', which are implemented by *methods* and *handlers*. The methods are, as usual, for reasoning. The handlers are for "doing", i.e., for provoquing side-effects. The combination and interaction of methods and handlers generates the behavior of the agent. Doing is not a goal in itself but brings the environment in such a state as to make reasoning work.

As it turns out there is finally not much going on in the method. If we think of the case model, then its shape (e.g., the heuristic classification horse-shoe [4]) is actually determined by the problem solving method that is underlying the reasoning. However, its role in reasoning is, in a sense, minimal. All that is required is some appropriate coordination between the different handlers that are working on it. If we take the bold step of viewing the problem solving method as the representation (which is not so bold in light of the notion of inference structure as used in [18]) then we can state that the role of representation is the coordination of behaviors.

So our next hypothesis is that representation is a means for coordination. An experiment by Ming Tan on cost-sensitive decision trees [17, 16] illustrates the same idea in a robotic context (Figure 12). Ming has developed a system for an approach and recognize task. A decision tree expresses the knowledge on how to recognize cups in a real world scene, observed with a scanner mounted on the arm of a mobile robot. The decision tree, which is illustrated in Figure 12 is used to determine the cheapest type of scan that will lead to additional information in order to classify an object as being a cup or not. Cost is determined by the cost of positioning the robot with respect to the object, and the cost of executing the scan. After every decision the robot moves to a location (i.e., distance to the object under investigation) where a particular scan has the best effect. This information is acquired during a training phase.

One can say that the decision tree representation regulates the coordination between the activities of the robot. Some of these activities are normally called actions (like moving to a location); other perceptions (the scan operations). But clearly the boundary is vague. They are just activities that change the state of the robot in some way and thus are equally valid to call them actions or perceptions. The decision tree representation coordinates all of these in a coherent and cost-effective approach and recognize behavior.

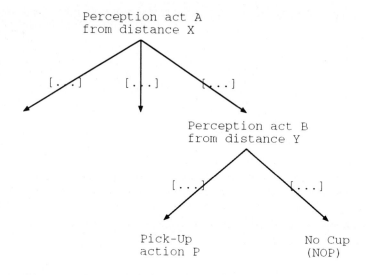

Fig. 12. A partial decision tree like those used in Ming Tan's experiment. Every decision node requires a particular perception from a particular distance to the object. Depending on the outcome of the test another such perception may be required, a pick up action is selected (the class) or it is decided not to pick up the object.

4.3 Ways of coordination

To more clearly define the issue of coordination let's take a multi-agent scenario. When do we say that two agents are behaving in a coordinated fashion (Figure 13).

Knowledge level coupling: whenever an observer rationalizes the behavior of multiple agents by ascribing goals and knowledge to them that, assuming their rational behavior, explains their behavior, we speak of knowledge level coupling between the agent processes.

Knowledge level coupling occurs in many situations. For example a good soccer team coordinates the different players in a rationalisable way. Also the coordination between the two teams is an example of knowledge level coupling (see [3] for more on this example). In a more traditional context phenomena of performance and explanation are knowledge level coupled: although a problem may be solved in a very different way that it is being afterwards explained, there has to be a rational relation between the two for the explanation to be convincing.

Symbol level coupling: whenever two systems coordinate by sharing or exchanging symbol level structures we call them symbol level coupled.

Symbol level coupling is a traditional trick in AI. The emphasis on knowledge and representation exerts its thorough influence here. For example,

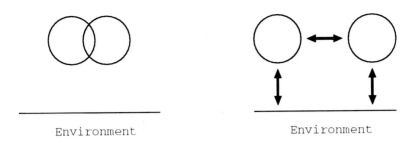

Fig. 13. Two forms of realizing coordinated behavior. In symbol level coupling, to the left, agents share or exchange representations. In structural coupling, to the right, their structures are such that, within the specific environment they share, their behavior is coordinated. Both forms of coordinating would seem like knowledge level coupling to an observer.

in machine learning focus has been for many years on the acquisition of knowledge representations and concept learning. Agent behavior, if considered at all, was being achieved via the representation of the knowledge underlying the behavior. With the rise of multi-agent scenarios, learning research has focussed on collaborative concept learning. Indeed, it was envisaged that coordination of multiple agents could be done only if they shared at least some aspects of each others goals and knowledge.

Structural coupling: whenever two systems coordinate without exchange of representation, but by being mutually adapted to the influences that they experience through their common environment we call them structurally coupled [11].

As extensively argued by Maturana and Varela (e.g. [11]) structural coupling can explain a variety of biological, cognitive and social phenomena.

4.4 External coordination

In a previous section we have explained the role of representation for coordination. To explain the nature of these representations, we first need to distinguish between internal and external representations (Figure 14).

External representations are physical structures in the environment that play a role in the mutual structural coupling of agents. The coordinated behaviors of agents are triggered by such external representations. A particular coordination of behavior must not be ever-present. It is only intelligent under certain situations. These situations are initially "represented" by external representations. For example, a side walk is not simply the thing that looks like what we are used to call a side-walk. It is the thing that plays a coordinating role in the behavior of pedestrians and drivers. Without that sidewalk no coordination, without the coordination no sidewalk. Or, to take up the

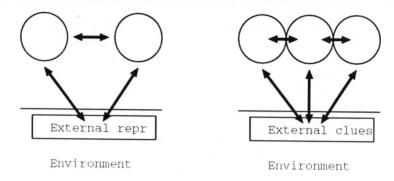

Fig. 14. External representations play a role in coordinating behaviors. External representations are internalized by having a third process that, in the presence of the right external clues, leads to a coordination as if the external representation were present.

soccer team again, the coordination of the players (within and across teams) is mediated primarily by... the ball. And, to go back to the problem solving example, the case model – which we viewed as the representation – plays the coordinating role between all the activities that gather information (actions and perceptions).[5]

4.5 Internalizing representation

An intelligent system is viewed as a collection of coordinated processes. The coordination happens, either by structural coupling through the environment, or by structural coupling with other mediating processes within the same agent. We envisage that the coordinating effect of an external structure can be taken over by a mediating process within the agent that, under the right circumstances (i.e., external clues in the environment) leads to the coordination of the different processes as if the external structure were present. How the processes actually get triggered is unclear but it does seem to rely on external clues or cues (Figure 14).

The hypothesis is that here cognition comes in. The coordinated behavior is as if mediated by representation, but it is achieved without representation. Representations originate in the physical world outside, but they are internalized as soon as processes within the agent take over the role of coordination. For example, the role of the hand that guides the child in its first steps is later taken over by an internal coordinating process. However, these

---

[5] This is a slight re-interpretation of Clancey's argument [5] for the usefulness of the metaphor of blackboard: it holds the external representation that regulates the coordination between multiple experts (knowledge sources).

coordinating processes are not structures as such but a process (with no special status and similar to the others) that influences the others in such a way that they become coupled in a particular way, i.e., to function in a seemingly coordinated fashion.

Why not do it with representation then? One might say that the agent, at the moments of behavior, might have the representation. Indeed, most of the distinctions that an agent seems to be making in the eyes of an observer must somehow be the result of a structural feature of that agent. In that way, however, just about anything within that agent that has a role in making that distinction is part of the representation and the notion of representation looses all value. Moreover, it could be that the structural elements start playing a completely different role in another situation (which is a feature of structural coupling, as argued by Maturana and Varela [11]). Thus, the representation looses its character of having an intrinsic meaning. Rather it acquires its meaning only when it is effectively used.

5   Conclusion

This chapter has investigated the relationship between intelligence and the agent-hood of an intelligent agent. Architectural proposals typically focus on one of these and the approaches are so dramatically different that one is inclined to believe in hybrid architectures as the only solution.

The purpose was not to give a complete overview of the different architectures that have been proposed. We have only highlighted some of the techniques and problems that have been used to apply cognitive architectures in the context of physical autonomous agents.

In the second part of the chapter we have speculated about another way to link cognition and behavior. We have taken the reader through a number of arguments that motivate the conclusion that coordination between processes is the key element in behavior and cognition. Representation, we have argued, originate in the external environment, and are internalized as coordinating processes. At best they may generate the idea of sensation of a representation but there is no structure in it that deserves the name, really.

Admittedly we are nowhere near to a full theory yet. However, to recapitulate from the introduction, the research direction brings together results from behavior-based and cognitive approaches, applies equally much to single agent as to multiple agent systems, and at least hints at an explanation of the (epi)phenomena of internal representation for cognition.

Acknowledgments

I wish to thank Luc Steels and Rodney Brooks for the opportunity to explain and develop these ideas. Many others have influenced these ideas over the last

5 years, during a stay at CMU (Tom Mitchell, Jeff Schlimmer, Ming Tan), at the CEAB in Blanes, Catalunya (Enric Plaza, Ramon Lopez de Mantaras, Carme Torras), and at the VUB AI-Lab. I recognize major influences from William Clancey, Tim Smithers, Rolf Pfeifer, Luc Steels, and Rodney Brooks, in no particular order. The author is presently a senior researcher for the Belgian National Science Foundation (NFWO).

References

1. J.R. Anderson. The Architecture of Cognition. Harvard University Press, Cambridge, MA, 1983
2. J. Blythe and T. Mitchell. On becoming reactive. In Proceedings 6th International Conference on Machine Learning, pp. 255–257. Morgan Kaufmann, 1989
3. P. Brazdil, M. Gams, S. Sian, L. Torgo, and W. Van de Velde. Learning in distributed systems and multi-agent environments. In Y. Kodratoff (ed.), Machine Learning–EWSL–91, pp. 412–423, Berlin, 1991. Springer-Verlag
4. W. J. Clancey. Heuristic classification. Artificial Intelligence, 27: 289–350, 1985
5. W. J. Clancey. Model construction operators. Artificial Intelligence, 53(1): 1–115, 1992
6. J.H. Gennari, P. Langley, and D. Fisher. Models of incremental concept formation. Artificial Intelligence, (40): 11–61, 1989
7. J. E. Laird, A. Newell, and P. S. Rosenbloom. SOAR: an architecture for general intelligence. Artificial Intelligence, 33: 1–64, 1987
8. J. E. Laird, E. S. Yager, M. Hucka, and C. M. Tuck. Robo-Soar: An integration of external interaction, planning, and learning using Soar. In W. Van de Velde (ed.), Toward Learning Robots, pp. 113–130. MIT Press, Cambridge, MA, 1993
9. Pat Langley, Kevin Thompson, Wayne Iba, John Gennari, and John Allen. An integrated cognitive architecture for autonomous agents. Technical report, Department of Information and Computer Science, University of California, Irvine, CA, 1991
10. H. Maturana. Biology of cognition. In F. Varela and H. Maturana (eds.), Autopoiesis and Cognition. Reidel, London, 1980
11. H.R. Maturana and F. J. Varela. The Tree of Knowledge: The biological roots of human understanding. Shambala, Boston and London, revised edition, 1992
12. T. Mitchell, J. Allen, P. Chalasani, J. Cheng, O. Etzioni, M. Ringuette, and J. Schlimmer. Theo: A framework for self-improving systems. In K. VanLehn, editor, Architectures for Intelligence. Erlbaum, Hillsdale, NJ, 1990
13. A. Newell. The knowledge level. Artificial Intelligence, 18: 87–127, 1982
14. A. Newell. Unified Theories of Cognition. Harvard University Press, Cambridge, MA, 1990
15. L. Steels. Components of Expertise. AI Magazine, 11(2): 29–49, 1990
16. M. Tan. Learing cost-sensitive decision trees. PhD thesis, School of Computer Science, Carnegie Mellon University, Pittsburgh, PA, 1992
17. M. Tan and J. Schlimmer. Cost sensitive concept learning of sensor use in approach and recognition. In Proceedings of the 6th International Machine Learning Workshop, 1989

18. W. Van de Velde. Inference stucture as a basis for problem solving. In Y. Kodratoff (ed.), Proceedings of the 8th European Conference on Artificial Intelligence, pp. 202–207, London, 1988. Pitman

19. W. Van de Velde. Reasoning, behavior and learning: a knowledge level perspective. In Proceedings of Cognitiva 90, pp. 451–463, November 1990

20. W. Van de Velde. Tractable rationality at the knowledge-level. In L. Steels and B. Smith (eds.), Proceedings AISB'91: Artificial Intelligence and Simulation of Behaviour, pp. 196–207, Springer, Berlin, 1991

21. W. Van de Velde. Issues in knowledge level modelling. In J-M. David, J-M Krivine, and R. Simmons (eds.), Second Generation Expert Systems, pp. 211 – 231. Springer, Berlin, 1993

22. K. VanLehn. Architectures for Intelligence. Lawrence Erlbaum, Hillsdale, NJ, 1989

Circle In The Round:
State Space Attractors for
Evolved Sighted Robots

Philip Husbands, Inman Harvey, Dave Cliff
School of Cognitive and Computing Sciences
University of Sussex, Brighton BN1 9QH, UK
email: philh or inmanh or davec@cogs.susx.ac.uk

Abstract. This paper presents an analysis of an artificially evolved dynamical network-based control system for a simulated autonomous mobile robot engaged in simple visually guided tasks.

Keywords. Genetic algorithms, recurrent dynamical networks, visually guided behaviours, dynamical systems

1 Introduction

After explaining our methodology for artificially evolving control systems for autonomous mobile robots, this paper presents and analyses recent results of experiments in concurrently evolving simulated robot control systems and sensor morphologies. The robots are engaged in simple navigation-based tasks in differing environments. Recurrent dynamical 'neural' networks make up the control system, and the primary sensory input is visual, via a pair of minimal bandwidth sensors. The structure and size of the networks are under evolutionary control as are properties of the visual sensors.

The work described here forms part of the early stages of a research program to thoroughly explore the potential of artificial evolution for developing autonomous agents. In order to progress most effectively, we think it is important to have a firm understanding of the evolutionary mechanisms and the systems they produce. General forms of analysis are necessary to throw light on, for example, the necessary conditions for the development of certain types of behaviours; whether or not there are underlying general behaviour-generating principles for classes of evolved agents[1]; whether or not different classes of artificial neurons result in significantly different evolutionary trajectories. They will also be needed to find out how robust any given evolved agent is, how general its behaviours are, and so on. Hence a major aim of this paper is to present some of our tools for analysis and to show how they can be used.

---

[1] A detailed elucidation of these may have interesting repercussions throughout AI and Cognitive Science.

Using a dynamical systems perspective, we give a thorough and general analysis of one of the successful evolved simulated robots. We show that it is robust in the face of noise and is highly fit over a large range of environments from a particular class. The controller is general: its dynamics are *not* the same in the different environments. The analysis of behaviour makes use of an appropriate state space. We are able to show that in some of these environments the entire state space is a single basin of attraction for a point attractor corresponding to a high scoring behaviour according to the task-based evaluation function used during evolution. The state spaces for the other environments have two attractors, both corresponding to highly fit behaviours. In other words, the robot has evolved to the point where it is 100% guaranteed to succeed at its given task – the visuo-motor couplings, via network dynamics and the visual structure of the robot's world, are in perfect harmony relative to the evaluation task. It is pointed out that these robots, being dynamical systems, use forms of animate vision.

Some of the problems of simulation work, particularly those involving vision, are briefly discussed. This leads into a description of ongoing work in which specialised visuo-robotic equipment is used to concurrently evolve visual morphologies and control networks *without* recourse to simulation of the agent environment coupling.

The structure of the paper is as follows. The early sections establish the scope of our work, justify the use of artificial evolution, explain its mechanisms, and discuss our reasons for basing the control systems on recurrent dynamical neural nets. The particular artificial neurons used in the work reported in later sections are then detailed. Next we move on to describe a series of experiments involving the concurrent evolution of control networks and visual morphologies. Techniques for analysing the evolved control systems are presented. These are then used to give a detailed analysis of one of the systems evolved in the experiments described earlier. There follows a discussion on how such analyses might be extended to tackle more complex cases. The paper ends with an introduction to current work in which artificial evolution is performed in the real world without the need for simulated sensing.

Lack of space means that a great many issues relating to the work are not dealt with or only very briefly covered. For further details, especially on methodological issues, see [22, 10, 16, 9, 23].

2 What sort of robots?

We are investigating mobile robots in environments with flat floors, where it is assumed that wheeled motion is relatively easy (subject to some slippage); and where there are obstacles and walls. So the focus is on navigation, using touch sensors and vision for sensory inputs, and motor outputs to wheels. One of the mobile robots built in our group is battery-powered, about 40 cms diameter and 30 cms high, running on left and right wheels with independent motors, and a third trailing castor for stability [29]. The sensors on this include the whiskers and bumpers as shown in the plan view of Figure 1. An onboard notebook 486 PC implements any desired control system. The signals to the motors can be represented as a real value in the range [-1.0,1.0]. This range is divided up into five more or less equal segments, and depending on which segment the signal falls into, the wheel will either remain stationary or rotate half/full speed forwards/backwards.

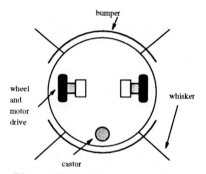

Fig. 1. Plan view of the mobile robot

Our simulation work has been based on this robot, and an earlier prototype, with the addition of vision. The simulations are at the physical level, with the characteristics of the motors, collisions with obstacles, slippage on the floor etc. modelled to include appropriate noise profiles. For vision, both in simulation and in reality, low bandwidth 'insect-like' eyes are assumed; in effect a small number of photoreceptors with genetically specified angle of acceptance and direction. The vision simulation uses ray-tracing with anti-aliasing via 16-fold super-sampling. Simulated vision rapidly becomes very expensive, and difficult to make accurate, as the required resolution increases and the environments become more complex. Towards the end of this paper we discuss our solution to this problem. However, the experiments described in this paper were of a minimalist enough nature that simulation was feasible.

The aim of the work described here was to explore some of the basic issues in artificial evolution, rather than to develop a practical controller for a particular robot. We regard realistic simulations as being suitable for these purposes. However, in other work we have used our evolutionary methodology to develop controllers for real robots [17, 24].

3 What model of cognition?

Inevitably anyone's approach to building autonomous robot control systems is determined, consciously or unconsciously, by their model of cognition. Our starting point is that we see a robot, and indeed any organism, as a physical dynamical system, with its own internal dynamics, coupled to a world (also a dynamical system) through sensory-inputs and motor-outputs. The sensory-inputs perturb the internal dynamics without constraining them [2, 30, 31]. We completely ignore the 'problem' of how internal models can accurately represent the 'real' world. We believe that such types of analysis, which may or may not be useful at the behavioural level of description, are positively harmful if taken uncritically as a blueprint for the construction of a physical dynamical system, a 'robot brain'.

A dynamical system is taken here to be any system that can be characterised by a finite number of state variables, and a dynamical law that specifies how those state variables change with time. Computers inhabit one very small corner of the space of all dynamical systems, and the all-too-common assumption that cognition is some form of computation is, we feel, a stultifying restriction on possible models for a robot control system, as well as in the rest of AI.

We are sympathetic to Brook's rejection of the dogma of functional decomposition [6, 7, 5]. However, we are sceptical of the alternative of behavioural decomposition advocated by Brooks, if it also is taken as a dogma (not that we are suggesting that he does). The methodology we advocate does not rely *a priori* on any form of decomposition, although control systems thus produced are open to analysis *a posteriori*; which may reveal in practice that some sort of decomposition is then possible or useful.

4 Why use evolution?

Brooks' approach very sensibly advocates an incremental approach to building up more sophisticated behaviours 'on top of' previous simpler ones. Each layer of behaviour is implemented from sensors through to motor-output largely independently of the other layers, the interactions between such layers being restricted to suppression and inhibition of lower layers. Through this method, the decomposition by activity of the behaviour is converted into decomposition by layer of the behaviour-producing mechanism. Each layer should be thoroughly debugged before adding the next layer, and the restrictions on interactions between the layers aids the designer's task.

Nevertheless, even with such restrictions, the addition of a new layer of mechanism has repercussions over and above any directly wired in links of suppression or inhibition — repercussions arising from the fact that the robot is coupled with its world. As the number of layers increases, the number of such repercussions, intended or unanticipated, can scale up horrendously;

as well as first-order repercussions, these in turn can stimulate higher-order consequences. In design, the purpose of modularity, or of decomposition of any kind, is to help the human designer minimise the unanticipated, but we believe that the practical limits of this approach are already being reached [22].

Evolution in the natural world does not suffer the same limitations that a human designer toils under, and each of us is an existence proof of the ability of evolution, given appropriate resources, to produce sophisticated control systems. Artificial evolution (subject to some caveats) allows us to specify *what* behaviour we desire without specifying *how* this should be achieved. Hence, as further discussed in [16, 9], we advocate artificial evolution for the development of control systems for autonomous mobile robots. Here sensors are taken to be an essential and integrated part of the control system. Hence they should be subject to evolutionary alteration along with the rest of the system. Ideally so should motor properties. However, the genetic specification of motor properties and general robot morphologies are currently beyond the scope of our work.

5   What is artificial evolution?

Artificial evolution involves the application of Genetic Algorithms (GAs) [14, 20], which are algorithms using a few ideas borrowed from natural evolution, and primarily used for function optimisation in highly complex domains where analytic solutions are not possible [27, 4, 13]. Standard GAs can be considered as search techniques operating in pre-defined search spaces, using populations of trial points in the spaces to guide where the next generation of such trial points should search.

A mapping is defined from a string of characters (the analogy with the genotype of an organism) to a trial point in the search space which can be evaluated (the 'phenotype'). One of the caveats alluded to above is that this mapping should be such that, in some sense, similar genotypes are associated with phenotypes that are also in general similar. In the present context, a mapping should be chosen such that any string of characters from some alphabet, of a predefined length, can be interpreted as specifying the control system for a robot.

In a standard GA[2], an initial population (perhaps of size 100) of randomly created genotypes would specify the initial population of control systems. One at a time, each of these is evaluated on the given task(s) to be tackled in the given environment; the result of evaluation is a real number. This number is translated into a measure of fitness. In an initial randomly generated population, of course, most will be useless; but any that score higher than average

---

[2] There are a number of variations on this basic algorithm, some involving asynchronous parallel distributed processes, but in all the basic elements of selection, reproduction, recombination and mutation are still at the core [21].

are given a higher than average chance of contributing genetic material to the next generation.

This is done by creating a reproductive pool of genotypes biased towards fitter members of the population. From this pool of 'parents' genotypes are taken in pairs; a randomly chosen crossover point is taken, and an offspring genotype is created by taking part, as defined by the crossover, from one of the parents, and the remainder from the other parent. A mutation genetic operator, applied at a given small rate of probability, mutates some of the characters in the genotype. The offspring thus created replace members of the previous generation and the evaluate-breed-replace cycle continues.

This process continued through successive generations, with the aim of seeing fitness rise over time. Although individual elements of the algorithm are random, the result is *not* random search, but something much more powerful. It can be shown [14, 20] that the algorithm exploits useful building blocks or schemata — roughly speaking, any fairly short lengths of a genotype which may 'code for' some useful part of the phenotype. In a population of size n, the number of such schemata that are usefully processed each generation is $\mathcal{O}(n^3)$.

6 Is our GA different?

GAs are usually used for optimisation, but evolution in the natural world is not optimisation in a pre-defined search space. There was no 'problem' posed several billion years ago for which present-day animals are some sort of 'solution'. The world that we live in was itself largely formed in co-evolution with our ancestors. What has happened in evolution, which is not normally permitted in standard GAs, is the development over the long term of more complex structures (associated in general with longer genotypes) from simpler ones. The notion of a pre-defined search space, whose dimensionality is associated with a fixed genotype length, becomes less useful when genotype lengths are allowed to increase to any arbitrary length.

A single well-defined task, or set of tasks, for a robot control system to tackle, can be taken as defining a restricted search space. Standard GAs are one possible tool for optimising a control system for the given problem. But looking ahead to the need to add new tasks for robots to do that were not originally anticipated — and on the inevitable demands for incremental adaptation of pre-existing systems — it follows that we need to be able to do evolution rather than optimisation. The SAGA framework [18] has been developed to deal with the rather different dynamics of GAs when genotype lengths must be allowed to increase to arbitrary lengths through evolution.

It turns out that when this happens, populations are inevitably largely genetically converged, as a *species* — SAGA stands for Species Adaptation Genetic Algorithms. The genetic operators must allow genotype lengths to change from one generation to the next, but it can be demonstrated that any such increases in genotype length will be restricted to gradual small steps[3]. Selection needs to be rank-based, and generally should be significantly stronger than in a standard GA (where people are usually trying to avoid premature genetic convergence; in SAGA, the population is always converged). Whereas in standard GAs, recombination is taken to be the dominant genetic operator, with mutation relegated to a background operator for ergodicity, in SAGA mutation is a much more significant force, and is applied at a higher rate, of the order of one mutation per genotype [18, 19, 15].

7   What building blocks for a control system?

We are relying on evolution for the design of a control system, but we must choose appropriate building blocks for it to work with (this is another of the caveats mentioned earlier). Some have advocated production rules [12] (Classifier Systems [14] are the GA version of these). Some propose Lisp-like programming languages [26]. Brooks [8] has proposed using Koza's ideas applied to a high-level behaviour language. Beer [3, 32] has used dynamical neural networks. It is only this last approach that we are in broad agreement with.

Our intuitions, supported by our simulation results and recent work with real robots [17], are that the primitives manipulated by the evolutionary process should be at a rather low-level. Any high level semantic groupings inevitably incorporate the human designer's prejudices, and will probably give rise to a more coarse-grained fitness landscape with more steep precipices. But evolution requires that the fitness landscape is in general not too rugged. It might be thought that the use of low-level primitives necessitates aeons of trials before any interesting high-level behaviour emerges, but our experience indicates otherwise.

Another factor concerning high-level languages is that the injection of noise into a system seems contrary to the rationale for such languages. The injection of noise into the lowest levels of a control system can be shown to have valuable effects on the dynamics; and has the additional benefit of blurring the fitness landscape and making it less rugged for evolution.

Our criteria for deciding on component primitives are:

- They are the primitives of a dynamical system.
- The system should operate in real time, and the timescales on which the components work must be in some sense appropriate for the world the robot inhabits.
- The system should be 'evolvable', not 'brittle'; in the sense that many of the possible small changes in the way components are bolted together should result in only small changes in resulting behaviour.

---

[3] What counts here as a small step is dependent on the ruggedness of the particular fitness landscape in genotype space. It is associated with the correlation length of such a landscape [25].

– Incremental change in the complexity of any structure composed of such primitives should be possible.

There may be many possible components and general architectures that meet these criteria. The particular choice we have focused on is that of recurrent dynamic realtime networks, where the primitives are the nodes in a network, and links between them. The nodes act much as many artificial 'neurons' in a neural network, though with particular characteristics. The links are unidirectional, have time delays between units, and may be weighted. There are no restrictions on network topologies, arbitrarily recurrent nets being allowed. When some of these nodes are connected to sensors, and some to actuators, the network acts as a control system, generating behaviours in the robot.

We are exploring a whole class of networks with these general properties. The next section details the particular type of net used in the experiments described later. It was chosen as a simple example of the sort of net we are interested in.

7.1 Artificial neuron mechanism

The artificial neuron mechanism employed to date has been designed for its usefulness in control applications rather than for biological plausibility or ease of analysis. Figure 2 represents the operation of a neuron. There are separate channels for excitation and inhibition. Real values in the range [0,1] propagate along excitatory links subject to delays associated with the links. The inhibitory (or veto) channel mechanism works as follows. If the sum of excitatory inputs exceeds a threshold, T_v, the value 1.0 is propagated along any inhibitory output links the unit may have, otherwise a value of 0.0 is propagated. Veto links also have associated delays. Any unit that receives a non zero inhibitory input has its excitatory output reduced to zero (i.e. is vetoed). Note that this means that a unit may be vetoed but still able to produce inhibitory output, as long as the sum of its excitatory inputs are high enough. In the absence of inhibitory input, excitatory outputs are produced by summing all excitatory inputs, adding a quantity of noise, and passing the resulting sum through a simple linear threshold function, $F(x)$, given in Equation 1 below. Noise was added to provide further potentially interesting and useful dynamics and to give an indication of the properties of a physical implementation of a unit which would be likely to include naturally occurring noise. The noise was uniformly distributed in the real range [-N,+N] (see Section 8 for further discussion of the role of noise). In the work described here all connections had unit weights.

$$F(x) = \begin{cases} 0, & \text{if } x \le T_1 \\ \frac{x-T_1}{T_2-T_1}, & \text{if } T_1 < x < T_2 \\ 1, & \text{if } x \ge T_2. \end{cases} \tag{1}$$

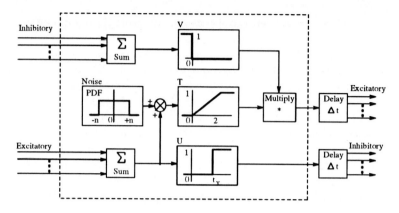

Fig. 2. Block diagram of neuron operation

In all our work to date the networks have been simulated in software. Their continuous nature is modelled by using a very fine time slice approach. At each time step of the robot kinematics simulation, sensor readings are fed into the network, the network is run through a number of cycles (with a variance to counter distorting periodic effects), and then the network outputs are converted into motor signals which allows the calculation of the new position of the robot. Time delays on the connections refer to network cycles, hence a unit delay is much smaller than the duration of a single robot step. In the work described in this paper unit delays are used throughout. The values of the other parameters mentioned above were set at N=0.1, T_v=0.75, T_1=0.0 and T_2=2.0, giving a slope of 0.5 to F.

7.2 Evolving the networks

In the experiments analysed later, a fixed number of nodes are identified as input nodes. These receive input directly from the sensors. There is one such node for each sensor. There are also a fixed number of output nodes, two for each motor. The outputs of each motor neuron pair are differenced to give the full motor signal range of [-1.0,1.0]. These signals are passed on to the relevant motors. There are an arbitrary number of internal nodes which fall into neither of the previous two classifications. The genetic encoding specifies the number of internal nodes, individual neuron properties, and the how the neurons are linked up. The size and topology of the network is unrestricted. The genotypes are strings of characters interpreted in a sequential manner from left to right. The encoding has been designed so that genetic operations (crossover and mutation) on 'parents' always produce 'children' coding for legal networks (see [16, 9] for full details).

As mentioned earlier, in much of our work to date, the weights and delays on network connections are set to a unit value, and the value of this delay implicitly determines the timescale on which the network operates. We are now moving towards having delays of widely ranging values. Learning (to use a behavioural level of description) can be considered as no more than the consequence of internal dynamics at different timescales; changing of weights in a network is just one of the ways in which this can be achieved.

8 Why do we use noise?

As will be shown later, the internal noise in the networks significantly alters their dynamics. In addition this helps to make the control system more evolvable and less brittle. Incidentally, it underlines the point that it is not very useful to try and interpret the networks as 'computing' outputs from inputs; rather, the network is a particular dynamical system perturbed by the inputs.

In addition, in our simulations at the physical level a significant amount of noise is deliberately added to sensor inputs, motor outputs, and movement in the world such as that arising from collisions. The noise profiles are based on measurements of the real system. Each robot is scored over a number of trials, and the *worst* of these scores is used as the final evaluation. In this way we wish to encourage robustness, and on transfer from simulation to the real world our goal is that the 'real' values of parameters should lie within the 'envelope of noise' we have created around the parameter values used in simulation. For details of recent work in which this transfer is successfully achieved see [24].

9 Experimental results: vision in an arena

Earlier results in simulation using navigation without vision in a cluttered environment, and with vision (two photoreceptors) in a single fixed circular arena, have been reported elsewhere [10, 9, 16]. This paper will report on new results using vision in differing circular arenas.

The arena has a black wall, and a white floor and ceiling. As well as the tactile sensors shown in Figure 1, the simulated robot is provided with two photoreceptors. These are placed at equal angles (the angle of eccentricity) either side of the the robot's forward facing midline. They each have the same angle of acceptance which specifies their cone of vision. Angle of eccentricity and angle of acceptance are illustrated in Figure 3. The simulated sensors produce a real number in the range [0.0,1.0] proportional to the brightness level in their field of view. As well as the control network size and topology, the photoreceptors' angles of eccentricity and acceptance are also under genetic control. The angle of eccentricity can vary between 0 and $\frac{\pi}{2}$ and the angle of acceptance between 0 and π. In this way the visual morphology is concurrently evolved along with the control network. The visual structure of

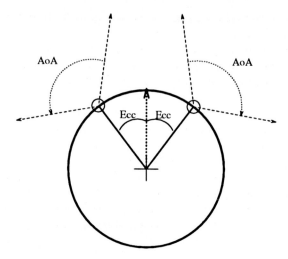

Fig. 3. Angle of acceptance and eccentricity for the two-photoreceptor robot. A top-down view of the robot, and the relevant angles

the robot's world varies markedly over the range of values of these two angles (see [9] for details).

The larger part of the genotype encodes a network, as described earlier. A genetic algorithm using SAGA principles was then used, with populations of size 60.

The evaluation of each robot involved decoding the genotype for the control network and visual morphology. The resulting robot was then given 8 trials. At the start of each trial it was placed in a random orientation and position (this position being restricted to a band near the wall). A fixed length of time was allowed (100 simulation timesteps), during which it was evaluated according to:

$$\mathcal{E} = \sum_{\forall t} \exp(-s \mid \mathbf{r}(t) \mid^2)$$

where $\mathbf{r}(t)$ is the 2-D vector from the centre of the arena to the robot's position at time t, and $\forall t$ denotes the entire (finite) lifetime of the robot. The sum is over the time steps of the simulation. The parameter s in this Gaussian is chosen so as to ensure that the robots collect virtually no score near the edge of the arena, and hence the evaluation function $\mathcal{E}$ implicitly sets the goal of: reach the centre of the arena as soon as possible and stay there. At the end of the 8 trials the *worst* (i.e. lowest) value of $\mathcal{E}$ is assigned as the fitness of the robot: this encourages robustness.

In the first set of experiments the cylinder radius was fixed at 20 units and the wall height at 15 units. The robot radius was 2 units. Success on this task was reached on a number of runs, within 100 generations each time. Two different successful control systems, termed C1 and C2, have been described elsewhere [11]. In both cases, the robots make a smooth approach towards the center of the arena, and then circle there, either on the spot or in a minimum radius circle. A typical behaviour for C1 is shown in Figure 4 and the network that generated it is illustrated in Figure 5. Compare these with the behaviours and networks of C2 given later; they are quite different in many respects but share some basic underlying principles revealed by the kind of deeper analysis given later. Before analysing the C1 and C2 results fully, we had speculated that the control system might be recognising that the centre of the arena had been reached by monitoring the absolute values of the photoreceptor inputs: at the centre of the world, the photoreceptor inputs would take on an absolute value determined by the ratio of wall-height to floor-radius of the particular circular arena used during evolution.

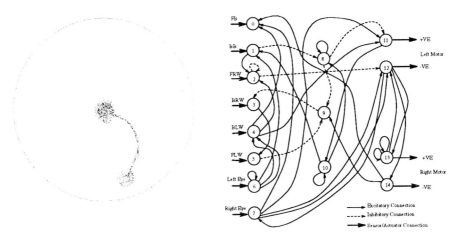

Fig. 4. Typical path of a successfully evolved robot, which heads fairly directly for the centre of the room and circles there, using input from 2 photoreceptors. The direction the robot is facing is indicated by arrows for each time step, largely superimposed

Fig. 5. The network diagram for the robot that produces the behaviour shown in the previous figure

With this in mind, we set up a more difficult task, where the height of the wall could vary over one order of magnitude, from 4 to 40 units; the diameter of the arena remaining at 40. The full range of possible heights was divided into 10 equal sections, and each robot was given 10 trials, with a height of wall chosen at random from each in turn of the 10 sections. In this way it could be ensured that it was tested across the full range. Thus no use could be made of absolute light values; as is usual in our experiments, the evaluation of the robot was based on the *worst* score it obtained across its trials.

Once again, on several runs success was reached within 100 generations. Furthermore, on analysing the successful robots it was found that they had evolved with either both photoreceptors pointing straight ahead, or at the minimum angle apart such that their cones of acceptance largely overlapped. In other words, for all practical purposes they had thrown one photoreceptor away and were succeeding with a single receptor of one pixel.

For the purposes of comparison, some robots that had earlier been successfully evolved for a single fixed wall-height, were tested across the full range of wall-heights used in the second experiment. Much to our surprise, although many failed, C2 succeeded, and in fact did better than any of those evolved under the varying conditions. On analysis, there were certain similarities among all the successful robots, including C2.

Whereas C2 had its photoreceptors, with a 45° angle of acceptance, placed at 60° to each side of the robot's centre line, a detailed analysis shows that it only ever makes use of a single receptor at any given time, although whether it is using the left or the right one varies with the immediate environment. Hence C2 is effectively monocular, although in a more subtle way than the others. Figure 6 shows the average performance of C2 over the range of wall heights. The average of 100 runs at each height are shown. The theoretical optimum score is between 75 and 80. But because the whole system is shot through with noise and the Gaussian tapers off very sharply from the centre, any score over 60 is highly fit, meaning the robot spent the vast majority of its lifetime very close to the centre.

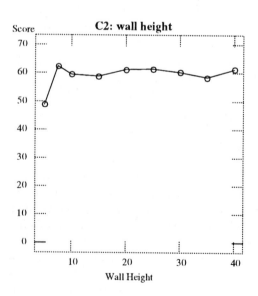

Fig. 6. C2 evaluation scores over different wall heights

Space does not allow a full analysis of more than one network, so because it has the largest number of interesting properties, many of them shared with one or more of the other successful controllers, and illustrates the analysis techniques particularly well, C2 has been chosen. Before it can be analysed in detail, the techniques used must be introduced.

10 Analysis

In the following sections we show that the evolved controllers can be thoroughly understood and we explain how they provide general solutions. The first two subsections deal with the main techniques used in the later full analysis of C2. These are an analysis of the control networks based around its feedback loops, and the use of a low dimensional state space for visualising how the robot behaviours vary with the visual signals being received. The crucial idea is that to fully understand how the evolved control system works, it is necessary to analyse in a general way the interplay between (i) the visual structure of the robot's world determined by its evolved visual morphology and (ii) the control network dynamics. This can be done if the visual structure of the robot's world can be mapped out, and it is understood how the motor outputs vary for all the possible left photoreceptor/right photoreceptor visual signal combinations in the world, while taking into account the internal network dynamics. With this information it should be feasible to construct vector flow fields, or phase portraits, showing how the robot behaviour varies across its world.

In the following analysis phase portraits are constructed in this way for the particular networks and class of worlds used in the experiments described earlier. The details would need to be a little different for other experimental setups, and the resulting state spaces may not be so easy to visualise, but the general idea should carry through. These issues are discussed more fully in Section 14.

11 Network analysis

Before going on to analyse particular controller networks, a commonly occurring sub-network configuration will be looked at in detail. This will enable a deeper understanding of the full networks described later. The details of the analysis are peculiar to the networks used in this work, but the general method could be used with many other types of noisy net. The use of more analytically contrived networks would probably entail a different, and presumably cleaner, analysis.

Nearly all the successful evolved networks we have studied involve one or more neurons with excitatory feedback loops. This arrangement often allows internal noise to rapidly build up the output from the unit to the upper threshold level without requiring any external input to the network. The unit then acts as a source of constant high output (unless vetoed) which typically contributes to the motor signals giving the basic behaviour of the robot. We will refer to such neurons as *generator units*. Their properties will fall out of the following general analysis of simple feedback loops. An understanding of such sub-networks can greatly simplify the analysis of full evolved controller networks. Their properties are not entirely obvious since internal noise is uniformly distributed with mean zero. In the following all connections have unit time delays, which is the case for all the experiments described in this paper.

(i)   (ii)

Fig. 7. Excitatory feedback loops

Consider the single neuron feeding back onto itself as shown in Figure 7(i). The input at time t, I_t, is related to the output at time $t-1$, O_{t-1}, by $I_t = F(O_{t-1} + n_t)$, where F is the transfer function of Equation 1 and n_t is the noise injected by the unit at time t. Ignoring the threshold terms for the moment, and concentrating on the linear part of the function, we can write:

$$O_t = k(O_{t-1} + n_t) \tag{2}$$

where k is the slope of the linear part of F. Hence,

$$O_0 = kn_0 \tag{3}$$
$$O_1 = k(kn_0 + n_1) \tag{4}$$
$$\cdots O_t = k^{t+1}n_0 + k^t n_1 + \cdots + kn_t \tag{5}$$
$$= \sum_{i=0}^{t} k^{t+1-i} n_i = \sum_{i=0}^{t} a_i(t) \tag{6}$$

where $a_i(t)$ is a substituted variable. Now the noise terms, n_i, can be considered as independent random variables uniformly distributed over the real range [-N,+N], giving them probability distribution function $P(x) = \frac{1}{2N}$. Using elementary probability theory, it can be shown that they each have mean $\mu_{n_i} = 0$ and variance $\sigma_{n_i}^2 = N^2/3$. The linear relationships between the $a_i(t)$s and n_is (for any fixed t and i) means that the $a_i(t)$s can be regarded as independent random variables with mean $\mu_{a_i(t)} = 0$ and variance $\sigma_{a_i(t)}^2 = k^{t+1-i} N^2/3$. In other words the output of the unit can be regarded as the sum of a series of independent random variables. To return to the temporarily ignored threshold conditions of F, because O_t can never become negative (lower threshold condition), we can use a version of the central limit theorem to guarantee that the distribution of the random variable Z given below,

$$Z = \frac{O_t - \sum_{i=0}^{t} \mu_{a_i(t)}}{\sqrt{\sum_{i=0}^{t} \sigma^2_{a_i(t)}}} \tag{7}$$

is closely approximated by the *positive* half of the standardized normal distribution $\mathcal{N}(0,1)$. Now, $\sum_{i=0}^{t} \mu_{a_i(t)} = 0$ and $\sum_{i=0}^{t} \sigma^2_{a_i(t)} = \sum_{i=0}^{t} k^{t+1-i} \frac{N^2}{3} = \frac{N^2}{3} \times \frac{k - k^{t+2}}{1-k}$, hence:

$$Z = \frac{O_t}{\frac{N}{\sqrt{3}} \times \sqrt{\frac{k - k^{t+2}}{1-k}}} \quad , (k \neq 1) \tag{8}$$

Clearly the properties of the feedback loop depend crucially on the value of k. Given that we have used a fixed slope of 0.5 for the neuron transfer function, k can be *effectively* altered by changing the number of feedback loops. This is because the transfer function involves a summation of inputs. With two loops $k = 1$, and $O_t = \sum_{i=0}^{t} n_t$. Because of the threshold conditions on the neurons, this means the output will follow a drunkard's walk between 0 and 1. With one loop $k = 0.5$ and, given that the maximum possible value for n_t is 0.1, the higher order terms become vanishingly small. Hence the output can be well approximated by:

$$O_t = k^3 n_{t-2} + k^2 n_{t-1} + k n_t \tag{9}$$

The expression for Z given in Equation 8 can be used to estimate the expected time for O_t to reach the upper threshold value of 1.0. Clearly this depends on the value of k, or at least the number of feedback loops as explained above. The larger k, the shorter the time. From Equation 2 and using the fact that the maximum negative amount of noise is -N, once the upper threshold has been reached it can only be maintained if $k(1 - N) > 1$, i.e. $k > 1/(1-N)$. Setting N=0.1, the value used in all our experiments to date, then $k > 1.11$. With three loops $k = 1.5$, and the probability of O_t reaching 1.0 by $t = 40$ is given by $2(1 - \Phi(0.0025))$, from the formula for Z, and where Φ is the cumulative distribution function of $\mathcal{N}(0,1)$. The value of this expression is greater than 0.999.

Hence we can conclude that with three feedback loops, and in the absence of other inputs, a neuron will easily reach the saturation condition within 40 network cycles, the minimum possible per *single* robot step, and therefore will act as a generator unit. With two feedback loops it will act as a noisy generator unit with its output performing a random walk between 0 and 1. This is exactly what is observed. In other words, these sorts of circuits rapidly settle down into stable dynamical regimes, although those regimes may involve stochastic elements. Since during this (extremely brief) settling down period changes in visual signals are ignored and motor outputs are not altered, networks built around circuits of this kind are amenable to the particular style of analysis developed over the following sections. This point will be covered in more detail later, where it will be seen to be crucial in allowing the use of the vector flow fields introduced in Section 13.2.

Surprisingly, the two unit mutual feedback loop shown in Figure 7(ii) has properties equivalent to the single unit loop. Using the same notation as above, $O_{1_t} = k(O_{2_{t-1}} + n_{1_t})$, with a symmetric expression holding for O_{2_t}. It is straightforward to show that,

$$O_{1_t} = k^{t+1}\eta_{1_0} + k^t\eta_{2_1} + \cdots + k^2\eta_{2_{t-1}} + k\eta_{1_t} \tag{10}$$

where,

$$\eta_t = n_{1_t} \quad \text{, even } t \tag{11}$$
$$= n_{2_t} \quad \text{, odd } t \tag{12}$$

Since n_1 and n_2 have exactly the same distribution this expression is identical to that in Equation 6 and the same analysis holds. Indeed, the interplay between the delays on the connections and the noise injected at each unit, means that a loop containing any number of units will have the same behaviour; any of its neurons can act as generator units.

Any external inputs to loops can be treated in a simple additive fashion, with the same sort of analysis as above holding. Specifically, if a unit has external input I_t and a single feedback loop,

$$O_t = \sum_{i=0}^{t} k^{t+1-i} n_i + \sum_{i=0}^{t} k^{t+1-i} I_i \tag{13}$$

When the loop has $k = 0.5$, this can be approximated by:

$$O_t = k^3(n_{t-2} + I_{t-2}) + k^2(n_{t-1} + I_{t-1}) + k(n_t + I_t) \tag{14}$$

If the loop from a unit involves m additional units the external input will be attenuated by k^m (assuming $k < 1$) before it feeds back into the unit. Because of the rapid settling down period of these circuits (over which time the visual inputs can be regarded as unchanging), visual inputs to networks built around these types of feedback loops can be regarded as perturbing the network dynamics in a straightforward manner.

Results from this section will be used later in analysing evolved control networks, where we will see that characterising a network in terms of its feedback loops can be a very useful tool in determining its stable state signal levels.

12 $r\phi$ Space

In order to understand the generation of the robot's behaviour, and in particular how the evolved network/visual sensor system is adapted to its simple environment, a global view of the visual structure of the (robot's) world would be useful. Coupled with an understanding of the control network's dynamics, particularly how motor signals change with visual inputs, in principle this global view would allow the construction of a vector flow field indicating the robot motion at each point in the world. How feasible this is for any given control system will depend largely on how noisy the network dynamics are. However, by definition, the dynamics of the most successful visually guided systems cannot be too noisy; otherwise there would be no correlations between visual signals and behaviours. Another factor in such an analysis is the dimensionality of the state space used. We can only usefully visualise low dimensional spaces. It turns out that the networks of interest here have properties which allow the use of a low dimensional state space and so analysis was feasible.

The uniformly lit cylindrical arena, with its uniformly coloured walls, floor and ceiling, affords some useful symmetries that can be exploited in such an analysis and visualisation of the robot control system dynamics. Cylindrical symmetry means that, for any given visual morphology, the visual inputs[4] are determined by the robot's distance from the centre of the arena (r) and the angle it makes with the arena radius passing through its centre (ϕ – in the following it is defined as the clockwise angle between the arena radius and the robot orientation arrow). This is illustrated in Figure 8: in terms of visual signals, situation A is identical to situation B. Hence the visual structure of the world for a particular robot can be mapped out by sampling visual signals for a set of values of ϕ between 0 and 2π at each of a set of distances (r) along a radius from the arena centre to the edge. Such a space can be used to individually map out signal levels for the left and right photoreceptors, or the combination of these two. The most global view is gained from mapping out the regions with different (left photoreceptor signal, right photoreceptor signal) combinations.

The later analyses using $r\phi$ spaces will be more easily understood after a brief examination of some of the geometric properties of this space. Three types of motion used in later analyses are covered. Note that in the ϕ dimension the 2π boundary wraps round to the 0 boundary, giving a cylindrical space.

- *Straight line*. Straight line motion in cartesian space does not map onto a straight line in $r\phi$ space. From Figure 9 it is readily seen that moving in a straight line backwards forces the following relationship:

$$rsin(\phi) = k \tag{15}$$

---

[4] Noisy sensors mean that average visual signals are being referred to.

When the robot starts from a position with $r\phi$ coordinates (r_0, ϕ_0),

$$k = r_0 \sin(\phi_0) \tag{16}$$

- *Rotation about right wheel.* Figure 10 illustrates a counter-clockwise rotation about the right wheel. Using the cosine rule, it can be seen that the following $r\phi$ equations hold:

$$k^2 = r^2 + rad^2 - 2\cos(\pi/2 - \phi) \tag{17}$$
$$= r^2 + rad^2 - 2\sin(\phi) \tag{18}$$

Hence,

$$r^2 - 2\sin(\phi) = k^2 - rad^2 = K \tag{19}$$

When the robot starts from a position with $r\phi$ coordinates (r_0, ϕ_0), and it has wheel base radius rad units,

$$K = r_0^2 - 2\sin(\phi_0) \tag{20}$$

- *Rotation about left wheel.* Similarly, for a clockwise rotation about the left wheel it can be shown that:

$$r^2 + 2\sin(\phi) = Q \tag{21}$$

where Q is a constant.

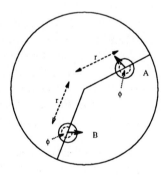

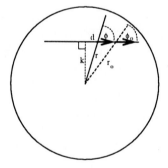

Fig. 8. Illustration of r and ϕ parameters. Robots at A and B, oriented along the thick arrows, have identical values for r and ϕ

Fig. 9. Relation between r and ϕ for straight line motion

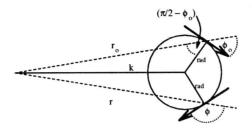

Fig. 10. Relation between r and ϕ for rotation about right wheel

13 C2 analysis

13.1 The network

The full C2 controller network is shown in Figure 11, its evolved visual morphology is: angle of eccentricity= 60^0, angle of acceptance= 45^0. This provides it with two fairly wide-angle photoreceptors with a large angular separation. A typical behaviour generated by C2, for arena wall height 15, is shown in Figure 13. As we shall see, at very low wall heights the behaviour is very similar except for the direction of rotation about the centre.

We will start our analysis of the network/visual sensor system by investigating some of the basic properties of the network. In this analysis we are interested only in the visually guided aspects of C2; unless it is started up against the wall, the tactile sensors are never active — it has evolved to make very good use of vision in successfully accomplishing its implicit task. Units with no outputs can be immediately eliminated. By studying time plots of neuron activities, motor outputs and sensor inputs, as shown in Figure 14, over a range of environmental conditions, it may be possible to highlight neurons that play no part in visually guided behaviours. In this case a small number of neurons can be eliminated. We can usefully redraw the slightly reduced C2 network to emphasise the visual pathways, as shown in Figure 12. Note that the tactile sensor input neurons have been taken over as internal units in the evolved network. Only those neurons referred to in the main argument have been included in the figure. The other neurons have no significant role in behaviour generation.

The first thing to note is that unit 6 (left photoreceptor input neuron) has only inhibitory outputs. These will only be active if its inputs sum to greater than T_v (0.75). Note that one of its inhibitory links goes to the left motor's negative input neuron (unit 13). Given all the excitatory links feeding into 13 from unit 7 (right photoreceptor input neuron), clearly, if active, the inhibitory link from 6 to 13 will have a major effect on the motor outputs. So, understanding the conditions under which the inhibitory links from 6 become active is one of the keys to understanding the behaviour generating mechanisms of the network.

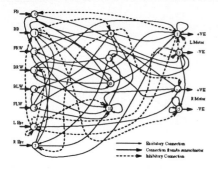
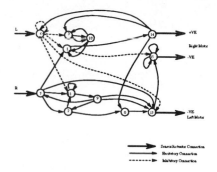

Fig. 11. Full C2 control network. The left-hand column are units originally designated as input units: FB=Front Bumper;
BB=Back Bumper; FRW=Front Right Whisker; BRW=Back Right Whisker; BLW=Back Left Whisker; FLW=Front Left Whisker. Right-hand column shows output units, which are paired and differenced to give two motor signals in the range [-1,1] from four 'neuron' outputs in the range [0,1]. Centre column shows 'hidden units'

Fig. 12. C2 visual guidance pathways. Note that, for the sake of clarity, the positions of the left and right motor outputs have been interchanged

Another important aspect of the architecture is the fact that unit 1 (with direct input from unit 7) has an inhibitory link to the right motor's negative input neuron (unit 15). Since 15 doubly feeds back onto itself it will act as a noisy generator node unless vetoed. Hence, understanding the conditions under which unit 1 produces inhibitory output is also important in elucidating the network operation. This will entail unravelling the series of interconnected feedback loops involving unit 7.

We will start by understanding the characteristics of unit 7's output. It has visual input from the environment (here referred to as E_r) and excitatory input from 9. If the inhibitory links from 6 are active it is clear that all the right photoreceptor visual pathways are rendered inactive, so in the following it is assumed that 6 is not vetoing. The tangle of network surrounding unit 9 is probably best understood by concentrating on unit 11. This unit is involved in two feedback loops sharing no common units other than itself; 11-9-2-11 and the direct feedback loop. From the arguments of Section 11, these two loops are enough to ensure that this neuron acts as a noisy generator unit with output varying between 0 and 1. In addition it has external visual input from 7 (direct and via 2). The effect of this input can be approximated by thinking of 11 as having a single direct feedback loop and two external inputs of $\frac{E_r}{2}$ and $\frac{E_r}{4}$. Approximating the feedback loop with $k^3 + k^2 + k$ (from Equation 9),

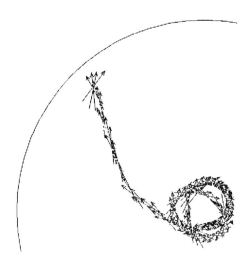

Fig. 13. Typical behaviour of the C2 controller, with noise. The robot starts near the edge of the arena, moves to the centre, and then rotates around the right wheel. As can be seen, the C2 controller drives the robot in reverse (backwards)

this gives an expected output of about $\frac{3E_r}{4}$. Given the additional loops 11 is involved in, and the attenuation argument of section 11.1, its visually-based output, in the stable state, should be approximately E_r. Adding this to its noisy generator behaviour, 11's output should vary approximately between E_r and 1. Unit 9's output will be about half this, hence 7's output will be approximately:

$$O_7 = \frac{3E_r}{4} + \frac{n_7}{4} \tag{22}$$

Where n_7 varies randomly between 0 and 1.

Returning to unit 1, we now understand its inputs from 7 and therefore shift the focus to unit 10. Unit 10 is involved in two feedback loops, 10-3-10 and 10-1-14-3-10. Since these share a common node other than 10, namely 3, this is *not* the situation of Equation 8 with $k = 1$. Rather, we have the case of one loop feeding into another giving an output close to $2 \sum_{i=0}^{t} k^{t+1-i} n_i$ with $k = 0.5$ (from Equation 6). Using Equation 9, the value of this expression varies randomly between 0 and approximately 0.2. Unit 10 also has input from 7 via 14 and 3. Hence 10's output is approximately given by,

$$O_{10} = n_{10} + \frac{O_7}{4} = n_{10} + \frac{3E_r}{16} + \frac{n_7}{16} \tag{23}$$

Where n_{10} varies between 0 and approximately 0.2.

Hence unit 1's input is approximately:

$$I_1 = O_7 + 2O_{10} = 1.125E_r + 0.38n_7 + 2n_{10} \tag{24}$$

Fig. 14. Record of observables and activity levels for the (with-noise) activity illustrated in Figure 13

The maximum values of the random parts of this expression, n_7 and n_{10}, are high enough to mean that unit 1 can act as a veto unit even in the absence of visual input. However in that case its inhibition will be very noisy. It is guaranteed to have inhibitory effect only if $E_r > 0.66$. So in general unit 1 will act as a noisy veto unit, but the higher E_r the more likely are its inhibitory links to be active, and the less likely the lower E_r.

As stated earlier, as soon as the inputs to unit 6 exceed the veto threshold, its inhibitory links become active and it closes down all the right eye visual pathways. Inputs to 6 are from the environment (here referred to as E_l) and from unit 14. Hence,

$$I_6 = E_l + O_{14} \approx E_l + \frac{O_1}{2} \approx E_l + 0.28E_r + 0.09n_7 + 0.5n_{10} \qquad (25)$$

Vetoing is guaranteed if $E_l > 0.75$. Given the value of n_{10}, it will occur in a noisy way for $E_l > 0.65$. At other values vetoing will be very noisy; if the above expression exceeds 0.75 the right eye pathways are closed down and their contributions to the expression go to zero, pulling the value below the threshold. It should be noted that the vetoing mechanism turns off excitatory output from a unit but *not* inhibitory outputs. Hence the veto feedback to unit 6 does not affect the above argument.

A crucial point to note at this juncture is that because the C2 network is built around the kinds of feedback loops described earlier, allowing the above analysis, then, by virtue of the 'settling down' argument of Section 11, the visual inputs can be seen to perturb the underlying network dynamics in a straightforward manner. This means that we can predict the motor behaviour given the visual input without explicitly having to take into account temporal factors (the robot's state history). It is this fact that lets us use the simple visualisable state spaces presented in the next section. More recent work with real robots has made use of the same type of network but with slightly altered timescales. Here state history has to be taken into account (the networks are no longer guaranteed to settle down between successive input samples) and slightly different higher dimensional state spaces, including a time dimension, have been used to analyse behaviours [24]. Networks operating on the same timescales as in this paper, and therefore amenable to exactly the same analysis, have also been successfully evolved to control real robots [17].

We can now summarise the motor output characteristics of the network. When E_l and E_r are both low, unit 15 acts as a noisy generator unit and unit 13 receives strong input from the noisy generator unit 11 via the double connection from 9. Hence we expect both motors to be activated in reverse, given a noisy straight line backwards behaviour. When E_r is high and E_l is low, 13 still receives strong input but 15 is now vetoed. Hence we expect a backwards rotation about the right wheel. The rotation will be noisy (jumping between rotation and straight line backwards) unless $E_r > 0.66$. When E_l is high and E_r is low, 13 is vetoed but 15 is left free to act as a noisy generator. Hence we expect a noisy backwards rotation about the left wheel. At intermediate combinations of E_l and E_r we expect a noisy alternation between backwards rotation about the left wheel and backwards rotation about the right wheel.

This network analysis can be combined with a use of $r\phi$ space to gain a global picture of the dynamic of the whole control system (network plus sensors). Only with this global picture can we really understand how the evolved controllers work.

13.2 The global picture

Figure 16 shows how $r\phi$ space is divided up into regions of different (E_l, E_r) visual signal combinations when the cylinder wall height is 15 units. Each bounded region has a constant combination over its area. This is different

Table 1. Table showing visual signals and dominant C2 behaviours in distinct visual regions of world for wall height 15. RrB= backwards rotation about right wheel; LrB= likewise about left wheel; SlB= straight line backwards; LrBsl= noisy oscillation between LrB and SlB

region	(lsig,rsig)	behaviour	region	(lsig,rsig)	behaviour	region	(lsig,rsig)	behaviour
1	(0.33,0.33)	SlB	2	(0.33,0.39)	RrB	3	(0.33,0.44)	RrB
4	(0.39,0.33)	RrB	5	(0.44,0.33)	RrB	6	(0.33,0.50)	RrB
7	(0.39,0.39)	RrB	8	(0.50,0.33)	LrBsl	9	(0.33,0.56)	RrB
10	(0.56,0.33)	LrBsl	11	(0.39,0.50)	RrB	12	(0.50,0.50)	LrBsl
13	(0.50,0.39)	LrBsl	14	(0.44,0.50)	RrB	15	(0.50,0.44)	LrBsl
16	(0.56,0.39)	LrBsl	17	(0.39,0.56)	RrB	18	(0.44,0.56)	RrB
19	(0.56,0.44)	LrBsl	20	(0.28,0.33)	RrB	21	(0.22,0.39)	RrB
22	(0.11,0.44)	RrB	23	(0.11,0.50)	RrB	24	(0.22,0.56)	RrB
25	(0.28,0.56)	RrB	26	(0.56,0.28)	LrB	27	(0.56,0.22)	LrB
28	(0.56,0.11)	LrB	29	(0.50,0.11)	LrB	30	(0.44,0.11)	LrB
31	(0.39,0.22)	SlB	32	(0.33,0.28)	SlB	33	(0.22,0.33)	RrB
34	(0.11,0.39)	RrB	35	(0.11,0.56)	RrB	36	(0.39,0.11)	LrB
37	(0.33,0.22)	SlB	38	(0.28,0.28)	SlB	39	(0.11,0.33)	RrB
40	(0.33,0.11)	LrB	41	(0.17,0.33)	RrB	42	(0.17,0.56)	RrB
43	(0.56,0.17)	LrB	44	(0.33,0.17)	SlB	45	(0.22,0.22)	SlB
46	(0.17,0.28)	SlB	47	(0.28,0.17)	SlB	48	(0.11,0.28)	SlB
49	(0.28,0.11)	SlB	50	(0.17,0.17)	SlB	51	(0.11,0.22)	SlB
52	(0.44,0.44)	RrB	53	(0.22,0.11)	LrB	54	(0.11,0.11)	LrB
55	(0.11,0.17)	SlB	56	(0.17,0.11)	SlB			

from the combinations in its neighbouring regions. Each region is labeled, along its right hand boundaries, with a unique integer. Figure 15 shows the actual grey level inputs (black=0, white=1) to the left and right photoreceptors represented in $r\phi$ space for this wall height. Table 1 gives the (E_l, E_r) combinations for each labelled region along with the dominant motor behaviour these will produce. These behaviours were deduced from the above network analysis. From this table the phase portrait of Figure 17 was produced. Using the geometrical properties of $r\phi$ space discussed earlier, a vector flow field was constructed across the regions. Where there is a noisy oscillation between two behaviours, a vector average was used. We can see that there appears to be a single attractor at $(rad, \frac{\pi}{2})$, where the robot wheel span radius is rad. This attractor[5] (marked with an X) corresponds to a rotation backwards about the right wheel, with the right wheel at the centre of the arena. Indeed the *whole* state space appears to be a basin of attraction for this single attractor. That is, from wherever, and at whatever orientation, the robot is started it will go to the centre and rotate in a very low radius circle. In other words, it cannot help but succeed at its implicit task. This is exactly the behaviour that was observed on countless runs of the simulated robot, as illustrated for a typical run in Figure 13. Some of these runs were translated

---

[5] The word attractor is being used a little loosely here, given the noise in the system, including the actuators.

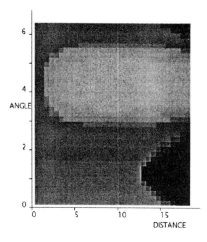
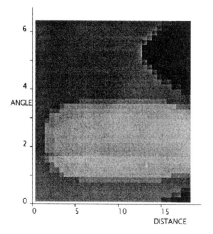

Fig. 15. Grey level 0-1 (black-white) $r\phi$ representations of left and right visual signals for C2 for wall height 15

into $r\phi$ space to see if the predicted phase portrait was confirmed. The resulting trajectories are shown in Figure 18, plotted over the right photoreceptor grey levels. Clearly the vector flow field is confirmed, with an attractor at $\left(rad, \frac{\pi}{2}\right)$.

Remember that C2 generalised very well over a wide range of wall heights. Figure 19 shows the grey level inputs for left and right photoreceptors at wall height 5 represented in $r\phi$ space. Figure 20 shows how the space is divided up into (E_l, E_r) combination regions. Clearly the visual structure is quite different. In a similar way to that described earlier, Table 2 was constructed. From this the vector flow field of Figure 21 was produced. It can be seen that this is different from the one constructed for wall height 15. There is an attractor at $\left(rad, \frac{3\pi}{2}\right)$ and what looks like a weak attractor at $\left(rad, \frac{\pi}{2}\right)$. The phase portrait is confirmed by Figure 22 which plots trajectories in $r\phi$ space from many runs of the simulated robot at this low wall height. Note that the existence of the two attractors, one much stronger than the other, is confirmed. The $\left(rad, \frac{3\pi}{2}\right)$ attractor corresponds to a backwards rotation about the left wheel with the left wheel at the centre. Again the controller acts very robustly, always moving to the centre and staying there.

C2 is an example of the power of artificial evolution. The visual morphology and network dynamics have evolved together to produce a very robust and general control system for the implicit task described by the evaluation function. In the face of noise, indeed by exploiting noise, the controller cannot help but succeed. Although the controller succeeds no matter where it is started in the arena, during evolution it was always started near the edge. Although space does not allow a full discussion of them, there are a number of other observations worth noting. The phase portraits show that at both wall

Table 2. Table showing visual signals and dominant C2 behaviours in distinct visual regions of world for wall height 5. LrB=backwards rotation about left wheel; RrB= likewise about right wheel; SlB= straight line backwards; LrRrb= noisy oscillation between LrB and RrB; LrBsl= noisy oscillation between LrB and SlB

region	(lsig,rsig)	behaviour	region	(lsig,rsig)	behaviour	region	(lsig,rsig)	behaviour
1	(0.56,0.56)	LrRrb	2	(0.56,0.61)	LrRrb	3	(0.56,0.67)	LrRrb
4	(0.61,0.56)	LrB	5	(0.67,0.56)	LrB	6	(0.56,0.72)	LrRrb
7	(0.61,0.61)	LrB	8	(0.72,0.56)	LrB	9	(0.56,0.78)	LrRrb
10	(0.78,0.56)	LrB	11	(0.61,0.72)	LrB	12	(0.72,0.72)	LrB
13	(0.72,0.61)	LrB	14	(0.67,0.72)	LrB	15	(0.72,0.67)	LrB
16	(0.78,0.61)	LrB	17	(0.61,0.78)	LrB	18	(0.67,0.78)	LrB
19	(0.78,0.67)	LrB	20	(0.50,0.56)	LrRrb	21	(0.50,0.61)	LrRrb
22	(0.39,0.67)	RrB	23	(0.33,0.72)	RrB	24	(0.39,0.78)	RrB
25	(0.50,0.78)	LrRrb	26	(0.78,0.50)	LrB	27	(0.78,0.44)	LrB
28	(0.72,0.33)	LrB	29	(0.67,0.39)	LrB	30	(0.61,0.44)	LrB
31	(0.56,0.50)	LrRrb	32	(0.44,0.56)	RrB	33	(0.33,0.61)	RrB
34	(0.33,0.67)	RrB	35	(0.33,0.78)	RrB	36	(0.44,0.78)	RrB
37	(0.78,0.33)	LrB	38	(0.67,0.33)	LrB	39	(0.61,0.33)	LrB
40	(0.56,0.44)	LrB	41	(0.39,0.56)	RrB	42	(0.78,0.39)	LrBsl
43	(0.56,0.33)	LrBsl	44	(0.50,0.50)	LrBsl	45	(0.33,0.56)	RrB
46	(0.83,0.44)	LrBsl	47	(0.56,0.39)	LrBsl	48	(0.44,0.50)	RrB
49	(0.33,0.83)	RrB	50	(0.44,0.83)	RrB	51	(0.50,0.89)	LrRrb
52	(0.50,0.83)	LrRrb	53	(0.56,0.83)	LrRrb	54	(0.83,0.50)	LrRrb
55	(0.89,0.50)	LrRrb	56	(0.89,0.44)	LrB	57	(0.83,0.33)	LrB
58	(0.50,0.44)	LrB	59	(0.44,0.44)	RrB	60	(0.39,0.50)	RrB
61	(0.39,0.89)	RrB	62	(0.44,0.89)	RrB	63	(0.83,0.56)	LrBsl
64	(0.89,0.39)	LrBsl	65	(0.50,0.39)	LrBsl	66	(0.33,0.50)	RrB
67	(0.28,0.56)	RrB	68	(0.22,0.61)	RrB	69	(0.11,0.67)	RrB
70	(0.11,0.72)	RrB	71	(0.22,0.78)	RrB	72	(0.28,0.78)	RrB
73	(0.28,0.89)	RrB	74	(0.33,0.94)	RrB	75	(0.50,0.94)	LrRrb
76	(0.56,0.89)	LrRrb	77	(0.94,0.50)	LrB	78	(0.94,0.33)	LrB
79	(0.89,0.33)	LrB	80	(0.89,0.28)	LrB	81	(0.78,0.22)	LrB
82	(0.78,0.11)	LrB	83	(0.72,0.11)	LrB	84	(0.67,0.11)	LrB
85	(0.61,0.22)	LrB	86	(0.56,0.28)	LrB	87	(0.50,0.28)	LrB
88	(0.50,0.33)	LrB	89	(0.39,0.39)	RrB	90	(0.28,0.44)	RrB
91	(0.28,0.50)	RrB	92	(0.22,0.56)	RrB	93	(0.11,0.61)	RrB
94	(0.11,0.78)	RrB	95	(0.22,0.89)	RrB	96	(0.28,0.94)	RrB
97	(0.39,1.00)	RrB	98	(0.44,0.94)	RrB	99	(0.89,0.56)	LrRrb
100	(0.94,0.44)	LrRrb	101	(1.00,0.39)	LrB	102	(0.94,0.28)	LrB
103	(0.89,0.22)	LrB	104	(0.61,0.11)	LrB	105	(0.56,0.22)	LrB
106	(0.44,0.28)	SlB	107	(0.33,0.33)	SlB	108	(0.22,0.50)	RrB
109	(0.11,0.56)	RrB	110	(0.11,0.89)	RrB	111	(0.22,0.94)	RrB
112	(0.33,1.00)	RrB	113	(0.94,0.22)	LrB	114	(0.89,0.11)	LrB
115	(0.56,0.11)	LrB	116	(0.50,0.22)	LrB	117	(0.28,0.28)	SlB
118	(0.22,0.39)	SlB	119	(0.17,0.44)	RrB	120	(0.11,0.50)	RrB
121	(0.11,0.83)	RrB	122	(0.17,0.94)	RrB	123	(0.28,1.00)	RrB
124	(0.39,0.94)	RrB	125	(0.83,0.61)	LrRrb	126	(1.00,0.28)	LrB
127	(0.94,0.17)	LrB	128	(0.83,0.11)	LrB	129	(0.50,0.11)	LrB
130	(0.50,0.17)	LrB	131	(0.44,0.22)	LrB	132	(0.22,0.22)	SlB
133	(0.17,0.33)	SlB	134	(0.11,0.44)	RrB	135	(0.11,0.94)	RrB
136	(0.17,1.00)	RrB	137	(0.22,1.00)	RrB	138	(1.00,0.33)	LrB
139	(1.00,0.22)	LrB	140	(1.00,0.17)	LrB	141	(0.94,0.11)	LrB
142	(0.44,0.11)	LrB	143	(0.33,0.17)	SlB	144	(0.11,0.33)	SlB
145	(0.11,1.00)	RrB	146	(1.00,0.11)	SlB	147	(0.33,0.11)	SlB
148	(0.17,0.17)	SlB	149	(0.11,0.22)	SlB	150	(0.67,0.67)	LrRrb
151	(0.94,0.39)	LrRrb	152	(0.39,0.11)	SlB	153	(0.22,0.11)	SlB
154	(0.11,0.11)	SlB	155	(0.11,0.17)	SlB	156	(0.11,0.28)	SlB
157	(0.11,0.39)	SlB	158	(0.28,0.11)	SlB	159	(0.17,0.11)	SlB

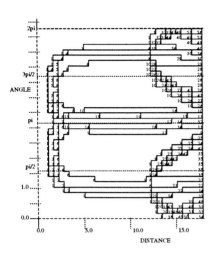

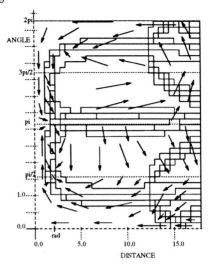

Fig. 16. Distinct visual regions for C2 for wall height 15, shown in $r\phi$ space

Fig. 17. Constructed phase portrait for C2 at wall height 15

heights for large r the robot will always rotate in the direction that will bring it normal to the wall in the shortest time; it then moves backwards towards the centre. The phase portrait in Figure 17 suggests that, at wall height 15, if the robot actually took a route directly to the centre (ϕ constant at 0), it may get stuck in a cyclical attractor. But the noise in the actuators means this is impossible (ϕ cannot be held at 0) and in practice the robot moves in slightly off-centre, getting drawn rapidly to the point attractor. Lastly, it is worth noting that C2's network architecture is somewhat reminiscent of a Brooksian Subsumption architecture [7], with the left and right visual pathways almost separate, and the left photoreceptor network being able to inhibit the right photoreceptor network under certain environmental conditions.

14 Extending the analysis

The network analysed has many interesting subtleties but is relatively simple. So were its environments and its sensory signals. So, although we were able to provide a thorough analysis and, in particular, were able to show how the network dynamics and visual morphology had evolved in tandem to provide a highly robust system for achieving the arena centring task, the question still remains as to how far this kind of analysis can be taken.

Of course it is not possible to give any sort of definitive answer to this question, but there are a number of issues that appear to be highly pertinent. A major concern is in finding the right level of abstraction to allow a meaningful analysis with manageable state spaces. Two techniques to help in this

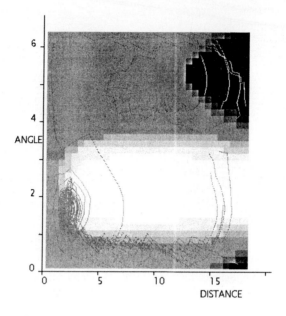

Fig. 18. C2 trajectories for wall height 15 shown in $r\phi$ space and superimposed over right eye grey levels

direction were used here: collapsing the network dynamics into feedback loop properties, and collapsing motor output dynamics into behaviour categories. The global picture afforded by using the $r\phi$ space to construct a visualisation of the robot's visual world was made possible by exploiting the simplicity and symmetries of the world. For more complex environments, especially dynamic environments with changing lighting conditions, that kind of global picture will just not be possible. Some kind of statistical categorisation will probably be needed to produce a manageable visual-inputs state space. This will get harder, and will probably necessitate more abstract spaces, as the number of visual receptive fields increases. However, work with real robots reported in [24] shows that it is certainly possible to extend the kind of analysis reported in the present paper to more complex environments and tasks. Useful *analyses* of existing mechanisms can often adopt higher levels of abstraction than can be tolerated in the *design* of the mechanisms. So although we may need artificial evolution to help develop complex systems, this does not necessarily mean that we will not be able to understand its products.

This paper has been concerned with ways of analysing visually guided behaviours from a dynamical systems perspective. Other researchers have been treading related paths, but have not been specifically concerned with visuo-motor behaviours. In particular, Randy Beer and his students have analysed a number of behaviour generating network controllers in dynamical

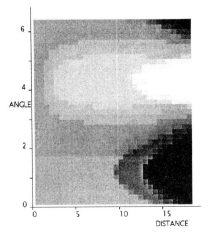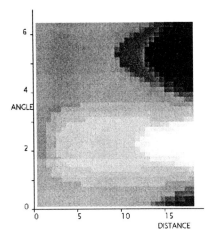

Fig. 19. Grey level 0-1 (black-white) $r\phi$ representations of left and right visual signals for C2 for wall height 5

systems terms. Beer and Gallagher analysed recurrent dynamical network locomotion controllers for an artificial insect [3]; while Yamauchi and Beer have shown how the dynamics of a network shaped by a genetic algorithm are capable of generating sequential behaviour and learning without recourse to an explicit lifetime learning algorithm and without an a priori discretization of states or time [33]. Tim Smithers has argued for agent environment systems to be viewed from a dynamical systems perspective [28].

15 Animate vision

Animate vision [1] is an approach that advocates that visual perception should be considered as a dynamic behaviour that an animal performs in order to achieve its goals in a dynamic environment. This is in contrast to traditional approaches to computer vision which tended to take snapshots of a stationary world, and then analyse them. It follows from this concept of animate vision that sensing and motor behaviour must interact closely with each other. In contrast, the advocates of functional decomposition assumed that vision was one task, and action something different; research on one, it was thought, should be done in a different room from research on the other.

From our viewpoint of cognition, that a robot or animal should be seen as a whole system with its own internal dynamics, coupled to a dynamical world, the tenets of animate vision appear perfectly natural. Furthermore, the behaviour of the successful robots in the circular arena supports an interpretation in terms of primitive animate vision. Any individual snapshot that the robot could be considered to take would consist of a single grey-scale

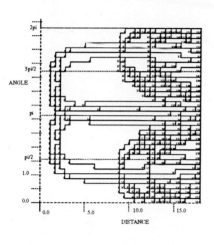 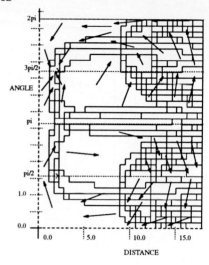

Fig. 20. Distinct visual regions for C2 for wall height 5, shown in $r\phi$ space

Fig. 21. Constructed phase portrait in $r\phi$ space for C2 at wall height 5

value for the one pixel of its one effective photorector; on its own, this could be no basis for creating an internal model of the world at that moment in time, and no basis for a decision whether to circle left, right, or go straight ahead. Yet the action that the robot takes of continually moving, switching between straight line motion and circular motions with a greater or lesser radius, allows the variation in light input to be used as a basis for 'deciding' when to increase or decrease the radius or move in a straight line; and this results in the robot reaching the centre and remaining there, as encouraged by the evaluation function. The dynamics of the overall system inevitably force it to the centre and then hold it there; if it strays away it is very quickly sucked back.

16 Further work: moving into the real world

The experiments discussed in this paper were intended to test the plausibility of our approach. However, the simulated visual environment was very simple and computational costs will increase dramatically as the visual environment becomes more complex. Indeed, even ignoring computational costs, the plausible modelling of visual inputs in such circumstances is highly problematic. Hence we have developed an experimental setup that allows the whole evolutionary process to work with a real robot, moving in the real world, and without recourse to simulated vision. This apparatus is briefly discussed below, see [17] for full details and an account of early experimental results achieved with it.

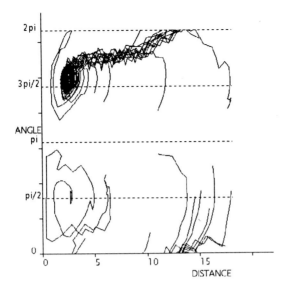

Fig. 22. C2 trajectories translated into $r\phi$ space for wall height 5

16.1 The gantry-robot

A gantry runs along the top of a frame about 1.5m by 1m. The gantry is moved in the X-direction by a stepper motor, and a platform is moved across the gantry in the Y-direction by a second stepper motor (see Figure 23). Supported below this, on the bottom of a vertical pole, is our 'part-real, part-virtual' robot. This has a camera facing in a vertical direction at an angled mirror which gives horizontal vision. The camera does not rotate, but rotation of the mirror, together with adjustment of the software subsampling the CCD image allows vision in any direction. Around the circumference of the camera is an array of whiskers (touch-sensors).

This robot is thus 'real' in the sense that it uses real light for its vision, real touch for its whiskers, and can move around in the Toytown-scale environments that we can build for it. It is partly 'virtual' in the sense that it has virtual left and right motors, modeling those of the original mobile robot, which are aligned in the direction of gaze given by the angled mirror; motion commands sent to these virtual motors are in practice converted to the appropriate movements of the X and Y stepper motors, for horizontal translation, and of the angled-mirror mounting, for rotation. It is also 'virtual' in the sense that the (genetically specified) virtual photoreceptors used for vision are implemented through sub-sampling the CCD image.

The genetically specified control networks for this robot are run off-board on a fast PC communicating via an umbilical cable. Through the use of stepper motors for X and Y motion, a supervisory program always knows where

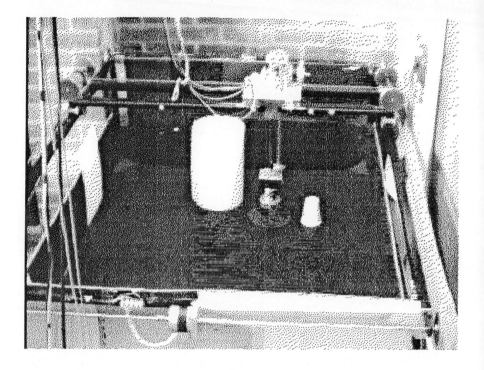

Fig. 23. The Gantry-Robot

the robot is, this knowledge being used to evaluate its performance — even though the control system for the robot has no such privileged knowledge. This setup, with off-board computing and avoidance of tangled umbilicals, means that the apparatus can be run continuously for long periods of time — making artificial evolution feasible. A top-level program automatically evaluates, in turn, each member of a population of control systems. A new population is produced by selective interbreeding and the cycle repeats. The gantry-robot is being used in this way to further our understanding of how to concurrently evolve visual morphologies and control networks for visually guided robots.

Another advantage of this apparatus is that it is possible to automatically collect many forms of data for use in analyses like those given earlier in this paper. See [24] for such an analysis of behaviours evolved on the gantry-robot.

17   Conclusions

We have shown it is possible to evolve dynamical network controllers plus visual morphologies that produce simple robust autonomous behaviours across a range of related environments.

We have shown that these systems can be thoroughly understood, and introduce one strand of our current research where the simulation of the agent environment coupling is abandoned.

Acknowledgements

Work on the Sussex evolutionary robotics project is funded by EPSRC grant GR/J18125. Thanks to Michael Wheeler and Tim Smithers for useful discussions on some of the material covered in this paper.

References

1. D. H. Ballard. Animate vision. Artificial Intelligence, 48: 57–86, 1991
2. R.D. Beer. A dynamical systems perspective on autonomous agents. Technical Report CES-92-11, Case Western Reserve University, Cleveland, OH, 1992
3. R.D. Beer and J.C. Gallagher. Evolving dynamic neural networks for adaptive behavior. Adaptive Behavior, 1(1): 91–122, 1992
4. R. Belew and L. Booker (eds.). Proceedings of the Fourth International Conference on Genetic Algorithms. Morgan Kaufmann, 1991
5. R. A. Brooks. Coherent behavior from many adaptive processes. In D. Cliff, P. Husbands, J.-A. Meyer, and S.W. Wilson (eds.), From Animals to Animats 3: Proceedings of The Third International Conference on Simulation of Adaptive Behavior, pp. 22–29. MIT Press/Bradford Books, Cambridge, MA, 1994
6. R.A. Brooks. Intelligence without reason. In Proceedings IJCAI-91, pp. 569–595. Morgan Kaufmann, 1991
7. R.A. Brooks. Intelligence without representation. Artificial Intelligence, 47: 139–159, 1991
8. Rodney A. Brooks. Artificial life and real robots. In F. J. Varela and P. Bourgine (eds.), Proceedings of the First European Conference on Artificial Life, pp. 3–10. MIT Press/Bradford Books, Cambridge, MA, 1992
9. D. Cliff, I. Harvey, and P. Husbands. Explorations in evolutionary robotics. Adaptive Behavior, 2(1): 73–110, 1993
10. D. Cliff, P. Husbands, and I. Harvey. Evolving visually guided robots. In H. Roitblat J. Meyer and S. Wilson (eds.), From Animals to Animats 2: Proceedings of the 2nd International conference on the Simulation of Adaptive Behaviour, pp. 374–383. MIT Press/Bradford Books, 1993
11. D. T. Cliff, P. Husbands, and I. Harvey. Analysis of evolved sensory-motor controllers. In Proceedings of Second European Conference on Artificial Life, ECAL93. Brussels, May 1993
12. M. Dorigo and U. Schnepf. Genetic-based machine learning and behavior-based robotics: A new synthesis. IEEE Transactions on Systems, Man, Cybernetics, 23(1): 141–154, 1993
13. S. Forrest (ed.). Proc. 5th Int. Conf. on GAs. Morgan Kaufmann, 1993
14. David E. Goldberg. Genetic Algorithms in Search, Optimization and Machine Learning. Addison-Wesley, Reading, MA, 1989

15. I. Harvey. The SAGA cross: The mechanics of recombination for species with variable length genotypes. In R. Manner and B. Manderick (eds.), Parallel Problem Solving from Nature,2, pp. 269–278. North-Holland, 1992

16. I. Harvey, P. Husbands, and D. Cliff. Issues in evolutionary robotics. In H. Roitblat J. Meyer and S. Wilson (eds.), From Animals to Animats 2: Proceedings of the 2nd International conference on the Simulation of Adaptive Behaviour, pp. 364–373. MIT Press/Bradford Books, 1993

17. I. Harvey, P. Husbands, and D. Cliff. Seeing the light: Artificial evolution, real vision. In D. Cliff, P. Husbands, J.-A. Meyer, and S. Wilson (eds.), From Animals to Animats 3, Proc. of 3rd Intl. Conf. on Simulation of Adaptive Behavior, SAB'94, pp. 392–401. MIT Press/Bradford Books, 1994

18. Inman Harvey. Species adaptation genetic algorithms: The basis for a continuing SAGA. In F. J. Varela and P. Bourgine (eds.). Proceedings of the First European Conference on Artificial Life, pp. 346–354. MIT Press/Bradford Books, Cambridge, MA, 1992

19. Inman Harvey. Evolutionary robotics and SAGA: the case for hill crawling and tournament selection. In C. Langton (ed.), Artificial Life III, pp. 299–326. Santa Fe Institute Studies in the Sciences of Complexity, Proceedings Vol. XVI, Addison-Wesley, Redwood City CA, 1994

20. John Holland. Adaptation in Natural and Artificial Systems. University of Michigan Press, Ann Arbor, USA, 1975

21. P. Husbands. Genetic algorithms in optimisation and adaptation. In L. Kronsjo and D. Shumsheruddin (eds.), Advances in Parallel Algorithms, pp. 227–277. Blackwell Scientific Publishing, Oxford, 1992

22. P. Husbands and I Harvey. Evolution versus design: Controlling autonomous robots. In Integrating Perception, Planning and Action, Proceedings of 3rd Annual Conference on Artificial Intelligence, Simulation and Planning, pp. 139–146. IEEE Press, 1992

23. P. Husbands, I. Harvey, D. Cliff, and G. Miller. The use of genetic algorithms for the development of sensorimotor control systems. In P. Gaussier and J-D. Nicoud (eds.), Proceedings of From Perception to Action Conference, pp. 110–121. IEEE Computer Society Press, 1994

24. P. Husbands, I. Harvey, N. Jakobi, D. Cliff. Evolved network controllers for mobile robots. In preparation

25. Stuart Kauffman. Adaptation on rugged fitness landscapes. In Daniel L. Stein (ed.), Lectures in the Sciences of Complexity, pp. 527–618. Addison Wesley: Santa Fe Institute Studies in the Sciences of Complexity, 1989

26. J. Koza. Genetic Programming: On the programming of computers by means of natural selection. MIT Press, 1992

27. J. D. Schaffer, editor. Proceedings of the Third International Conference on Genetic Algorithms, San Mateo, CA, 1989. Morgan Kaufmann

28. Tim Smithers. On why better robots make it harder. In D. Cliff, P. Husbands, J.-A. Meyer, and S. Wilson (eds.), From Animals to Animats 3, Proc. of 3rd Intl. Conf. on Simulation of Adaptive Behavior, SAB'94, pp. 54–72. MIT Press/Bradford Books, 1994

29. A. Thompson. Evolving electronic robot controllers that exploit hardware resources. In Proc. of Third European Conference on Artificial Life, Springer-Verlag, 1995 In Press

30. Tim van Gelder. What might cognition be if not computation. Technical Report 75, Indiana University Cognitive Sciences, 1992
31. F. Varela, E. Thompson, and E. Rosch. *The Embodied Mind*. MIT Press, 1991
32. B. Yamauchi and R. Beer. Integrating reactive, sequential, and learning behavior using dynamical neural networks. In D. Cliff, P. Husbands, J.-A. Meyer, and S. Wilson (eds.), From Animals to Animats 3, Proc. of 3rd Intl. Conf. on Simulation of Adaptive Behavior, SAB'94, pp. 382–391. MIT Press/Bradford Books, 1994
33. Brian M. Yamauchi and Randall D. Beer. Sequential behavior and learning in evolved dynamical neural networks. Adaptive Behavior, 2(3): 219–246, 1994

Animal and Robot Navigation

Ulrich Nehmzow

Department of Computer Science
Manchester University
Manchester M13 9PL
United Kingdom
u.nehmzow@cs.man.ac.uk

Abstract. It is argued that the following three properties are foundations of robust robot navigation:

- The use of landmarks (and, in particular, the use of a compass sense),
- the use of canonical paths, and
- the use of topological rather than geometrical maps.

Some examples of successful animal navigation are presented that support this view. We have performed initial experiments with mobile robots to investigate mechanisms suitable to implement such navigational architectures. Experiments concerning navigation by dead reckoning are presented, and a differential light compass is introduced to aid robot navigation.

1 Introduction

Most of the work to date concerning navigation of mobile robots uses internal geometrical representations of the robot's environment to perform navigational tasks. MOBOT III, for example, constructs such a geometrical representation autonomously from sensor data ([8]), other robots use maps supplied by the designer ([6]). Usually the maps are altered in the course of operation, depending on sensory perception.

This "classical" approach has the advantage that the resulting map is intelligible to the human operator, which means that a special interface between the robot's representation of its environment and the way a human sees it is not necessary. This allows for easy supervision of the robot operation by the designer. On the other hand, the approach is time consuming and requires large amounts of memory. Furthermore the resulting maps contain information not necessarily needed for the immediate task ahead. If, for example, the robot's task was to move ahead, avoiding one obstacle in front of it, information concerning objects *behind* the robot as well as in the *far distance* to the sides or ahead is not needed. To store such information puts an additional burden onto memory and processing requirements.

Alternatively, one could merely use information concerning the neighbourhood relationships between certain landmarks the robot perceives. Such topological mappings occur, for example, in animals and humans in the mapping of sensors onto the cortex ([9, 17], see also [3]). As they only represent the topological relationship between locations, but not the distances, they require less time to build and less memory to store.

Topological maps have been used for robot navigation, too. Examples are *Toto* ([12]), *Alder* and *Cairngorm* ([14]). In Toto's case landmarks are defined by an environmental property such as "wall" (predefined by the designer) and a compass heading, in Alder's and Cairngorm's case landmarks are identified by the excitation pattern of a self-organising feature map (regarding self-organising feature maps, see [7]).

In this paper we argue that successful and robust robot navigation can be achieved by using (computationally cheap) topological maps, if two further restrictions are satisfied: firstly, the environment contains features which can serve as landmarks detectable by the robot (in particular, that reference landmarks — a compass sense — are used); and secondly, the robot follows canonical rather than arbitrary paths in order to disambiguate sensor signals and allow for reliable landmark recognition.

The paper is structured as follows: Section 2 discusses the various types of landmarks suitable for robot and animal navigation and gives examples of how animals use them. Section 3 discusses the role of canonical paths in navigation, section 4 describes human navigational skills using the example of navigation between Hawaii and Tahiti, and section 5 presents experiments in dead reckoning navigation with mobile robots. The use of a *differential light compass (DLC)* and its effect on the reliability of navigation is discussed in this section as well.

2 Landmarks

2.1 Local Landmarks

Local landmarks we define as those landmarks whose relative position to the robot or animal changes as a result of motion. Typically such landmarks are trees, rocks, rivers, buildings *etc.* in the animate world, and beacons, reflective strips, colour markings *etc.* in the robot environment.

Ants

Wehner and Räber ([20]) report that apparently desert ants (*Cataglyphis bicolor*) use visual stimuli to identify a home location. Their experiment (see figure 1) suggests that in order to move to a particular location ants try to match the current retinal image with a stored one. Figure 1 (top) shows the initial situation: two identical markers at equal distances to the left and the right of the mound serve as landmarks. The small dots indicate homing positions of individual ants. If these markers are moved apart to twice the original distance (thus halfing their angular extent) the ants return to either of the two markers (figure 1, middle). If, then, the size of the markers is doubled (returning their angular extent to the original size) the ants return to the original home location (figure 1, bottom).

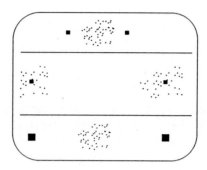

Fig. 1. Landmark use in *Cataglyphis bicolor* (after [20])

Cosens reports that wood ants *Formica aquilonia* change their paths in his laboratory in relation to large rocks (which are part of the laboratory environment) being shifted ([2]). Again, this indicates that ants use local landmarks for navigation.

Bees

Gould and Gould ([5, p.148]) report von Frisch's findings concerning the use of landmarks in the navigation of bees: bees who have been trained on a dog-leg route with the turn near a prominent tree will continue to fly to the food by way of the tree, whereas bees trained on a similar dog-leg route in the open country will fly along the shortest route. Prominent landmarks such as forest edges will be used predominantly for navigation by honey bees, the sun is merely used as a reference point in the dance, indicating the direction to the food source ([5, p.148ff],[18, p.177]).

Cartwright and Collett ([1]) report experiments with bees that show a striking similarity to the experiments with ants by Wehner and Räber ([20]). As in the ants, bees seem to use the angular extent of local landmarks for navigation, displacement of artificial landmarks leads to a corresponding displacement in return position. Cartwright and Collett: "We conclude that bees learn the angular size of the landmark when viewed from the position of the food source and use this to guide their return." They also report that "the compass bearings of the landmarks on the retina are of more importance than their apparent size". This suggests that the actual physical features of a landmark, although being taken into account once near the desired location, provide only a secondary source of information.

In summary of their experiments they say that "the conclusion to be drawn from the experiments ... is that bees do not find their way using anything analogous to a floor plan or map of the spatial layout of landmarks and food source. The knowledge at their disposal is much more limited, consisting of no more than a remembered image of what was present on their retina when they were at their destination".

Waterman ([18, p.176]) confirms this by reporting that honey bees stop foraging if artificial landmarks are introduced (or removed) into their territory and make orientation flights before they continue to forage; similarly they make reconnaissance flights in unknown territory before they begin to forage.

2.2 Reference Landmarks

Reference landmarks we define as those landmarks that do not change their relative position due to robot or animal motion. Typically these are distant forests or mountains, the sun and stars, the polarisation of the sky, the magnetic north pole *etc.*

Ants

Desert ants (*Cataglyphis bicolor*) are able to return to their nest directly after completion of a foraging trip. They do not retrace their outward journey (by, for instance, following a pheromone trail). Once near the nest, they start searching for the nest in spiral-like motions of ever widening circles ([4, p.61]). Experiments in featureless planes show that the ants' navigation is, in those cases, based purely on dead reckoning ([19]). In order to achieve this, they orientate their direction according to polarisation patterns of the sky ([20]). Such patterns can be regarded stationary over the relatively short foraging time of desert ants and therefore provide reliable directional cues. As an ant does not perform pitching and rolling movements during foraging and homing, the sky pattern changes only in relation to rotations about the vertical axis.

Bees

Gould and Gould's experiments ([5, p.126]) show that honey bees *Apis mellifera* orientate their flight paths with the help of the sun (and, as Waterman reports ([18]), also using the polarisation of the sky).

The results relevant for the discussion here are depicted in figure 2. They show that bees' paths in relation to the hive are determined by the position of the sun. If the hive is moved by several kilometers over night (figure 2, top), the bees will leave the hive in the morning in the same direction as they did the previous day. If the hive is moved while the bees are away feeding (figure 2, middle), they will return to the old location, discover their error and fly to the new location of the hive. In experiments with the *Edinburgh R2* and *Grasmoor* (see section 5.1 for a description of these robots) a similar effect

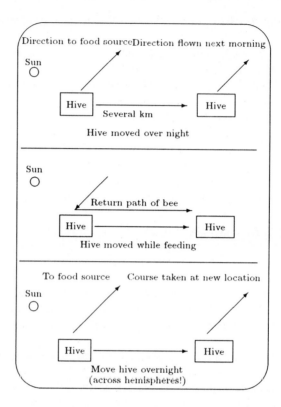

Fig. 2. The use of reference landmarks in *Apis mellifera*

can be observed: if the position of the main light source is moved laterally on the return leg of the robot's journey, the robot will return to a location which is deflected in relation to the apparent movement of the light source[1]. If, thirdly, the hive is moved over whole continents, i.e. the apparent movement of the sun is a different one to what it was at the original location, bees will fly off in the same angle they used to fly at the original location, but get confused in the course of the day (figure 2, bottom).

Birds

Planetarium experiments with indigo buntings (*Passerina cyanea*) about to migrate show that birds can use the stars for navigation. In the planetarium the birds orientated themselves in the same direction they would have done in the open country. If the artificial sky was shifted, their flight orientation changed accordingly ([18, p.109]). Through another set of experiments it was also shown that the birds *learn* to interpret the star patterns for navigation.

---

[1] This experiment is described in section 5.2 of this paper.

3 Canonical Paths

The use of unchangeable (canonical) paths improves navigation reliability greatly: the task of navigating to a goal location is simplified to one of staying on a (possibly learned) path. Gallistel ([4, p.38]) reports that human navigators prior to the latter part of the eighteenth century used to "sail the parallels", that is sail due North or South and then due East or West[2](even at the expense of making detours), thus reducing the error introduced by trigonometry and furthermore exploiting trade winds and ocean currents (which are both parallel to latitudes). If landmarks are known that describe the desired path piloting (navigating with respect to internal representation and observed landmarks) can be used instead of the more unreliable dead reckoning. Preliminary experiments with *Grasmoor* have been conducted that demonstrate this type of navigation: the robot navigates using local landmarks and a differential light compass (see section 5.2). Canonical paths are indeed used by insects for successful navigation between (stationary) food sources and the nest.

Wood ants (*Formica rufa*), for example, use the same paths between their food sources and their nest so intensely that the forest floor shows these paths clearly: in the course of many years the covering vegetation has been removed and black trails indicate the usual way of the ants ([2]).

A further example for the use of canonical paths in animal navigation are Canvasback ducks. They have, for example in the Mississippi valley, very narrow flight paths which they follow each year during migration ([18, p.176]). The knowledge of these paths seems to be *acquired* in some cases (bees), in others they must be genetically *preprogrammed*. First-year juvenile terns, for example, do not depend upon the leadership of an adult to find their migratory paths ([18, p.177]).

4 Polynesian Navigation

Without using compasses or geometrical maps the inhabitants of the Polynesian islands were able to navigate successfully over several thousand kilometers of open sea (the journey between Tahiti and Hawaii — 6000 km long — has been performed recently by the *Hokulea*, using traditional navigation methods ([10, 11],[18, ch.3])). Such navigation was successful, because all available sources of information (local and reference landmarks as well as special properties of the environment) were exploited by the navigators.

To determine an initial course, the navigators used the stars, moon and sun as reference landmarks. The direction of setting or rising of a particular star, for example, would give the direction for a particular island (reference landmark,

---

[2] We have used a similar behaviour for some of our experiments with the *Edinburgh R2* — the robot first moves parallel to the x-axis, then parallel to the y-axis of its coordinate system.

see figure 3). They also use local landmarks of the island they are departing from to adjust their initial course ([18, ch.3]).

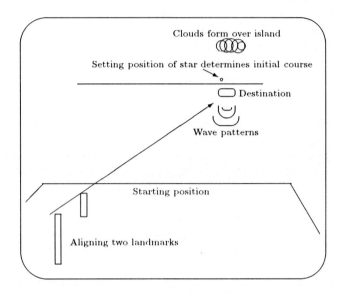

Fig. 3. The use of reference landmarks and local landmarks

During the journey wind and wave patterns (which are usually unique for certain parts of the ocean, because they are influenced by the position of islands) were used as local landmarks; in fact the weather patterns in many parts of the world are so reliable that they are used for navigation not only by humans, but also by animals. Sparrows and warblers, for example, fly 2000 to 3000 km from Canada and New England to South America's north coast. Computer modelling suggests that, provided the departure time is closely timed (which indeed seems to be the case, according to observations), the birds will be directed to their destinations by the repeating wind patterns ([18, p.27]).

Although a navigator in a canoe has a view of only 15 km, islands can be detected from a much greater distance: because land is warm in comparison to the sea, air masses rise over land, cool down in the process and form clouds (see figure 3). These clouds are stationary over an island. Furthermore, atholls reflect a green light onto the underside of these white clouds. Certain land birds fly out over the sea for a certain distance and can therefore be used as an indication of nearby land.

In addition to the intelligent use of reference and local landmarks, however, successful navigation would not be possible if the environment wasn't "benign", that is if certain properties of the environment didn't help to achieve navigation. The Hawaiian islands, for example, are stretched out in a chain 750 km long (but only 150 km wide). This means that the navigator has to get the bearing only roughly right; as soon as one island is detected, local landmarks can be used to sail to the actual destination of the journey. This principle applies to robotics as well: we believe that robot performance (and therefore navigation as one aspect of robot performance) always has to be seen with respect to the robot's environment. A general purpose robot does not exist.

5 Experiments in Robot Navigation

Robot navigation based on pure dead reckoning is unlikely to succeed if the path of the robot is complicated and long. This is due to inevitable sensor and actuator inaccuracy (e.g. variation of sensor readings due to surface, colour and position variation as well as electrical noise, or the fact that turning angles cannot be determined more accurately than a few degrees), as well as effects stemming from the interaction with the environment (e.g. wheels turning, although the robot is pressing against a wall: the robot is *not* moving, yet seems to measure forward motion!).

Instead of using pure dead reckoning, we propose a robot navigation system that uses local landmarks such as passive or active beacons, walls and corners, *and* reference landmarks (a compass sense), canonical paths and a topological representation of the environment. We have already shown that such a system can be used for successful location identification and limited navigation ([15]). The robots used in those early experiments were equipped with tactile sensors only, whereas the robots used here have far more sensors (see section 5.1).

For short travelling distances (up to four infrared sensor ranges, approximately 4 m) we propose a navigation system that is based on both navigation in relation to reference landmarks (see section 5.2) and local landmark recognition and identification. How local landmarks can be identified has been described earlier ([14]), experiments regarding navigation by means of reference landmarks are presented here.

5.1 The Robots

Initial experiments were conducted by us at Edinburgh University (Psychology Department), using the *Edinburgh R2* ([16]). This robot is shown in figure 4, it is a mobile robot of roughly quadratic shape (25cm x 25cm). It has a gripper with two degrees of freedom. Eight infrared proximity sensors are mounted around the robot, 25 cm above ground. White objects are detected as far away as 40 cm, objects coated with reflective tape can be seen from distances of 1 m and more. In addition to the infrared sensors the robot has six light sensors (five around the perimeter, one pointing upwards),

Fig. 4. The *Edinburgh R2*

two break beam sensors and one colour sensor in the gripper. Mounted at a height of 5 cm are five tactile sensor pads, two additional tactile sensor pads are fitted inside the gripper. Through wheel encoders the robot can also sense the velocity of each of its two wheels (passive caster wheels are mounted at the front). The speed of the robot can be controlled by software and goes up to 40 cms^{-1}.

Subsequent experiments, confirming the suitability of the differential light compass for robot navigation, and additionally using a series of local landmarks, were conducted at Manchester University (Department of Computer Science), using the mobile robot *Grasmoor* (see figure 5). This robot is equipped with four infrared range-finding sensors, two tactile sensors and four light sensors. The infrared sensors detect local landmarks at a distance of approximately 30 cm.

Fig. 5. *Grasmoor*

5.2 Navigation with the Differential Light Compass

The Differential Light Compass

The *Edinburgh R2* is equipped with six light sensors, of which five are used for the light compass. *Grasmoor* has four light sensors[3]. The arrangement of the sensors of these two robots is shown in figure 6.

The differential light compass is based on the assumption that a continuous gradient of light intensity in the environment exists (provided, for instance, by the sun or a distant light source), and that the robot's light sensors all have similar sensitivity. The current heading of the robot (dx,dy) is then estimated by the following equations[4]:

$$dx = \frac{S1 - S4}{|S1 - S4| + |S2 - S3|}$$
$$dy = \frac{S2 - S3}{|S1 - S4| + |S2 - S3|}$$

where S1 ...S4 denote the signal values of light sensors one to four, dx the heading in x-direction and dy the heading in y-direction (scalars). During

---

[3] The light sensors are based on cadmium sulphide light dependent resistors (dark resistance greater than $20M\Omega$, resistance at 100 lux typically $5k\Omega$, rise time at 10 ft candle 45ms, fall time at 10 ft candle 55ms).

[4] Equations are given for *Grasmoor's* case.

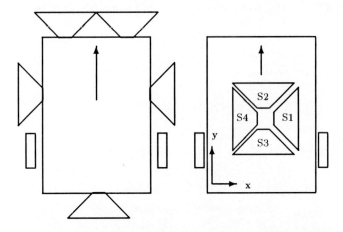

Fig. 6. Arrangement of light sensors on the *Edinburgh R2* (left) and *Grasmoor* (right)

the outward journey the current position of the robot is estimated in discrete time steps according to the following equation, assuming constant travelling speed:

$$x(t+1) = x(t) + dx$$
$$y(t+1) = y(t) + dy.$$

During the return journey, the *Edinburgh R2* moves such that x and y coordinate decrease[5]. The robot stops once they lie below a predefined threshold. *Grasmoor* acts similarly, but because the robot also uses sequences of several landmarks for navigation it initiates a landmark searching behaviour once the differential light compass indicates the robot is close to the desired location. The robots' behaviour is, in fact, not unlike that of navigating desert ants ([19]).

Performance with and without Differential Light Compass

To assess the reliability of navigation based on dead reckoning, we have conducted two experiments with the *Edinburgh R2* that compare a navigational system that uses a reference landmark (in this case a light gradient) with one that uses proprioception alone.

In the latter case the robot's current position is constantly computed, using the wheel encoders and an estimate of the current heading. The latter is determined by the fact that the robot is programmed to perform only 45 degree turns when avoiding obstacles.

---

[5] The origin is, by definition, at the starting location.

It turns out that there are many sources of perturbation that render such a system unreliable for complex navigational tasks. For once, the supposedly 45 degree turns of the robot are never precise, furthermore, the speed of each individual wheel varies slightly in between measuring intervals. For long paths the error accumulates beyond the level where dead reckoning alone can still successfully be used for navigation.

Dead reckoning, using the DLC, uses basically the same algorithm as the previous experiment, but instead of *estimating* the current heading it is *measured*, using the DLC. Again, while the robot is moving through its environment, the current position is constantly updated. On returning to the starting location, the robot first moves in the "correct" direction along the x-axis, then along the y-axis. It then repeats this procedure until the total error lies below a predefined threshold. Moving along both axes instead of returning home on a straight path is computationally cheaper and avoids rounding errors due to the fact that for all computations integer arithmetic has to be used.

Results

In the first experiment the robot had to avoid two obstacles, and then return to the starting location. Figure 7 shows the results of this experiment (Numbers indicate the final returning positions if the DLC was used, letters those where the DLC was not used). Both approaches work satisfactorily; as the robot has a infrared sensor range of ca. 100 cm (for especially prepared passive beacons) the actual starting location was visible from all returning positions.

If, however, a more complicated path as shown in figure 8 is followed, pure dead reckoning quickly becomes so unreliable that it can no longer be used. Figure 8 shows that without using the DLC the robot ended up too far from the actual starting location to be able to see a beacon, whereas when using the DLC a successfull return was achieved.

A further experiment to measure the navigational error was conducted as follows: the robot followed a path as shown in figure 9. At the location marked 'R' the robot was manually instructed to return to the starting location (marked 'S'). The current heading of the robot was determined, using the DLC, the current position was computed using dead reckoning. Figure 10 shows the return position of the robot in x/y coordinates in cm (eleven runs), as well as the distance of the center of the robot from the starting location in cm (in the run marked '*' the robot crossed the starting location several times). Also shown is the displacement in per cent of total distance travelled. To interpret figure 10, it should be born in mind that the sensors of the robot are 12 to 25 cm away from the center, and have a range of another 100cm for passive, reflective beacons. In other words: a beacon would have been visible from all return positions in the eleven runs of the experiment. Another interesting point to make is that the typical navigational error of a human navigator is 5 to 10 per cent of the total distance travelled, the same

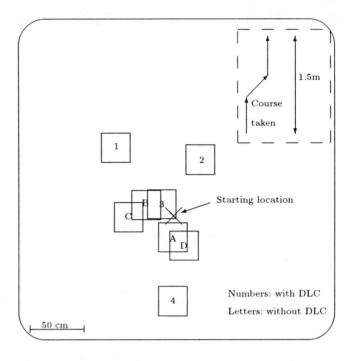

Fig. 7. Navigation with and without DLC. 1st Experiment.

can be observed in desert ants ([4, p.41]). The robot's navigational error of about 8 % falls within this range.

Subsequent experiments with *Grasmoor* have confirmed the findings presented above.

The Influence of a Reference Landmark on Robot Navigation

In section 2.2 we described an experiment that shows that bees orient their paths according to the position of the sun.

A similar experiment was conducted with the *Edinburgh R2*. Our robotics laboratory has two windows, next to each other. The robot's task was to move in a straight line, and upon the experimenter's command return back to the starting location, using the differential light compass. On the outward journey (straight line of about 150 to 200 cm) the left window was bright, whilst the right window was darkened by blinds. For the return journey, the left window was darkened and the right one was bright. The resulting end position of the robot was shifted in relation to the shift of reference landmark, as in the experiment with the bees (see figure 11). A control experiment was conducted in which the lighting remained unchanged throughout the experiment, the results of this experiment are also shown in figure 11.

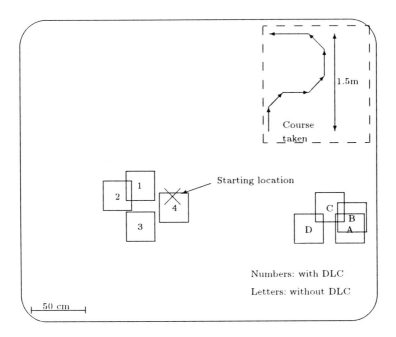

Fig. 8. Navigation with and without DLC. 2nd Experiment

5.3 Local Fluctuations in Light Gradient

Local fluctuations in light gradient occur for instance if the robot moves into shadows, or approaches a window off-center. As long as a gradient exists at all this does not affect navigation, provided the robot returns along the exact same path it used during the outward journey. However, as this is impossible, local fluctuations in light gradient do introduce an error. We have found, though, that if the starting location is marked by a passive beacon, and journey distances do not exceed the range of about four sensor ranges (4 m), the robot reliably returns to the starting location, despite the fact that local fluctuations in light gradient are encountered. We have not only conducted experiments in the laboratory, but also in the open, using a natural light gradient for navigation. At the time of the experiment the sky was still bright, but the sun had set. The robot twice returned to the starting location after two journeys of about 3 m each, travelling over rough terrain and thus being forced to realign repeatedly towards the intended travelling direction.

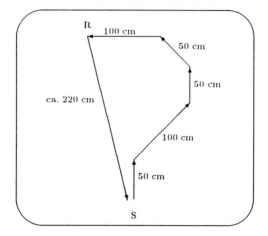

Fig. 9. Course steered to measure navigational error

x [cm]	70	0	20	-30	-10	0	-5	-25	-25	-25	-30	-5,5
y [cm]	30	60	35	0	25	75	-10	-10	0	60	-50	19,5
d [cm]	76	60	40	30	27	75*	11	27	25	65	58	44,9
Error [%]	13	10,5	7	5,3	4,7	13,2	1,9	4,7	4,4	11,4	10,2	7,9

Fig. 10. Navigational displacement, using the DLC and dead reckoning

6   Summary and Conclusion

There is good evidence that animals use local landmarks such as prominent trees, rocks, forest edges *etc.* as well as reference landmarks such as sun, stars and magnetic senses to navigate successfully. It is also observed that in order to navigate between two known locations animals seem to prefer well defined and constant paths, even if this means longer travel distances.

As animal navigation is proven to be reliable and successful it is suggested that a robot navigation system should be based on the following three principles:

- It should use local and reference landmarks,
- it should use canonical paths, and
- it should be based on topological rather than geometrical maps.

First experiments with two mobile robots are presented and the reliability of navigation based on dead reckoning is investigated. The experiments show that for simple navigational tasks pure dead reckoning is sufficient, but that

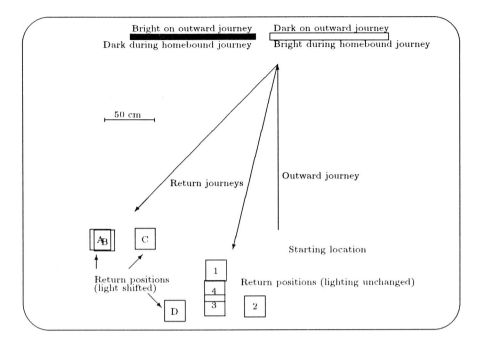

Fig. 11. The use of reference landmarks in the *Edinburgh R2*

this method becomes unreliable if paths are complicated and long. Using reference landmarks (such as, in our case, the sun) greatly improves navigational competence in the latter case. This finding leads us to conclude that for reliable robot navigation a compass sense is *essential*. In the experiments presented here this compass sense has been implemented by means of a differential light compass that uses a light intensity gradient to define a reference direction.

References

[1] B.A. Cartwright and T.S Collett, Landmark learning in bees, J. Comp. Physiol. (1983) 151: 521-543

[2] Derek Cosens, Department of Zoology, Edinburgh University, personal communication

[3] Patricia Smith Churchland, Neurophilosophy, MIT Press Cambridge and London, 1986

[4] C.R. Gallistel, The organisation of learning, MIT Press, 1990

[5] James L. Gould and Carol Grant Gould, The honey bee, Scientific American Library 1988

[6]   Peter Kampmann and Günther Schmidt, Indoor navigation of mobile robots by Use of learned maps, in G. Schmidt (ed.), Information Processing in Autonomous Mobile Robots, Springer Verlag, Berlin, 1991

[7]   Teuvo Kohonen, Self organization and associative memory, Springer Verlag, Berlin, 2nd edition, 1988

[8]   T. Knieriemen and E. von Puttkamer, Real-time control in an autonomous mobile robot, in G. Schmidt (ed.), Information Processing in Autonomous Mobile Robots, Springer Verlag, Berlin, 1991

[9]   E.I. Knudsen, Auditory and visual maps of space in the optic tectum of the owl, Journal of Neuroscience, 2, 1177-1194 (after [4])

[10]   Will Kyselka, An ocean in mind, University of Hawaii Press, Honolulu 1987

[11]   D. Lewis, We, the navigators, University of Hawaii Press, Honolulu, 1972

[12]   Maja Mataric, Navigating with a rat brain: A Neurobiologically-Inspired Model for Robot Spatial Representation, in [13]

[13]   Jean-Arcady Meyer and Stewart Wilson (eds.), From animals to animats, Proc. 1st Intern. Conf. on Simulation of Adaptive Behaviour, MIT Press, Cambridge and London, 1991

[14]   Ulrich Nehmzow and Tim Smithers, Mapbuilding using self-organising networks, in [13]

[15]   Ulrich Nehmzow, Experiments in competence acquisition for autonomous mobile robots, Ph.D. Thesis, Edinburgh University, 1992

[16]   Ulrich Nehmzow and Brendan McGonigle, Robot navigation by light, European Conference on Artificial Life, Brussels, 24.-26.5.1993

[17]   D.L. Sparks and J.S. Nelson, Sensory and motor maps in the mammalian superior colliculus, Trends in Neuroscience, 10, 312-317 (after [4])

[18]   Talbot H. Waterman, Animal navigation, Scientific American Library, New York 1989

[19]   R. Wehner and M.V. Srinivasan, Searching behaviour of desert ants, genus cataglyphis (formicidae, hymenoptera), Journal of Comparative Physiology 1981, 142, 315-338

[20]   R. Wehner and F. Räber, Visual spatial memory in desert ants Cataglyphis bicolor, Experientia 35 (1979)

From Local Interactions
to Collective Intelligence

Maja J. Mataric

Volen Center for Complex Systems, Computer Science Department, Brandeis University, Waltham MA 02254-9110, USA

maja@cs.brandeis.edu

Abstract This paper describes a research program with the goal of understanding the types of simple local interactions which produce complex and purposive group behaviors[1]. We describe a synthetic, bottom–up approach to studying group behavior, consisting of designing and testing a variety of social interactions and scenarios with artificial agents situated in the physical world. We propose a set of basic interactions which can be used to structure and simplify the process of both designing and analyzing group behaviors. We also demonstrate how these basic interactions can be simply combined into more complex compound group behaviors. The presented behavior repertoire was developed and tested on a herd of physical mobile robots demonstrating avoiding, following, dispersing, aggregating, homing, flocking, and herding behaviors.

Keywords Group behavior, collective intelligence, social interaction, behavior-based control, mobile robots

1 Introduction

Intelligence is a social phenomenon. Most intelligent animals exist in a society of kin, obey its rules, and reap its benefits. Societies vary in size and complexity, but have a key common property: they provide and maintain a shared culture.

To be understood, individual intelligence must be observed and analyzed within its social and therefore cultural environment. In contrast to traditional AI, which addresses intelligence as an isolated phenomenon, our work is based on the belief that intelligent behavior is inextricably tied to its cultural context and cannot be understood in isolation. This emphasis is similar to the founding principles of ethology, the study of animal behavior. Unlike the behaviorist branch of biology, ethology studies animals in their natural

---

[1]This paper is a combination of previously published work from Mataric (1992) and Mataric (1993). The work reported here was done at the MIT Artificial Intelligence Laboratory.

habitats. Similarly, this research attempts to understand intelligent behavior in its natural habitat: situated within a culture.

The complexity of culture results from the local interactions among individuals. This research will focus on exploring simple local social interactions which result in purposive group behaviors.

Nature abounds with systems whose complex global behaviors emerge from simple local interactions at all scales, from the subatomic (Gutzwiller 1992), to the semantic (Minsky 1986), to the social (Deneubourg, Goss, Franks, Sendova-Franks, Detrain & Chretien 1990). In contrast to these bottom–up effects, most of AI relies on top–down modularity. As expressed by Simon (1969), a system is well designed if it is analyzable through decomposition into non–interacting modules. However, imposing such modularity minimizes precisely the type of interactions that seem to generate complexity in nature. The global behavior of complex systems, such as groups of social agents, is determined by the local interactions of their constituent parts. These interactions merit careful study in order to understand the global behavior of the society. In natural systems, such interactions resulted in the evolution of complex and stable behaviors that do not lend themselves to traditional, top–down style of analysis.

In order to reach that level of complexity synthetically, such behaviors must be generated through a similar, interaction–driven, incrementally refined process. In this paper we describe such an approach, a combination of bottom–up experiments and theory with the purpose of of designing, observing, and formalizing group behaviors. In particular, our work deals with the problem of controlling a group of autonomous agents, such as a colony of robots foraging and collecting objects, or a group of agents learning from each other. Beni & Wang (1989) refer to such "swarm intelligence" as "the interplay of computation and dynamics." The goal of our work is to explore that interplay and determine how to best, and most simply, design the computational component in order to take advantage of the dynamics.

2   Related Work

Interactions between artificial agents have been studied almost entirely in simulation. For example, Steels (1989) describes a simulation of simple robots using the principles of self organization to perform a gathering task. Brooks, Maes, Matarić & Moore (1990) report a set of simulations in a similar domain, with a fully decentralized collection of non–communicating robots. Arkin (1992) describes a schema–based approach to designing simple navigation behaviors to be extended to multiple physical agents.

The field of Artificial Life (Alife) focuses on bottom–up modeling of biological complex systems. In particular, Alife features much work on simulating ant colonies (Colorni, Dorigo & Maniezzo 1992, Beni & Hackwood 1992, Drogous, Ferber, Corbara & Fresneau 1992, Travers 1988) and many others. Deneubourg et al. (1990), Deneubourg & Goss (1989), and their other work describes experiments with real and simulated ant colonies and examines the role of simple control rules and limited communication in producing trail formation and task division. Deneubourg, Theraulax & Beckers (1992) define some key terms in swarm intelligence and discuss issues of relating local and global behavior of a distributed system. Assad & Packard (1992), Hogeweg & Hesper (1985) and other related work also report on a variety of simulations of simple organisms collectively producing complex behaviors emerging from simple interactions.

In contrast to the majority of Alife research, this work is concerned with fewer but individually more intelligent agents, and the behavior emerging from their interactions.

3 Emergent Behavior

Emergent behavior is one of the main topics of research in the field of swarm intelligence. It is manifested by global states or time–extended patterns which are not explicitly programmed in, but result from local interactions among the system's components. Emergent behavior can be observed in any sufficiently complex system, i.e. a system which contains local interactions with spatial and/or temporal consequences.

Emergent phenomena are appealing because they appear to provide something for nothing, such as interesting collective behaviors emerging without *a priori* design or analysis of the system. In reality, any purposive behavior must be obtained either analytically or empirically. In this paper we will describe how a particular type of emergent collective behavior can be reliably obtained empirically, then analyzed and used as a tool. The next section addresses the difficulties of the alternative, analytical approach.

3.1 Analyzing Complex Behavior

Analyzing and predicting the behavior of a single situated agent is an unsolved problem in robotics and AI. Even highly constrained domains are intractable, and realistic worlds usually do not contain the structure, determinism and predictability necessary for strict formal analysis (Canny 1988, Brooks 1990 b, Brooks 1991).

Predicting the behavior of a multi–agent system is generally even more complex than the single–agent case. In general, no satisfactory solution exists for predicting the behavior of a system with nontrivial interacting components. Statistical methods used for analyzing particle systems do not directly apply as they constrain the type and amount of interactions among the components (Weisbuch 1991). While systems with large numbers of simple and simply interacting components can be analyzed this way (Wiggins 1990), no tools are available for systems consisting of comparatively few but more complicated components with complex interactions.

Instead of attempting to analyze arbitrary complex behaviors, this work focuses on providing a set of behavior primitives that can be used for synthesizing and analyzing a particular type of complex multi–agent systems. We will describe a set of local interactions which produce cohesive collective behaviors in this domain, and can be combined into variety of more complex global behaviors. The behaviors we describe emerge from local interactions, but rather than being unexpected, are repeatable and well understood.

4 Basic Principles

The following principles form the basis for the methodology used in this work.

- Agents are homogeneous in terms of hardware and software. There are no *a priori* leaders or followers.
- Agents do not use any explicit one–to–one communication. Instead, all communication is based on sensing the external state of nearby agents. This is akin to stigmergic communication in nature, which is based on modifications to the environment rather than direct message passing.
- Agents do not engage in any explicit cooperation. Instead, cooperation is implicit, and occurs through the world rather than through directed communication. Agents affect one another largely by means of their distribution and external state.
- Agents have no hidden goals. This work does not deal with the agents' underlying motivations, the possibility of cheating, competition, and ulterior motives. Agents are assumed to have a set of common, similar, or at least overlapping goals.
- Agents are able to detect other agents of the same kind. The ability to categorize the perceptible objects in the world into at least two classes: "others like me" and "everything else" is a necessary condition for intelligent collective behavior. With this ability, which is innate and ubiquitous in nature, even the simplest of interactions can produce purposive collective behavior, as we will demonstrate in the following sections.

5 Basic Interactions and Combinations

Interactions between individual agents need not be complex to produce complex global consequences. Here we describe some of the simplest types of interaction that serve as a basis for a rich behavior repertoire.

Table 1 defines the interaction primitives, or basic behaviors, we have discovered to be functionally sufficient building blocks for designing a variety

Collision Avoidance	the ability of an agent to avoid colliding with anything in the world. Two distinct strategies can be devised; one for other agents of the same kind, and another for everything else that the robot might encounter.
Following	the ability of two or more agents to move while staying one behind the other.
Dispersion	the ability of a group of agents to spread out over an area in order to establish and maintain some predetermined minimum separation.
Aggregation	the ability of a group of agents to gather in order to establish and maintain some predetermined maximum separation.
Homing	the ability to reach a goal region or location.

Table 1: The basic interaction primitives

of complex behaviors in our domain. In this work, a primitive is any group behavior, constructed with simple local interactions, that can be used as a building block and cannot be reduced to one or more of the other primitives. The primitives are combined by concurrent execution, or by switching between behaviors (Matarić 1994 a). For example, *herding*, a group behavior involving collective motion guided by a subset of the agents, could be constructed with a combination of *flocking* and *homing*. The agents on the fringes and at the front of the herd engage in homing behavior while the rest flock. Other examples of more complex behaviors, such as *foraging*, can also be implemented from a subset of the above primitives, as will be demonstrated shortly.

6 Methodology

Since complex interactive behavior depends strongly on the particular parameters of the system, in order to duplicate any such behavior it is necessary to either construct a very accurate model of the system, or to build the system and experimentally reproduce the behavior. Our work relies on the latter approach, a bottom–up experimental methodology, which drives and refines our theories. Although there is a great tendency to design behaviors around

Figure 1: Some of the twenty physical experimental agents used to demonstrate and verify our group behavior work. Each of the robots is a wheeled base with a two–pronged forklift for picking up, carrying, and stacking pucks. The robots are equipped with bump sensors, infra–red (IR) sensors on the forklift, and a radio transceiver for inter-robot communication and data collection

solving a problem, or simulating a naturally occurring system, it is necessary to go beyond finding a working solution to evaluating the effects of various possibilities.

Rather than attempt to to simulate biological systems, we try to elucidate the phenomena of social interaction by synthesizing, observing, and analyzing similar phenomena. Consequently, our evaluation criteria are based on performing and analyzing a variety of experiments in our synthetic domain.

Our group behavior experiments are implemented and tested on a collection of twenty physically identical mobile robots. Each robot is a 12"–long four–wheeled vehicle, equipped with one piezo–electric bump sensor on each side and two on the rear of the chassis. Each robot has a two–pronged forklift for picking up, carrying, and stacking pucks. For this purpose, the forklift contains six infra–red sensors: two pointing forward and used for detecting objects, two break–beam sensors for detecting a puck within the "jaw" and "throat", and two down–pointing sensors for aligning the fork over an object (figure 1).

The robots are equipped with radio transceivers, which use two radio base stations to triangulate the robots' position, and transmit and receive one byte of data per robot per second. The radio system is used for data gathering and for simulating additional sensors. In particular, radios are used to distinguish robots from other objects in the environment, an ability which cannot be implemented with the on–board IR sensors. The flexibility of the radio system also allows for testing a variety of sensing parameters.

The robot experiments are run fully autonomously with all of the processing and power on board. The robots are programmed in the Behavior Language, a parallel programming language based on the Subsumption Architecture (Brooks 1990 a, Brooks 1987). Their control systems consist of collections of parallel, concurrently active behaviors implementing the setf of interaction primitives.

6.1 Hardware Implications

The types of experiments that can be implemented on robots are strongly constrained by the sensory, mechanical, and computational limitations of the hardware. In addition to the expected sensory and mechanical error, our robots suffer from inaccurate steering and radio transmission. Further, the infra–red sensors have long and varying ranges. Consequently, not only do different robots have different sensing regions, but the sensitivity of sensors on a single robot varies as well.

Hardware variability between robots is reflected in their group behavior. Although programmed with identical software, the robots behave differently due to their varied sensory and actuator properties. Small differences between individuals become amplified as many robots interact over extended time. As in nature, individual variability creates richer, more complex interactions and thus demands more robust and adaptive behavior. The described interaction primitives were designed to be general enough to span a variety of mechanical and sensory variations. As with any physical system, the particular parameter values were tuned to the specifics of the system. The resulting interaction primitives were robust and repeatable within the allowable margins of the hardware.

7 Experimental Results

This section describes the robot implementation of the basic interaction primitives, and illustrates their performance. The results are plotted as the robots' position over time, based on radio data, displayed from a top view, and using trails to indicate time history. The sizes of the robots and the testing area shown in the display are scaled to the exact dimensions of the physical environment. The robots are shown as black rectangles aligned in the direction of their heading, with the ID numbers in the back, and white arrows indicating the front. The corner display shows actual elapsed time, in seconds, for each snapshot of the experiment. The data is also available on video tape.

7.1 Collision Avoidance

Moving while avoiding collisions is perhaps the most studied topic in mobile robotics. Finding a guaranteed general–purpose collision avoidance strategy for an agent situated in a dynamic world is difficult. In a multi–agent world the problem can become intractable. Observation of biological systems suggests that insects and animals do not have precise avoidance routines. Instead, they are equipped with a simple basic avoiding behavior which is effective most of the time, and a small number of special-purpose behaviors for the few important specific cases (Wehner 1987). Inspired by those data, we designed the following avoidance strategies:

```
Avoid-Other-Agents:
If an agent is within d_avoid
   If the nearest agent is on the left
      turn right
      otherwise turn left.
```

```
Avoid-Everything-Else:
If an obstacle is within d_avoid
   If an obstacle is on the right only, turn left.

   If an obstacle is on the left only, turn right.
   After 3 consecutive identical turns, backup and turn.

   If an obstacle is on both sides, stop and wait.
   If an obstacle persists on both sides,
      turn randomly and back up.
```

The `Avoid-Other-Agents` behavior takes advantage of group homogeneity. Since all agents execute the same strategy, the behavior can rely on the resulting spatial symmetry and take advantage of it. If an agent fails to recognize another with its other–agent sensors (in this case radios), it will subsequently detect it with its collision–avoidance sensors (in this case IRs), and treat it as a generic obstacle, using the `Avoid-Everything-Else` behavior.

To increase robustness and minimize oscillations, our strategies take advantage of the unavoidable noise and errors in sensing and actuation, which result in naturally stochastic behavior. This stochastic component guarantees that the an avoiding agent will not get stuck in infinite cycles and oscillations. In addition to the implicit stochastic nature of the robots' behavior,

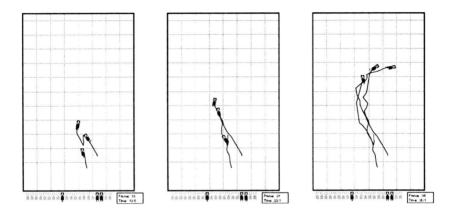

Figure 2: Continuous following behavior of 3 robots. The entire time history of the robots' positions is plotted

Avoid-Everything-Else also utilizes an explicit probabilistic strategy by employing a randomized move.

7.2 Following

Following can be implemented as an inverse of avoidance:

```
Follow:
If an agent is within d_follow
    If an agent is on the right only, turn right.

    If an agent is on the left only, turn left.
```

Our robots use the same sensors for both behaviors. They convert binary IR sensors into "range" sensors through the use of time. Distance from the object being followed in terms of time to collision is estimated from the time an object has continuously remained within the IR sensor range. Figure 2 illustrates *following* on 3 robots.

Following is analogous to osmotropotaxis exhibited by ants (Calenbuhr & Deneubourg 1992). While ants use the differential in pheromone intensity perceived by the left and right antennae to decide in which direction to turn, our agents turn based on the binary state of the front two IR sensors.

Under conditions of sufficient density, *collision avoidance* and *following* can produce more complex global behaviors. For instance, chemotropotaxic

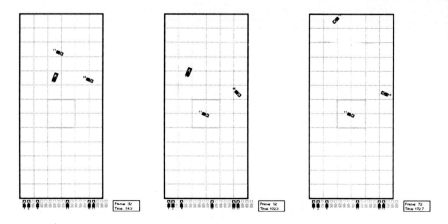

Figure 3: Dispersion behaviors of 3 robots

ants exhibit emergence of unidirectional traffic lanes. The same lane–forming effect could be demonstrated with robots executing *following* and *avoiding* behaviors. However, more complex sensors than IRs must be used in order to determine which direction to follow. If using only IRs, the robots cannot distinguish between other robots heading toward and away from them, and are thus unable to select whom to follow.

7.3 Dispersion

In order to balance goal–directed behavior against interference, the agents must be able to not only avoid but also escape congested situations. A robust *dispersion* behavior can be designed as an extension of the existing collision avoidance. While collision avoidance relies on the presence of a single agent, *dispersion* uses the local distribution (i.e. the locations of other agents within the range of the agent's sensors) in order to decide in which direction to move. The algorithm computes the local centroid to determine where most of the nearby agents are, and then moves away from that direction. Figure 3 illustrates dispersion with 3 robots.

```
Disperse:
If one or more agents are within d_disperse
    move away from Centroid_disperse.
```

In situations of high density, the system can take a long time to achieve a dispersed state since local interactions propagate far, and the motion of an individual can disturb the state of many others. Thus, *dispersion* is best viewed as an ongoing process that maintains a desired distance between agents while they are engaged in other behaviors.

7.4 Aggregation

Aggregation is the inverse of dispersion. Its goal is to bring all of the agents within a prespecified distance from each other.

```
Aggregate:
If nearest agent is outside d_aggregate
    turn toward the local Centroid_aggregate, go.

Otherwise, stop.
```

Dispersion and *aggregation* are similar to behaviors exhibited by army ants in the process of stabilizing the temperature of a bivouac. Individual ants aggregate and disperse by following the local temperature gradient (Franks 1989). Temperature control is a side–effect of these two interactions. Similarly, density control could be used as a means of collecting and distributing objects in the environment, as in the foraging and sorting behaviors we described earlier.

7.5 Homing

The simplest *homing* strategy is a greedy local one:

```
Home:
If at home
  stop.
  otherwise turn toward home, go.
```

Figure 4 illustrates the homing behavior of five robots. The data illustrates that the actual trajectories are far from optimal, due to mechanical and sensory limitations, in particular due to the error in the sensed position.

Individual *homing* is effective as long as the density of agents is sufficiently low. If enough agents are trying to home, they begin to interfere with each other. Interference worsens if the agents have non-zero turning radii, unequal velocities, or are subject to sensor and control errors. All of the above conditions are common in robots, suggesting the need for some form of group

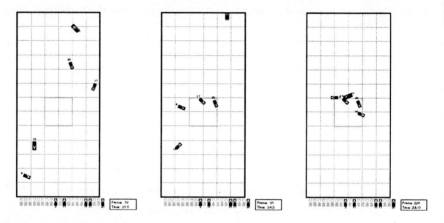

Figure 4: Homing behaviors of 5 robots. Four of the five robots reach home quickly and the fifth joints them about 60 second later

navigation. Flocking is an example of such navigation.

7.6   Flocking

Flocking is an essential means for a group[2] of agents to move together. Instead of constituting a new interaction primitive, flocking can be implemented as a combination of other primitives: avoidance, aggregation, dispersion, and homing. Each of these constituent behaviors commands a velocity to the effector that results in a stop, go, or turn. *Flocking* weights and combines these outputs.

```
Weight the inputs from
   avoid, aggregate, disperse, and home
   then compute a turning vector.
If in the front of the flock,
   slow down.
If in the back of the flock,
   speed up.
```

The choice of weights on the behavior outputs depends on the dynamics and mechanics of the agents, and the ranges of the sensors. A flocking behavior can be implemented with avoidance, aggregation and dispersion alone, but the flock would not move to a particular location but instead follow whatever

---

[2]A group is defined to be a collection of size three or more.

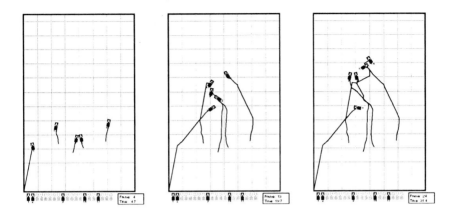

Figure 5: Flocking behavior of 5 robots. The robots are started out in a nearly lined–up dispersed state. They quickly establish a flock and maintain it as the positions of the individual robots within the flock fluctuate over time

direction the robots in the front are taking. Homing provides the flock with a consistent direction or a final goal location.

The flocking algorithm was inspired by an elegant graphics simulation of bird flocking implemented by Reynolds (1987). However, the robot implementation required many more details than the simulation, due to the complex robot interaction dynamics. Figure 5 demonstrates two runs of the flocking behavior with five robots.

7.7 Foraging

Foraging is a classical task for both animals and mobile robots. In both cases, the agents search for objects, collect them, and take them to a prespecified destination. Foraging can be implemented as a combination of the described basic interaction primitives. Unlike *flocking*, which combines the outputs of concurrently active constituent behaviors, the controller in *foraging* switches between the different behaviors under appropriate sensory conditions.

The high-level goal of a foraging group is to collect objects from the environment and deliver them home. In addition to having the basic social interaction repertoire, individual agents are also able to recognize and manipulate objects. However, they have no model of the environment, nor a global view of it.

The internal behavior structure of the each of the agents is identical, and specifies the conditions triggering each of the basic social primitives (Figure 6). Foraging is initiated by *dispersion*, followed by a search for objects. Finding an object triggers *homing*. Encountering another agent with a dif-

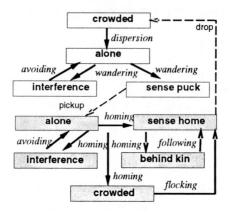

Figure 6: The behavior structure of the foraging agents. Shaded states indicate carrying an object. Basic interaction primitives are italicized

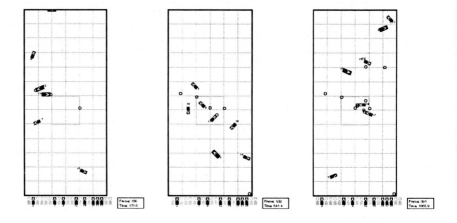

Figure 7: An example of the foraging behavior of 6 robots. About eight pucks have been delivered to the home region, marked with a grey box. Two of the robots are dropping off pucks while the others are wandering in search of additional pucks to pick up and deliver home

ferent immediate goal (as manifested by its external state, e.g. not carrying an object), induces *avoidance*. Conversely, encountering kin, another agent with the same external state (carrying something, or approaching an area with free objects) triggers *following*. Three or more followers initiate *flocking*. Reaching home and depositing the object restarts dispersion.

Foraging demonstrates how basic interaction primitives can be combined into a higher-level compound behavior. The combination is simple in that conflicts between two or more interacting agents, each potentially executing a different behavior, are resolved uniformly due to agent homogeneity. Since all of the agents share the same goal structure, they all respond to the environmental condition consistently. For example, if a group of agents is *flocking* toward home and it encounters a few agents *dispersing*, the difference in t' e agents' external state will either induce kin *following* or non–kin *avoidan* thus dividing the group again.

Figure 7 illustrates an example foraging run with six robots. Initially 'i pucks are located in a pile in the lower left hand of the workspace. Over time the robots gradually move the pucks into the home region, marked with grey bounding box.

8 Heterogeneous Groups

In the experiments described so far the agents were fully homogeneous. As a control study, as well as an attempt to address a larger variety of social interactions, we introduced dominance hierarchies into our agent societies.

We tested the performance of dominance hierarchies using hierarchical control strategies on two of the basic behaviors described above: aggregation and dispersion. These two behaviors were chosen because simple performance evaluating criteria could be applied. Specifically, given sufficient space, aggregation and dispersion can reach a static state, i.e. obtain the desired distance between the agents. Consequently, it is relatively simple to compare the performance of different aggregation and dispersion algorithms based on the time each required to reach static termination conditions. In contrast, following and flocking are not as simply evaluated due to their dynamic nature.

In this set of experiments the collection of agents was classified into a total order, based on a randomly assigned unique ID number, simulating an established pecking order in the group (Chase 1982, Chase & Rohwer 1987). Unlike the homogeneous algorithms, in which all agents moved simultaneously according to identical local rules, in the hierarchical case the ID number determined which agents were allowed to move while others waited. (In all cases, a simple precedence order was established such that within a small radius the agent with the highest ID got to move.) As in the homogeneous case, we tested multiple hierarchical algorithms for each of the group behaviors.

Using the software environment, we conducted 20 experiments with each

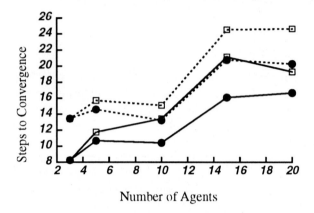

Figure 8: The performance of two different aggregation algorithms based on time required to reach static state. Two termination conditions were tested: a single aggregate (data points shown with boxes) and a few stable groups (data points shown with dots). The performance of hierarchical algorithms is interpolated with solid lines while the homogeneous ones are interpolated with dotted lines

group size (3, 5, 10, 15, and 20 agents) and each of the algorithms. Additionally, we tested the algorithms on two different degrees of task difficulty. For aggregation we tested two terminating conditions: a single aggregate containing all of the agents, and a small number of stable aggregates. The former terminating condition is more difficult. Similarly, for dispersion we tested two initial conditions: a random distribution of initial positions, and a packed distribution in which all of the agents start out in one half of the available space. The latter condition is more difficult.

We found that, in the case of aggregation, hierarchical strategies performed slightly better than totally homogeneous ones. Figure 8 plots the average number of moves an agent takes in the aggregation task against the different group sizes and the two different terminating conditions: a single aggregate and a few stable groups. Both hierarchical and homogeneous algorithms behaved as expected, improving on the simpler of the two terminating conditions. Their performance declined consistently with the growing group size.

Unlike aggregation, in the case of dispersion, homogeneous strategies outperformed hierarchical ones. Figure 9 plots the average number of moves an agent makes in the dispersion task for the different group sizes on two different initial conditions: a random distribution, and a packed initial state. Again, both hierarchical and homogeneous algorithms improved with the easier initial conditions.

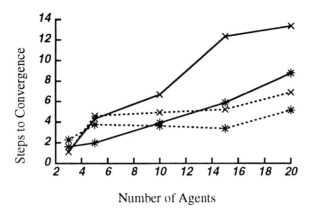

Figure 9: The performance of two different dispersion algorithms based on time to reach static state. Two initial states were tested: a random distribution (data points shown with stars) and a packed distribution (data points shown with crosses). The performance of the hierarchical algorithms is interpolated with solid lines while the homogeneous ones are interpolated with dotted lines

Although the performance difference between the homogeneous and hierarchical algorithms was repeatable and consistent, it was small, and its magnitude barely surpassed the standard deviation among individual trials for each of the algorithms and group sizes. The standard deviation was particularly significant in the case of small (3 and 5) group sizes. We believe that the difference between the two strategies would be negligible on physical agents.

The experiments comparing simple hierarchical and homogeneous algorithms demonstrate that for the described domain simple hierarchical strategies do not improve the global performance. In our experiments, hierarchies were assigned randomly since all of the agents are identical. More complex hierarchical strategies could be devised, but would require an increased perceptual and cognitive overhead. From our data we can hypothesize that for simple spatial domains 1) the simplest, homogeneous solution works well, and 2) quite a bit more knowledge and processing is required to significantly improve it.

9 Extensions

This paper has described our work on a methodology for principled design of group behaviors based on simple local interactions. Besides demonstrating basic interactions and their combinations, we have also implemented a method for automatic generation of higher–level behaviors through an adapted process of unsupervised reinforcement learning (Matarić 1994 a). We applied this method to learning to forage and generated a performance comparable to that of the hard–wired algorithm within fifteen minutes of real–time learning (Matarić 1994 c). Furthermore, we have successfully applied the learning methods to acquiring not only higher–level behaviors but also social rules such as *yielding* and *communicating* (Matarić 1994 b).

Our current work applies the basic behavior paradigm to different types of robots and tasks, as well as to heterogeneous groups. Finally, our future work extends the learning paradigms to allow for detecting sequences of states and actions, and storing the observed information. This ability to **imitate** can be used to propagate local interactions globally, both spatially and temporally, in a form of *social learning*. This type of learning through imitation is ubiquitous in nature (McFarland 1985) and appears to be innate. Our continuing work utilizes the described behavior primitives as the basis for social interactions learnable through imitation.

10   Summary

This paper has described a research program with the goal understanding the types of simple local interactions which produce complex and purposive group behaviors. We described a synthetic, bottom–up approach to studying group behavior, consisting of designing and testing a variety of social interactions and scenarios with artificial agents situated in the physical world. Specifically, we introduced a set of basic interactions or primitives which can be used to structure and simplify the process of both designing and analyzing group behaviors. We also showed that for simple spatial behaviors no dominance hierarchies are necessary, or indeed helpful.

We hope that synthesizing and observing group behavior will offer insights into understanding social interactions in biology, as well as help derive methods for principled behavior control in robotics and AI.

Acknowledgements

Many thanks to Matthew Marjanović, Stanley Wang, and Owen Wessling for making the unruly robot herd behave, and helping its collective intelligence emerge in spite of many hardware problems. Matt wrote and debugged the flocking software as well as the software for recording and displaying robot data.

The research reported here was done at the MIT Artificial Intelligence Laboratory. Support for this research was provided in part by the Jet Propulsion Laboratory contract 959333 and in part by the Advanced Research Projects Agency under Office of Naval Research grant N00014–91–J–4038.

References

Arkin, R. C. (1992), Cooperation without communication: Multiagent schema based robot navigation, Journal of Robotic Systems

Assad, A. & Packard, N. (1992), Emergent colonization in an artificial ecology, in F. Varela & P. Bourgine (eds.) Toward A Practice of Autonomous Systems: Proceedings of the First European Conference on Artificial Life, MIT Press, pp. 143–152

Beni, G. & Hackwood, S. (1992), The maximum entropy principle and sensing in swarm intelligence, in F. Varela & P. Bourgine (eds.) Toward A Practice of Autonomous Systems: Proceedings of the First European Conference on Artificial Life, MIT Press, pp. 153–160

Beni, G. & Wang, U. (1989), Swarm intelligence in cellular robotic systems, in P. Dario, G. Sandini & P. Arbischer (eds.) NATO ASI Series F, Vol. 102, Springer–Verlag, Berlin, pp. 703–712

Brooks, R. A. (1987), A hardware retargetable distributed layered architecture for mobile robot control, in IEEE International Conference on Robotics and Automation, Raleigh, NC, pp. 106–110

Brooks, R. A. (1990 a), The behavior language; user's guide, Technical Report AIM-1127, MIT Artificial Intelligence Lab

Brooks, R. A. (1990 b), Challenges for complete creature architectures, in J. A. Meyer & S. Wilson (eds.) From Animals to Animats: International Conference on Simulation of Adaptive Behavior, MIT Press

Brooks, R. A. (1991), Intelligence without reason, in Proceedings, IJCAI-91

Brooks, R. A., Maes, P., Matarić, M. J. & Moore, G. (1990), Lunar base construction robots, in IEEE International Workshop on Intelligent Robots and Systems (IROS-90), Tokyo, pp. 389–392

Calenbuhr, V. & Deneubourg, J. L. (1992), A model for osmotropotactic orientation (i), Journal of Theoretical Biology

Canny, J. F. (1988), The Complexity of Robot Motion Planning, MIT Press, Cambridge, MA

Chase, I. D. (1982), Dynamics of hierarchy formation: The sequential development of dominance relationships, Behaviour 80, 218–240

Chase, I. D. & Rohwer, S. (1987), Two methods for quantifying the development of dominance hierarchies in large groups with application to harris' sparrows, Animal Behavior

Colorni, A., Dorigo, M. & Maniezzo, V. (1992), Distributed optimization by ant colonies, in F. Varela & P. Bourgine (eds.) Toward A Practice of Autonomous Systems: Proceedings of the First European Conference on Artificial Life, MIT Press, pp. 134–142

Deneubourg, J.-L. & Goss, S. (1989), Collective patterns and decision-making, in Ethology, Ecology and Evolution 1, pp. 295–311

Deneubourg, J. L., Goss, S., Franks, N., Sendova-Franks, A., Detrain, C. & Chretien, L. (1990), The dynamics of collective sorting, in From Animals to Animats: International Conference on Simulation of Adaptive Behavior, MIT Press, pp. 356–363

Deneubourg, J. L., Theraulax, G. & Beckers, R. (1992), Swarm-made architectures, in F. Varela & P. Bourgine (eds.) Toward A Practice of Autonomous Systems: Proceedings of the First European Conference on Artificial Life, MIT Press, pp. 123–133

Drogous, A., Ferber, J., Corbara, B. & Fresneau, D. (1992), A behavioral simulation model for the study of emergent social structures, in F. Varela & P. Bourgine (eds.) Toward A Practice of Autonomous Systems: Proceedings of the First European Conference on Artificial Life, MIT Press, pp. 161–170

Franks, N. R. (1989), Army ants: A collective intelligence, American Scientist 77, 139–145

Gutzwiller, M. (1992), Quantum chaos, Scientific American

Hogeweg, P. & Hesper, B. (1985), Socioinformatic processes: Mirror modelling methodology, Journal of Theoretical Biology 113, 311–330

Matarić, M. J. (1992), Designing emergent behaviors: From local interactions to collective intelligence, in J.-A. Meyer, H. Roitblat & S. Wilson (eds.) From Animals to Animats: International Conference on Simulation of Adaptive Behavior

Matarić, M. J. (1993), Kin recognition, similarity, and group behavior, in Proceedings of the Fifteenth Annual Conference of the Cognitive Science Society, Boulder, Colorado, pp. 705–710

Matarić, M. J. (1994 a), Interaction and intelligent behavior, Technical Report AI-TR-1495, MIT Artificial Intelligence Lab

Matarić, M. J. (1994 b), Learning to behave socially, in D. Cliff, P. Husbands, J.-A. Meyer & S. Wilson (eds.) From Animals to Animats: International Conference on Simulation of Adaptive Behavior, pp. 453–462

Matarić, M. J. (1994 c), Reward functions for accelerated learning, in W. W. Cohen & H. Hirsh (eds.) Proceedings of the Eleventh International Conference on Machine Learning (ML-94), Morgan Kauffman Publishers, Inc., New Brunswick, NJ, pp. 181–189

McFarland, D. (1985), Animal Behavior, Benjamin Cummings

Minsky, M. L. (1986), The Society of Mind, Simon and Schuster, NY

Reynolds, C. W. (1987), Flocks, herds, and schools: A distributed behavioral model, Computer Graphics 21(4), 25–34

Simon, H. (1969), The Sciences of the Artificial, MIT Press

Steels, L. (1989), Cooperation between distributed agents through self-organization, in Workshop on Multi-Agent Cooperation, North Holland, Cambridge, UK

Travers, M. (1988), Animal construction kits, in C. Langton (ed.) Artificial Life, Addison–Wesley

Wehner, R. (1987), Matched filters – neural models of the external world, Journal of Computational Physiology A(161), 511–531

Weisbuch, G. (1991), Complex system dynamics, in Lecture Notes Vol. II, Santa Fe Institute Studies in the Sciences of Complexity, Addison–Wesley, NY

Wiggins, S. (1990), Introduction to Applied Nonlinear Dynamical Systems and Chaos, Springer–Verlag, NY

The Mobile Robot of MAIA: Actions and Interactions in a Real Life Scenario

Giulio Antoniol, Bruno Caprile, Alessandro Cimatti & Roberto Fiutem

Istituto per la Ricerca Scientifica e Tecnologica, I-38050 Povo, Trento, Italy

Abstract. Over the years, automated vehicles have evolved from reliable yet rigidly constrained AGVs to the fairly flexible ones of the HelpMate generation. It is interesting to observe that while such systems exhibit quite respectful autonomous navigation capabilities, their capacity of interacting with the users and the environment is still limited.

In this paper, the research work done at IRST in the framework of the Experimental Platform of MAIA is presented. In particular, a transport mission scenario (MAIA '94) is considered, in which autonomous navigation, speech recognition and planning represent three aspects of the same capacity that the system has of interacting with the external world in a reliable and autonomous fashion.

1   Introduction

Design and realization of artificial systems able to autonomously accomplish transport missions in relatively unstructured indoor environments, such as hospitals or office buildings, is an area of research that has witnessed a growing and growing interest from the AI and Engineering communities; yet, it is only since recent times that the subject has established as a technology mature enough to find its way through the marketplace [1, 2, 3].

In this respect, the joint effort that has been carried out at IRST[1] is to be seen from two, somewhat complementary, perspectives: the first, pragmatic, concerns the actual realization of a system able to perform transport missions, by establishing and autonomously carrying on a number of interactions with the surrounding environment and the humans; the second, more conceptual, concerns the use of the system itself as a flexible experimental setting on which different models of intelligent behaviour can be elaborated and tested.

A system (The *Experimental Platform of MAIA*) has therefore been defined which consists of two basic entities: the Mobile Robot and the Supervision Station. The Mobile Robot is endowed with the capability of navigating through the environment, autonomously dealing with obstacles or other events that may affect its motion; sensors, actuators and interface devices

---

[1] Research described in this paper is part of MAIA (Italian acronym for *Advanced Model of Artificial Intelligence*), the leading AI project presently developed at IRST [4, 5].

allow it to interact with people and the external world. From the Supervision Station, operations of the Mobile Robot can be scheduled, planned and supervised. Albeit the Experimental Platform is designed to deliver mail or office items, no obstacle in principle exist to apply its capabilities to other scenarios, such as surveillance, or operations in hostile environments. In this respect, that of moving along corridors or hallways can be seen as a particular one of a much wider class of interactions that we wish a well educated robot be able to perform – noteworthy among them all those based on acoustic and visual abilities. In this respect, what we are working towards is a system acting more like a butler rather than a porter, and like a butler hopefully able to satisfy our requests without needing specification of too many details concerning the execution.

The experience of the past years has clearly shown the realization of a system like this cannot be accomplished on the bare ground of abstract intuitions – regardless of how deep or intriguing they may be. Rather, it is crucial that an unrelenting process be carried out in which novel ideas and solutions could be proposed and thoroughly tested. The Experimental Platform that will be described in the following sections, can therefore be regarded as the framework in which the experimental work is carried out, as well as the very outcome of the experimental work itself.

In Sec. 2 the Experimental Platform of MAIA is be presented, and its two main components described. A few considerations are also reported about the general ideas which the system is most inspired to. Section 3 contains a description of the application scenario in which the system is going to be tested and a rather detailed presentation of the functionalities it can exhibit. In the last section, a series of open issues and ideas for future developments is finally collected and discussed.

2 The Experimental Platform

The *Experimental Platform* is designed to satisfy requests of missions (consisting in delivery tasks) that enabled users issue through the computer network. Two its main components: the *Mobile Robot* and the *Supervision Station* (see Fig. 1). The Mobile Robot is an autonomous vehicle equipped with a sensorimotor apparatus which make it possible, for the robot, to move within a dynamic environment; aboard the robot, speech processing primitives are also implemented which are able to support simple vocal interactions, such as receiving/confirming orders and recording and playing speech messages. The Supervision Station processes mission requests, plans the robot activities needed for the achievement of the tasks being requested, supervises the activities of the mobile robot and interacts with a specialized human operator. The communication flow among the two subsystems is generally the following: the Supervision Station sends to the Mobile Robot control sequences, and the Mobile Robot returns monitoring information.

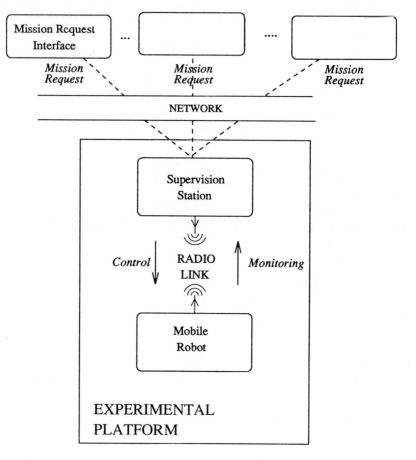

Fig. 1. The Architecture of the Experimental Platform

2.1 The Supervision Station

Generally speaking, the Supervision Station is the assembly of devices and modules by means of which operations of the Mobile Robot can be planned, executed and monitored (see Fig. 2). A Supervision Operator supervises the operations that the Supervision Station automatically carries out, and can take over the system whenever he finds it necessary or simply advisable (see Sec. 3.2).

Mission requests are first checked for validity, and then scheduled by the Supervisor module. The Supervisor takes into account the time constraint specified by the user in the request, also comparing them with statistical information (such as, for example, average mission times, "costs" of different paths, and so on). At present, statistical information about missions is not available (indeed, this is one of the by-products we expect to obtain from the

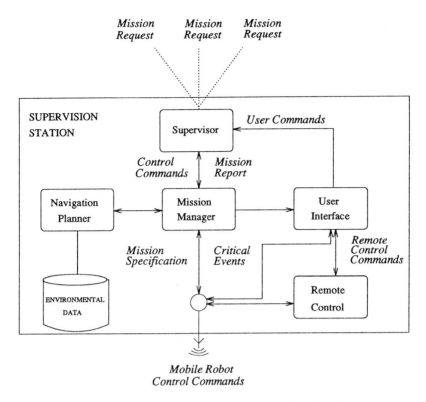

Fig. 2. The Architecture of the Supervision Station

experiments), and the Mission Scheduler simply handles a queue of missions, trying to satisfy queued mission requests as soon as the mobile robot becomes available.

When the scheduled mission is to be executed, the Supervisor module instantiates a Mission Manager for that mission. By the time being only one robot is employed, but the architecture is open to having more than one robot managed from the Supervision Station: multiple instances of the Mission Manager can be instantiated – one for each robot executing a specific mission. When the Mission Manager receives a mission request, it plans – at a strategic level – the activities needed to achieve the goal. The result is a combination of navigation and interactions that can be carried out by the Mobile Robot. During the strategic planning phase, the planning of navigation tasks is performed by a dedicated module, the Navigation Planner, which determines the most appropriate path to be followed in order to navigate between different locations; the Navigation Planner exploits a topological description of the navigation environment. Once that the needed activities are determined, the Mission Manager module starts to execute the mission, sending navigation and interaction requests to the Mobile Robot, and waiting for the monitoring information describing the result of the execution. If a problem occurs which cannot be solved by applying Mobile Robot's skills, the Mission Manager is responsible for planning an alternative course of actions; for instance, if the path planned to reach the destination cannot be traveled because of some obstacle, the Mission Manager may plan an alternative path.

The Supervision Station is designed to interact with a specialized operator via a User Interface. The interface shows a map of the mission environment and continuously displays the position of the robot and the current state of execution of the mission. Furthermore, it displays sensors data coming from the microphones and cameras set aboard the robot. The interface also allows the operator to communicate additional information about the environment, as, for example, that a certain corridor should not be used for navigation. Via the interface, the user can take control over the mission: for instance, the operator can suspend the current navigation, resume, and tell the system to change the planned path. An important feature of the User Interface is the possibility of activating the Remote Control module. This module implements functionalities for the direct control of the robot in terms of navigation and interaction commands that can be issued by voice or keyboard. For example, a remote control command could be: *"go on straight for ten meters"*, while an example of remote interaction could be the simple utterance of a warning message on the part of the robot. Vocal remote control commands are issued in continuous speech. The User Interface incorporates a continuous speech recognizer and a semantic interpretation module that translates the user sentences into the navigation or interaction primitives.

Design and realization of User Interface module is a critical and delicate matter: the Supervision System can impose a severe load on the Supervision Operator, especially in the case of multiple robots. It is therefore important that all the available man-machine interaction modalities and techniques be fully exploited. Advanced graphics and the use of voice when the operator is hands-busy or eyes-busy can help the operator in the accomplishment of complex supervision tasks. Their introduction into the system has therefore to be carefully considered and experimentally evaluated.

2.2   The Mobile Robot

From a purely functional point of view, the Mobile Robot consists of three main entities (see Fig. 3). The Robot Supervisor module receives the specification of a mission as a sequence of navigation, interaction, actuation and monitoring actions. It then checks the mission for validity and activates the proper subsystems. Whenever an activity is over, the Supervisor sends execution reports to the Supervision Station. In the present realization, functionalities of the system are still limited: the Robot Supervisor acts as an intermediate layer between the Supervision Station modules and the Mobile Robot subsystems. In the future, its functionalities (and responsibilities) will grow bigger to include the handling of certain critical situations now managed at a higher level.

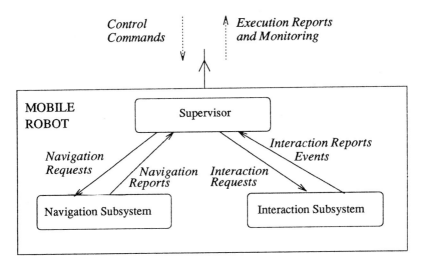

Fig. 3. The Architecture of the Mobile Robot

The Interaction Subsystem (Fig. 4) of the Mobile Robot receives high-level interaction requests, as, for example, *Verify-User-Presence* or *Load-Interaction*; it then tries to satisfy such requests by activating the underlying subcomponents, and finally returns the results to the Robot Supervisor. The Interaction Subsystem uses a continuous speech recognizer with a limited vocabulary for the interpretation of the vocal command issued by the user. Also, it communicates by voice through a speech synthesizer. The Interaction Subsystem can also detect asynchronous events of potential interest occurring in the environment as, for example, screams.

The Navigation Subsystem is organized in three main layers (see Fig. 5). At the top, the *symbolic layer*, being devoted to the processing of global information about the environment and the state of the system, deals with the large-scale aspects of navigation; the *sensorimotor layer* lies at the bottom and consists of all the modules and devices devoted to the control of actuators and the acquisition of data from the sensors; in the middle, the *behaviour level* bridges the gap between the two, linking the abstract descriptions of actions needed to achieve the navigation tasks with the modules that such actions physically implement. A detailed presentation of the Navigation Subsystem in terms of its component modules and capabilities is beyond the scope of the present paper and can be found in [6]. It may nevertheless be interesting to outline it briefly.

A navigation command consists of a *path* to be traveled that the Navigation

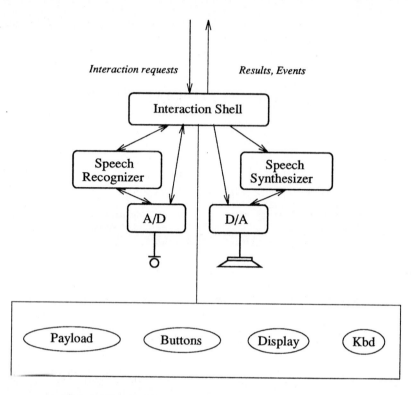

Fig. 4. The Architecture of the Interaction Subsystem

Subsystem receives through the Robot Supervisor. The path is a sequence of *virtual doors* expressing the set of geometrical constraints that the robot is expected to comply with when moving towards its destination. Using global information about the environment, the Navigation Shell translates path into a sequence of elementary actions (*basic behaviours*) whose successful execution would cause the robot to reach the goal. Basic behaviours as *"follow the corridor, until a corridor opens on the left"* are to be considered as functional modules able to autonomously deal with a given set of events or occurrences. A basic behaviour consists in a motor action typically performed in a reactive fashion whose execution is started and ended upon detection of certain events (*predicates* and *perceptions*). The agent that establishes which combination of processes implementing reflexes, predicates and perceptions is best suited to realize the basic behaviour to be performed is the Action Manager. The Action Manager also manages the activation and disactivation of such processes, and reports back to the Navigation Shell whenever the execution of a basic behaviour is completed – no matter whether successfully or not.

As far as the robot actually executes all the basic behaviours, the path is traveled to its end and the navigation command thereby accomplished.

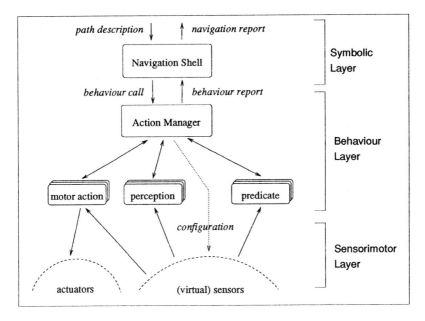

Fig. 5. The Architecture of the Navigation Subsystem

Should instead something go wrong the Navigation Shell, on the basis of the knowledge it holds on the environment, may try to apply suitable recovery strategies. When, finally, the sequence of basic behaviours implied in the path has been executed completely or it has unrecoverably failed, the Navigation Shell finally issues the navigation report.

2.3 A Highly Interacting System

The Experimental Platform we have presented in the previous sections shows features which allow for a high degree of interaction with the environment. The interaction ability arises at different levels in the architecture. At the lower level, the mobile robot is equipped for complex interaction with the environment: from the point of view of navigation, behaviour-based approach based on fusion of different sensing devices guarantees the ability to deal with complex and dynamic environments; from the point of view of communication with users, the vehicle is equipped with devices such as microphones, loudspeakers and speech modules which allow to perform complex tasks. At a higher level, the Supervision Station and its operator can interact in a fairly complex way with the external environment; the system can finally modify its behaviour according to the information acquired from the interaction.

The Mobile Robot and the Supervision Station are connected to the IRST internet Local Area Network (LAN) so that they can access all the network resources, share data and exchange information in an easy and effective way.

The architecture of the Robot itself is based on a LAN: robot hardware comprises three PC-486 boards connected through an Ethernet network. The main advantage of this choice consists in the possibility of relying on a large set of standard and reliable tools for communication and development. This architecture and the set of related tools allows us to concentrate on the conceptual and experimental issues rather than on aspects that could always be optimized should efficiency become a stringent issue.

3   MAIA '94

In this section the experimental setting is presented in which the capabilities of Experimental Platform of MAIA are tested. Before entering a more detailed description, it may be worth to preliminarily remark how our chief interest in this context will not be much that of measuring which level of performance single functionalities taken as isolated have attained, but rather that of understanding potentialities, limitations and possible flaws of the system as a whole. As we have already said, crucial for the system we have been developing is the capability of interacting with the external world in a wide variety of reasonably simple, yet realistical situations; to interact with the system should therefore be as "natural" and real-life as possible; indeed we expect that interactions with untrained personnel may yield valuable indications about the level of safety, ease of use and overall robustness of the system – beside suggesting extensions that may further challenge the flexibility of the architecture, or the working range of modules and devices.

An experimental framework (MAIA '94) has consequently been defined which can encompass most of the conditions in which the system is expected to work, while allowing large amounts of experimental data to be easily collected. In this framework, an experiment is a *mission* whose execution involves many of the skills the system is supposed to be endowed with – notably, autonomous navigation, and interaction with various kinds of users.

3.1   The Missions

Two are the kinds of missions which the MAIA '94 scenario consists of: (1) the *routine mission*, that is a repetitive duty to be performed one or more times during the day, such as the delivery of mail coming from outside the Institute; (2) the *spot mission*, that is, an unscheduled duty consisting in point-to-point delivery of documents or other office items, like faxes or internal mail. Both kind of missions are planned and executed on top of a suitable collection of modules designed to perform specific tasks, as, for example, point to point autonomous navigation or load/unload interaction with senders/receivers. Beside these, special functionalities have been implemented that are related to particular missions or situations, such as the recording of messages for the receiver or the system start-up procedure.

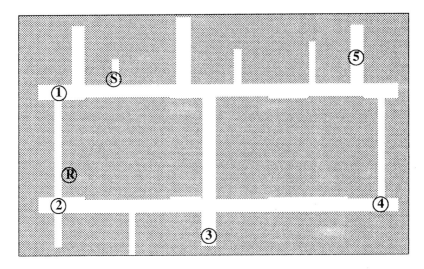

Fig. 6. The first floor of the building of our Institute. The *Delivery Stations* for the routine mission are labeled with numbers from 1 to 5, while label **S** indicates the place nearby the Supervision Station where the robot waits for commands. The "garage" of the robot is labeled with **R**.

The Routine Mission Routine missions consists in delivering mail coming from outside the Institute to the offices of the respective addressees. At present the number of delivery stations is limited to five (see Fig. 6), although more will be added as the experimentation progresses. Of course, the case may well happen that at some of the stations no mail is to be delivered. In order for the routine mission to be correctly scheduled, all the intermediate stops and the interactions that are expected to take place have therefore to be appropriately specified.

The mission starts with the robot placed nearby the supervision station, where the Supervision Operator provides to manually load the mail addressed to the various delivery stations into different, suitably labeled drawers. A *load interaction* starts then with the robot telling the operator to be ready and waiting for a "yes"/"no" answer from the operator. In case of negative answer (or no answer at all) the mission is aborted and the robot places itself in *stand-by* modality. When a positive answer is obtained, the robot waits until the operator pushes a button (a green one) placed atop the robot itself – thereby notifying that loading has been taken to completion. The robot is now ready to go; before moving, however, it asks for confirmation one last time. If a negative or no answer is issued, the robot waits again for the green button to be pushed; if no answer is received before a reasonable interval of time has elapsed (typically 10 seconds), the load interaction is discontinued, and the mission consequently aborted. When a positive answer is received, the robot starts navigating toward its first destination. The path for the routine mission is precompiled, and can be modified only when a critical situation occurs for which replanning is needed (for example, an obstruction along the route). In case of replanning, the order in which delivery stations are reached can be modified accordingly. Once a delivery station is reached, an *unload interaction* is activated.

The unload interaction goes pretty much as a load interaction does, except that in case of failure (negative or no answer) navigation towards the next destination is reestablished. Once that the last destination is reached and the mail delivered, the robot goes back to the supervision station, where it briefly reports about the mission – telling, for example, whether the mission was successful or not, or whether any object still lies undelivered in the drawers.

The Spot Mission   Unlike routine missions, spot missions cannot be scheduled in advance, as cannot operations implied in their execution be planned in detail; in this respect, spot mission can be considered as customized services that the system offers to enabled users. By means of a simple program available from the Unix shell of the network, the user contacts the Supervision Station and is requested to fill – through a suitable graphic interface – a sort of *spot mission request form*. Sender, receiver (receivers, in case of multiple destinations) and other information are therefore specified and collected. Requests of spot missions is accepted only during fairly limited intervals of time (typically, every day from 10am to 12am), and they are going to be actually executed only after the routine mission has been successfully completed.

Start-Up – The start-up procedure requires that a human operator manually activate the on-board systems and devices, and disconnect the robot from the power supply. A software primitive is available which allows to check whether the start-up procedure has been successfully completed. Once the system is ready, the robot autonomously moves to an appropriate location nearby the Supervision Station (see Fig. 6), where it waits for commands.

Batteries Recharge – The charge level of batteries is an information available to the Supervision Station, and a recharge mission can therefore be automatically scheduled whenever a need for it exists. Since the total autonomy of the vehicle is estimated in 4–6 hours, no need for recharging the batteries typically occurs during the period of time in which the robot is operative. When such period ends, a recharge mission is always executed before shutting the system down.

Mission Confirmation – This interaction is typical of any spot mission, and actually represents its first phase. Before leaving the supervision station to execute the spot mission, the robot asks the Supervision Operator whether the mission is confirmed. In case of negative answer (or no answer within a certain interval of time) the mission is aborted, and the robot stands by waiting for other spot missions.

Messages Recording – When loading the robot the user can record vocal messages that will then reported to the receiver once that destination is reached. In this way, the robot can deliver objects and messages that the sender may for any reason wish to associate with the objects themselves – for example, instructions, words of thank, and so on.

3.2   The Supervision Station

Missions are scheduled, planned, and their execution autonomously managed from the Supervision Station. Albeit many operations take place automatically, a human operator (the Supervision Operator) can intervene on the system whenever he thinks it appropriate – typically to modify the way operations are being executed, to remotely control the robot, or to monitor the environment by means of the robot's sensors. In particular, it is possible for the Supervision Operator to select between two different interactions modalities:

– *Automatic Handling of Critical Situations*: the system tries to cope with the critical situations notifying the Supervision Operator of any action consequently undertaken; the Supervision Operator may take over the system, should he judges that his intervention is necessary;
– *Manual Handling of Critical Situations*: whenever a critical situation occurs, the system switches automatically to *telecontrol* modality.

System-Operator Interface   The Supervision Operator interacts with the system through a graphic interface featuring a map of the environment, and reporting certain information pertaining missions. In particular it is possible:

– to visualize the path synthesized by the Planner and currently being traveled;
– to visualize the position and orientation of the robot;
– to visualize images taken from the cameras set aboard the robot;
– to listen to the acoustic input of the microphones set aboard the robot;

From the Supervision Station, pending missions can also be rescheduled, postponed or altogether cancelled.

Mission Management   As soon as the robot becomes available, the first pending mission is executed. During the execution, the Supervision Operator can perform a small set of operations impacting on the mission itself. They are:

– *Stop* – The mission is asynchronously interrupted and the status of the robot becomes *stopped*. This is the only command that is possible to issue during the execution of the mission; all the other commands can be issued only when the robot's status is stopped;

- *Resume* – The status of the Robot is resumed from stopped; the mission is reestablished maintaining the original path.
- *Abort* – The current mission is aborted, and the robot goes back to the supervision station;
- *Remote Control* – The status of the Robot switches to *telecontrolled*, and navigation commands as complex as, for example, *"enter the next corridor on the right and keep on going for ten meters"* become available to the Supervision Operator. Commands are issued via voice [7] and their execution can be monitored through the graphic interface. Remote control modality will be more exhaustively described in the next section;
- *Change Path* – The current path is superseded and a new one is defined which may hopefully better account for obstructions along the route, or other reported problems.

Telecontrol Possibility of remotely control operations performed by robots or other automated devices may have a crucial impact on the functioning of many systems designed to interact with the external world. Typically, remote control is based on a sensor/effector chain able to dispatch to the remote device the commands that the operator wants to impart, and to report back to the operator about the effects that the commands themselves had. Albeit traditionally successful, very "direct" ways of imparting commands, as those based on joystick-like devices and monitors, do not always prove to be the most appropriate ones. Contol of a planetary probe from earth, for example, would be virtually impossible in this way – because of the time that radio signals would take to travel the distance. In other cases, information the robot sends back to the operator is incomplete or insufficient, typically due to some limits or constraints on the sensorial apparatus of the robot; sometimes, finally, one may not simply want to spend much time and attention in manually performing fine control operations unless it is absolutely necessary. The only way out one is left in all these cases is to endow the robot with levels of autonomy that may allow the operator to interact in more "synthetic" fashion.

Vocal commands as those reported in the previous section are good examples of what we mean with synthetic interaction: all the details inherent the execution are left to the actual ability of the robot to autonomously face the events that may occur. Likewise, information the robot send back to the operator is of synthetic nature: whether it was able to successfully complete the command or not, and, in case of failure, a brief report on what has presumably gone wrong. The telecontrol system that has been implemented on our system allows both direct and synthetic interactions.

The customary way to access telecontrol during navigation consists in first stopping the system by means of the *Stop* command, and in selecting then the *Remote Control* option. Neither hitting robot's frontal bumper, nor pressing the emergency (red) button forces the system to switch to telecontrol modality; this is automatically set when one of the two following errors occur:

- *inconsistent position* – the Navigation System has failed to maintain the position within acceptable limits of accuracy; the robot is "lost", and the navigation has consequently to be discontinued.
- *impossibility to replan*: this situation occurs when all possible paths to reach the destination appear to be occluded.

Back to Autonomous Navigation – When operated in telecontrol modality, the vehicle is disconnected from the modules devoted to maintain information of global nature about the navigation. In order for autonomous navigation to be reestablished at the end of a telecontrol session, it is therefore necessary that position and orientation of the robot be estimated. In order to do this and to propagate the information throughout the system, a suitable set of primitives is available to the Supervision Operator which can take advantage of the sensors the robot is equipped with.

3.3 The Robot

The mobile robot has been completely designed and assembled at IRST. It spans 50 centimeters in both length and width, and 90 centimeters in height. Kinematically, the vehicle is based on two motorized front wheels and a non motorized, pivoting rear wheel. Steering is thus obtained by appropriately setting the speeds of the two front wheels. The lowest section of the vehicle hosts all the electro-mechanical devices (motors, batteries, converters), while control and computing hardware finds place in the central section; sensors, devices for user interface and the drawers for the objects to be transported are lodged at the top.

Computing power is guaranteed by three on board Intel 486 microprocessors, connected to one another through an Ethernet network. The software environment is Unix-like: Linux operating system (standard and public domain) on two boards and LynxOS operating system (real-time and proprietary) on the third. The choice of a real-time operating system as the software environment for the sensory-motor layer stems from the consideration that low-level processing may require a time-sharing management more predictable than that typically featured by standard operating systems. Sixteen sonars able to detect reflecting surfaces in the range [0.2 m, 4.0 m] and a motorized b/w camera form the sensorial apparatus of the robot. An odometer has also been developed which provides estimates of robot's position and orientation.

4   Final Considerations

In this paper we have presented the Experimental Platform of MAIA, a system of robot navigation and delivery tasks. The system can be used as a testbed for experimental models and to perform delivery missions. We have first described the system architecture and some of the design principles and features, and presented an example of application where the system can be effectively used and its behaviours tested. In the following we discuss some open problems and limitations, and present some extensions to the system and further applications which we would like to investigate.

Several are the open problems related to autonomous navigation. At present the most stringent one concerns the possibility of making the robot navigate inside unstructured environments. This would require a much more accurate processing of sensorial information – especially of visual nature. However, some ideas are currently under development, and we hope to introduce some of the most promising results as soon as they attain a sufficient degree of reliability. Indeed, visual skills are related not only to the ability of moving, but also to the capability of interacting with the surrounding environment.

Although very natural, using speech as an input channel raises a whole new category of problems that are not encountered when using keyboard or mouse devices. Automatic speech understanding systems for robotic applications should be speaker-independent, cover spontaneous speech phenomena, and provide fault-tolerant semantic interpretation. These issues can be partially fulfilled by using robust approach to semantic interpretation, and a rejection criterion for unreliable recognized sentences [8]. Indeed, two other important issues should be thoroughly studied: all existent speech recognition/understanding systems give good recognition rate when the signal they are presented with is reasonably "clean"; mumbles, pauses, hesitations, so frequent in spontaneous speech, decrease system performance. Moreover, talking about speech recognition we usually refer to ideal situations in which the user is required to direct utterances into the microphone. A speech recognition system able to "track" the voice of the user under less constrained conditions would be of great avail to make interaction more user-friendly. An approach to this problem based on a microphone array is presently investigated at our Institute [9].

As for possible extensions in the functionalities of the system, we are strongly interested in the use of the mobile robot for data acquisition and system configuration. When the system has to be installed in a new environment, soon the problem arises of formalizing and providing the system with a great deal of information. For the system to work, it needs both the geometric and symbolic information about the environment in which it is going to operate. This implies that several levels of knowledge be integrated, for example by mapping the metric information concerning the environment in the data structures used by the planning modules. At present, all this information is inserted manually by specialized operators. This requires time and the costs are not negligible when compared with the costs of the whole transport system. It would be possible to design semi-automatic support tools that allow trained people to start from an electronic description of the environment (obtained from engineering design tools such as Autocad) to derive a suitable description in the format used by the navigation an planning subsystems. A more interesting approach has to do with the automatic learning of maps, through a systematic exploration of the environment by the mobile robots.

Elements of learning from past experience could also be introduced in order to extend the overall autonomy of the system. Suppose, for instance, that a mobile robot has to deliver some objects starting from a certain place and has to reach another place for delivery. Had the system only a limited knowledge of the environment, a classical planning approach would be bound to repeated failures. However, if the robot were able to acquire more information from the external environment (for example, reading the written text signs typically posted on doors), it could communicate it to the higher levels, in order to fill the possible gaps in the map and therefore in the path.

Acknowledgements – Many people have contributed to the work described in this paper. Let our grateful thanks go to the members of the Mechanized Reasoning, Speech Recognition, System Integration and Vision groups at IRST for several helpful discussions and suggestions. Authors feel also indebted to Gianni Lazzari, Tomaso Poggio and Luigi Stringa for the constant support provided.

References

1. Evans, J., Krishnamurthy, B., Barrows, B., Skewis, T., Lumelsky, V.: Handling real-world motion planning: a hospital transport robot. Proc. of IEEE Cont. Sys. Feb. (1992) 15–20
2. Lob, W.S.: Robotic transportation. Clin. Chem. 36(9) (1990) 1544–1550
3. Robertson, G.I.: HelpMate delivery robot operates safely amongst the general public. Proc. of ISIR '91, Detroit, 24 (1991) 1–12
4. Poggio, T., Stringa, L.: A project for an intelligent system: vision and learning. Int. J. Quant. Chem. 42 (1992) 727–739
5. Caprile, B., Lazzari, G., Stringa, L.: Autonomous navigation and speech in the mobile robot of MAIA. Proc. of SPIE, Boston (USA), 1831 (1992) 276–284
6. Cattoni, R., Di Caro, G., Aste, M., Caprile, B.: Bridging the gap between planning and reactivity: a layered architecture for autonomous indoor navigation. Proc. of IROS '94, Munich, (1994) 878–885
7. Antoniol, G., Cattoni, R., Cettolo, M., Federico, M.: Robust speech understanding for robot telecontrol. Proc. of ICAR '93, Tokyo, (1993) 205–209
8. Antoniol, G., Cettolo, M., Federico, M.: Techniques for robust recognition in restricted domains. Proc. of 3rd ECSCT, Berlin, 3 (1991) 2219–2221
9. Omologo, M., Svaizer, P.: Acoustic event localization using a crosspower-spectrum phase based technique. Proc. of ICASSP, Adelaide, 2 (1994) 273–276.

Of Elephants and Men[*]

Johan M. Lammens[1], Henry H. Hexmoor[2], and Stuart C. Shapiro[2]

[1] Advanced Robotics Technology and Systems Laboratory, Scuola Superiore S. Anna, 56127 Pisa, Italy; e-mail lammens@arts.sssup.it
[2] Computer Science Department, State University of New York at Buffalo, NY 14260, USA; e-mail {hexmoor|shapiro}@cs.buffalo.edu

Abstract. In the elephant paper, Brooks criticized the ungroundedness of traditional symbol systems and proposed physically grounded systems as an alternative. We want to make a contribution towards integrating the old with the new. We describe the GLAIR agent architecture that specifies an integration of explicit representation and reasoning mechanisms, embodied semantics through grounding symbols in perception and action, and implicit representations of special-purpose mechanisms of sensory processing, perception, and motor control. We present some agent components that we place in our architecture to build agents that exhibit situated activity and learning, and some applications. We believe that the Brooksian behavior generation approach goes a long way towards modeling elephant behavior, which we find most interesting, but that in order to generate more deliberative behavior we need something more.

1 Introduction and Overview

In the elephant paper [11] appearing in the proceedings of the predecessor of the current workshop, Brooks criticizes the ungroundedness of traditional symbolic AI systems, and proposes physically grounded systems as an alternative, particularly the subsumption architecture. Subsumption has been highly successful in generating a variety of interesting and seemingly intelligent behaviors in a variety of mobile robots. As such it has established itself as an influential approach to generating complex physical behavior in autonomous agents. In the current paper we explore the possibilities for integrating the old with the new, in an autonomous agent architecture that ranges from physical behavior generation inspired by subsumption to classical knowledge representation and reasoning, and a new proposed level in between the two. Although we are still struggling with many of the issues

---

[*] The research reported in this paper was carried out while the first author was a member of the SNePS Research Group at the department of Computer Science, SUNY at Buffalo, and was supported in part by Equipment Grant No. EDUD-US-932022 from SUN Microsystems Computer Corporation, and in part by NASA under contract NAS 9-19004. First author's current address: Hewlett-Packard Española S.A., R&D Department, Barcelona, Spain.

involved, we believe we can contribute to a solution for some of the problems for both classical systems and physically grounded systems mentioned in [11], in particular:

- The ungroundedness of symbolic systems (referred to as "the symbol grounding problem" by [19]): our architecture attempts to ground high level symbols in perception and action, through a process of embodiment.
- The potential mismatch between symbolic representations and the agent's sensors and actuators: the embodied semantics of our symbols makes sure that this match exists.
- Our symbolic representations do not have to be named entities. The knowledge representation and reasoning (KRR) system we use in our implementations allows the use of unnamed intensional concepts.
- We have some ideas about how to automate the construction of behavior generating modules through learning, but much remains to be done.

We agree with the requirement of physically implemented (as opposed to simulated) systems as the true test for any autonomous agent architecture, and to this end we are working on several different implementations. We will present both our general multi-level architecture for intelligent autonomous agents with integrated sensory and motor capabilities, GLAIR[3], and a physical implementation and two simulation studies of GLAIR-agents.

By an *architecture* we mean an organization of components of a system, what is integral to the system, and how the various components interact.[4] Which components go into an architecture for an autonomous agent has traditionally depended to a large extent on whether we are *building a physical system, understanding/modeling behaviors of an anthropomorphic agent*, or *integrating a select number of behaviors*. The organization of an architecture may also be influenced by whether or not one adopts the *modularity* assumption of Fodor [17], or a *connectionist* point of view, e.g. [46], or an *anti-modularity* assumption as in Brooks's subsumption architecture [9]. The *modularity* assumption supports (among other things) a division of the mind into a *central system*, i.e., cognitive processes such as learning, planning, and reasoning, and a *peripheral system*, i.e., sensory and motor processing [13]. Our architecture is characterized by a three-level organization into a Knowledge Level (KL), a Perceptuo-Motor Level (PML), and a Sensory-Actuator Level (SAL). This organization is neither modular, anti-modular, hierarchical, anti-hierarchical, nor connectionist in the conventional sense. It integrates a traditional symbol system with a physically grounded system, i.e., a *behavior-based* architecture. The most important difference with a behavior-based architecture like Brooks's subsumption is the presence of three distinct levels with different representations and implementation mechanisms

---

[3] Grounded Layered Architecture with Integrated Reasoning
[4] Our discussion of architecture in this paper extends beyond any particular physical or software implementation.

for each, particularly the presence of an explicit knowledge level. Representation, reasoning (including planning), perception, and generation of behavior are distributed through all three levels. Our architecture is best described using a resolution pyramid metaphor as used in computer vision work [6], rather than a central vs. peripheral metaphor.

Architectures for building physical systems, e.g., robotic architectures [3], tend to address the relationship between a physical entity, (e.g., a robot), sensors, effectors, and tasks to be accomplished. Since these physical systems are performance centered, they often lack general knowledge representation and reasoning mechanisms. These architectures tend to be primarily concerned with the *body*, that is, how to get the physical system to exhibit intelligent behavior through its physical activity. One might say these systems are not concerned with *consciousness*. These architectures address what John Pollock calls *Quick and Inflexible* (Q&I) processes [49]. We define consciousness for a robotic agent operationally as being aware of one's environment, as evidenced by (1) having some internal states or representations that are causally connected to the environment through perception, (2) being able to reason explicitly about the environment, and (3) being able to communicate with an external agent about the environment.[5]

Architectures for understanding/modeling behaviors of an anthropomorphic agent, e.g., cognitive architectures [4, 49, 42], tend to address the relationships that exist among the structure of memory, reasoning abilities, intelligent behavior, and mental states and experiences. These architectures often do not take the *body* into account. Instead they primarily focus on the *mind* and *consciousness*. Our architecture ranges from general knowledge representation and reasoning to body-dependent physical behavior, and the other way around.

We are interested in autonomous agents that are embedded[6] in a dynamic environment. Such an agent needs to continually interact with and react to its environment and exhibit intelligent behavior through its physical activity. To be successful, the agent needs to reason about events and actions in the abstract as well as in concrete terms. This means combining situated activity with acts based on reasoning about goal-accomplishment, i.e., deliberative acting or planning. In the latter part of this paper, we will present a family of agents based on our architecture. These agents are designed with a robot in mind, but their structure is also akin to anthropomorphic agents. Figure 1 schematically presents our architecture.

---

[5] A machine like a vending machine or an industrial robot has responses, but it is *unconscious*. See [15] for a discussion of independence of consciousness from having a response. Also, intelligent behavior is independent of consciousness in our opinion.

[6] "Embedded agents are computer systems that sense and act on their environment, monitoring complex dynamic conditions and affecting the environment in goal-oriented ways." ([31] page 1).

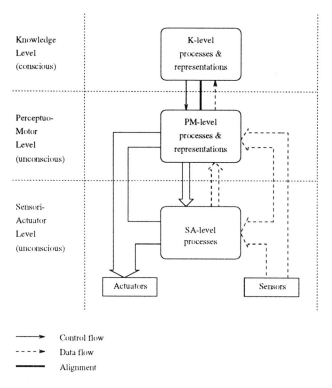

Fig. 1. Schematic representation of the agent architecture. Width of control and data paths suggests the amount of data passing through (bandwidth). Sensors include both world-sensors and proprio-sensors.

There are several features that we hope contribute to the robustness of our architecture. We highlight them below (an in-depth discussion follows later):

- We differentiate conscious reasoning from unconscious Perceptuo-Motor and Sensori-Actuator processing.[7]
- The levels of our architecture are semi-autonomous and processed in parallel.[8]

---

[7] We consider body-related processes to be *unconscious*, but that is not meant to imply anything about their complexity or importance to the architecture as a whole. Indeed, we believe that the unconscious levels of our architecture (the Perceptuo-Motor level and the Sensori-Actuator level) are at least as important to the architecture as the conscious one (the Knowledge level). We reserve the term *sub-conscious* for implicit cognitive processes such as category subsumption in KRR systems. See [56] for a discussion of sub-conscious reasoning.

[8] This autonomy is similar to Brooks's subsumption architecture [9], but at a more macroscopic level. Brooks does not distinguish between the three levels

- Conscious reasoning takes place through explicit knowledge representation and reasoning. Unconscious behavior makes use of several different mechanisms.
- Conscious reasoning guides the unconscious behavior, and the unconscious levels, which are constantly engaged in perceptual and motor processing, can *alarm* the conscious level of important events, taking control if necessary. Control and generation of behavior are layered and not exclusively top-down.
- Lower level mechanisms can pre-empt higher level ones. This is kind of subsumption on its head, but everything depends on the placement of behaviors in the hierarchy of course. We haven't quite decided yet whether inhibition should work the other way around as well.
- There is a correspondence between terms in the Knowledge Representation and Reasoning system on one hand, and sensory perceived objects, properties, events, and states of affairs in the world and motor capabilities on the other hand. We call this correspondence *alignment*.
- Our architecture should be appropriate both for modeling elephants and for modeling chess-playing agents.[9]

2   The GLAIR Architecture

In this section we discuss in detail our autonomous agent architecture for integrating perception and acting with grounded, embodied, symbolic reasoning.

2.1   Related Work

Architectures proposed in the literature do not fall into neatly separable classes, mainly because the scope of the models and the motivations vary widely. However, we can divide a review of related work into theoretical issues of agent architectures, on the one hand, and implemented architectures, on the other.

Theoretical Issues. We believe that behavior-based AI has adopted the right treatment of every day behavior for agents that function in the world. However, this has been done at the expense of ignoring cognitive processing such as planning and reasoning. Clearly, what is needed is an approach that allows for both. We believe that our architecture meets this need. As in behavior-based AI, GLAIR gains validity from its being grounded in its interaction with the environment, while it benefits from a knowledge level

---

we describe, as his work is solely concerned with behaviors whose controlling mechanism we would situate at the Perceptuo-Motor level.

[9] Though not necessarily for chess-playing elephants.

that, independent of reacting to a changing environment, performs reasoning and planning.

The Model Human Processor (MHP) is a cognitive model [12] that suggests the three components of *perception, cognition,* and *motor.* Cognition consists of working memory, long-term memory, and the cognitive processor. Perception is a hierarchy of sensory processing. Motor executes the actions in the working memory. This is a traditional symbol-system decomposition of human information processing. This type of decomposition has shown only limited success in building physical systems. Despite this, systems like SOAR adhere to this model. In our architecture, we deliberately avoid this kind of top-down problem decomposition by allowing independent control mechanisms at different levels to take control of the agent's behavior, and pre-empt higher level control while doing so. It may be necessary to allow higher level mechanisms to selectively inhibit lower-level ones as well, but we have found no good reason to do so yet.

A situated agent, at any moment, attends to only a handful of entities and relationships in its immediate surroundings. In this type of setting, the agent often does not care to uniquely identify objects. It is sufficient to know the current relationship of the relevant objects to the agent, and what roles the objects play in the agent's activities. Agre and Chapman in [2] proposed indexical-functional representations (which [1] refers to as deictic representations) to be the more natural way agents refer to objects in common everyday environments. They called entities and relationships of interest *entities* and *aspects,* respectively. With respect to its current activities, the agent needs only to focus on representing those entities and relationships. Although the objects in the environment come and go, the representations of entities and relationships remains the same. For example, the-cup-that-I-am-holding[10] is an indexical-functional notation that abstracts the essentials of what the agent needs to know in its interaction. These representations serve to limit the scope of focus on entities. For example, if the agent wants to pick up a cup, it does not need to know *who owns the cup* or *how much coffee the cup can hold;* only the relevant attributes of the cup apply. We believe that systems endowed with general KRR abilities can and should generate deictic representations to create and maintain a focus on entities in the world, but we have not yet designed an implementation strategy.

Implemented Architectures. Brooks's *subsumption* architecture, [9, 10, 11], clusters behaviors into layers. Low-level behaviors, like deciding the direction of motion and speed, can be inhibited (subsumed) by behaviors determined at higher levels, such as avoiding obstacles. Subsumption behaviors are written as finite state machines augmented with timing elements. A compiler is used to simulate the operation of finite state machines and parallelism.

---

[10] This kind of designation is merely a mnemonic representation intended to suggest the entity and aspect under consideration, for the purpose of our exposition. It is not the actual representation that would be used by an agent.

This architecture is implemented on a variety of mobile robots. Frequently used behaviors are placed at a lower level than less frequently-used behaviors. This organization of behaviors gives the system fast response time and high reactivity. Our architecture is similar to Brooks's in our intra-level implementations of behaviors. However, the subsumption architecture lacks the separation we have made into conscious and non-conscious spheres. In anthropomorphic terms, Brooks's agents are all non-conscious. We believe that the off-line specification and compilation of behavior modules is too inflexible for autonomous agents that can adapt to a wide range of circumstances, especially if they have to learn from their interactions with the environment. Pattie Maes has experimented with a version of a behavior-based architecture, ANA [44], which consists of competence modules for action and a belief set in a network relating modules through links denoting successors, predecessors, and conflicts. Competence modules have activation levels. Activations are propagated and the competence module with the highest activation level is given control. Maes has explored learning and has applied her architecture to robotic systems.

In the subsumption architecture, sensations and actions are abstracted by giving them names like "straightening behavior" in order to make things easier to understand for human observers. Much in the spirit of [1], we believe that behavior modules should more naturally emerge from the interaction of the agent with its environment. In contrast to hand coding behaviors and in order to facilitate embodiment, in GLAIR we are experimenting with (unnamed) emergent behavior modules that are learned by a robot from scratch. An (unnamed) behavior module can be thought of as a set of tuples (P,A) where P is a set of grounded sensations and A is an instance of an act. For instance, reaching for an object might be a set of tuples (vision/sonar data, wheel motor actuation). After learning, this new behavior module will become active only if the grounded sensations match any of the grounded sensations experienced before. As a measure of abstraction and generalization, we may allow near matches for sensations. To bootstrap the learning process, we need a set of primary or first-order ("innate", for the philosophically inclined) sensations and actions. We will return to this point briefly in Sect. 3.3.

The Servo, Subsumption, Symbolic (SSS) architecture [14] is a hybrid architecture for mobile robots that integrates the three independent layers of servo control, Brooksian behavior based modules, and a symbolic layer. Our architecture is similar to this in its general spirit of identification and integration of three distinct levels corresponding to levels of affinity-of-interaction (i.e., the rate at which it is in real-time contact with the world) with the outside world. This similarity also constitutes a point of departure, however, in that SSS is defined with respect to specific (and different) implementation techniques. For example, the symbolic layer in SSS seems to be a decision table versus a general KRR system as intended in GLAIR. Unlike GLAIR, SSS assigns particular tasks for each layer and uses a hard-wired interconnection channel among layers.

Albus et al's hierarchical control architecture [3] is an example of a robotic architecture; we would say it is *body centered*. This architecture proposes abstraction levels for behavior generation, sensory processing, and world modeling. By descending down the hierarchy, tasks are decomposed into robot-motion primitives. This differs from our architecture, which is not strictly top-down controlled. Concurrently, at each level of Albus's hierarchy, feedback processing modules extract the information needed for control decisions at that level from the sensory data stream and from the lower level control modules. Extracted environmental information is compared with the expected internal states to find differences. The differences are used for planning at higher levels.

Payton in [48] introduced an architecture for controlling an autonomous land vehicle. This architecture has four levels: mission planning, map-based planning, local planning, and reflexive planning. All levels operate in parallel. Higher levels are charged with tasks requiring high assimilation and low immediacy. The lower levels operate on tasks requiring high immediacy and low assimilation. Our architecture is similar in this respect. The reflexive planning is designed to consist of pairs of the form $\langle virtual sensor, reflexive behavior \rangle$. Each reflexive behavior has an associated priority, and a central blackboard style manager arbitrates among the reflex behaviors. Some of the problems with the earlier implementation due to using the blackboard model were solved in [53].

Rosenschein and Kaelbling's work [30, 31] describes tools (REX, GAPP, RULER) that, given task descriptions of the world, construct reactive control mechanisms termed *situated automata*. Their architecture consists of perception and action components. The robot's sensory input and its feedback are inputs to the perception component. The action component computes actions that suit the perceptual situation. We should note that unlike Brooks's behavior modules, situated automata use internal states, so their decisions are not Markovian (i.e., they are not ahistoric). They are mainly intended to produce circuits that operate in real time, and some properties of their operation are provable. The mechanism for generating situated automata, although impressive, seems too inflexible for autonomous agents that have to operate in a wide variety of (possibly unknown) circumstances. Perhaps the operation of our Perceptuo-Motor level could be modeled by a situated automaton, but we are not convinced that this is the right formalism to use, due to its inflexibility.

Gat in [18] describes ATLANTIS, an architecture for the control of mobile robots. This architecture has three components: control, sequencing, and deliberation. The control layer is designed as a set of circuit-like functions using Gat's language for circuits, ALPHA. The sequencing is a variation of Jim Firby's RAP system [16]. The deliberation layer is the least described layer. As with situated automata, we are not convinced that this is the right kind of formalism to use, for the same reasons.

An architecture for low-level and high-level reactivity is suggested in [22]. High-level reactivity is reactivity at the conceptual level. This architecture suggests that an autonomous agent maintains several different types of goals. High-level reactivity is charged with noticing impacts of events and actions in the environment on the agent's goals. Subsequently, high-level reactivity needs to guide the agent's low-level reactivity. Low-level reactivity is at the sensory, perceptual, and motor level. The mechanism for low-level reactivity is similar to other reactive systems that have components for perception and action arbitration. The novelty of this architecture is the incorporation of high-level reactivity and a supervisory level of planning and reasoning, which guides the choice of low-level reactive behaviors. In our present conception of agent architecture, we avoid a sharp separation between the two types of reactivity. We also relax the top-down nature of interaction between levels. Reactivity may be initiated at any level of our architecture either due to interaction with other levels or in direct response to external stimuli.

SOAR [38] was designed to be a general problem solving architecture. SOAR integrates a type of learning known as *chunking* in its production system. Recently, SOAR has been applied to robotic tasks [37]. In this framework, planning and acting is uniformly represented and controlled in SOAR. This approach lacks the ability of our architecture for generating behavior at non-conscious levels as well as the conscious level (or at different levels in general), and for having different-level behaviors interact in an asynchronous fashion. It also lacks our multi-level representations.

Simmons's Task Control Architecture (TCA) [61] interleaves planning and acting by adding *delay-planning constraints* to postpone refinement of planning until execution. For example, a plan for a robot to collect used cups for trash is decomposed into: navigate to the cup; pick it up; navigate to trash bin; deposit the cup. Since the robot does not have sensory information about the cup yet, the plan to pick it up is delayed until the robot gets close enough. Selectively delaying refinement of plans allows for reactivity. This type of "stepwise refinement" follows effortlessly from our architecture, without the need to explicitly implement it. Since conscious planning which goes on at the Knowledge level uses a more coarse-grained world model, there is simply no possibility to express fine details of planning and execution. These can only be represented and/or computed at the lower Perceptuo-Motor level and Sensori-Actuator level. Planning and execution in our architecture may proceed in a lock-step fashion, but they need not be. (see the discussion of engaged vs. disengaged reasoning in Sect. 2.6). TCA uses a message-passing scheme among modules that allows concurrent execution of tasks. It has been used to control the six-legged walking robot Ambler and a cup-collecting robot.

2.2 Architecture Levels

We now proceed to discuss one of the distinguishing characteristics of GLAIR: its three levels.

Motivation. The three levels of our architecture are of organizational as well as theoretical importance. Organizationally, the layered architecture allows us to work on individual levels in a relatively independent manner, although all levels are constrained by the nature of their interactions with the adjoining level(s). The architecture is hierarchical, in that level i can only communicate with levels $i-1$ and $i+1$, if any.

The levels of our architecture are semi-independent. While control flows mainly top-down and data mainly bottom-up, local control mechanisms t any level can preempt higher-level control, and these local mechanisms ter the data stream for their own purpose, in parallel with higher-level one Representations become coarser-grained from bottom to top, while con' data becomes more fine-grained from top to bottom. The terms in the Know¹ edge Level's KRR system model conscious awareness of the world (and tl body), and the perception and motor capabilities in the other levels provide the grounding for an *embodied semantics* of the former. Routine, reflex-like activities are controlled by close coupling of perception with motor actions at the (unconscious) Perceptuo-Motor and Sensori-Actuator levels. This close coupling avoids having to exert control over these activities from the conscious level, as in purely top-down structured architectures with a symbol level at the top of the hierarchy. In the latter kind of system, signals must first be transformed to symbols and vice versa. The low-level coupling provides for better real-time performance capabilities, and relieves the Knowledge level of unnecessary work.

In general, we have multi-level layered representations of objects, properties, events, states of affairs, and motor capabilities, and the various levels are *aligned*. By alignment we mean a correspondence between representations of an entity at different levels. This organization contributes to the robustness and computational efficiency of implementations. The semi-autonomous nature of the levels allows for graceful degradation of system performance in case of component failure or situation-dependent incapacitatedness. Lower levels can function to some extent without higher-level control, and higher levels can function to some extent without lower-level input.[11]

Our architecture allows us to elegantly model a wide range of behaviors: from mindless, spontaneous, reflex-like, and automatic behavior, e.g., "stop if you hit an obstacle," to plan-following, rational, incremental, and monitored behavior, e.g., "Get in the car now, if you want to go to LA on Friday."[12]

---

[11] For instance, in the context of autonomous vehicles, if obstacle avoidance or returning to the base is a lower-level behavior than planning exploration strategies, then a failure of the hardware implementing the latter does not necessarily prevent the former.

[12] The plan is to get in the car to go to the travel agency to get a ticket to fly to LA

In anthropomorphic terms, we identify the Knowledge level with consciously accessible data and processing; the Perceptuo-Motor level with "hardwired," not consciously accessible processing and data involved with motor control and perceptual processing; and the Sensori-Actuator level with the lowest-level muscular and sensor control, also not consciously accessible. The substrate of grounding and embodiment [19, 39, 62] of actions, concepts, and reasoning is mainly the Perceptuo-Motor level and to some extent the Sensori-Actuator level.

We will now explore representation and computation at the individual levels in more detail.

The Knowledge Level. The *Knowledge level* contains a traditional KRR and/or planning system like SNePS [58, 59], using a relatively course-grained representation of objects, events (including actions), and states of affairs. For instance, objects are represented at this level as unique atomic identifiers, typically without further detail about their physical characteristics or precise locations. It is possible to represent such detail explicitly at this level, but not required. Only if the detail becomes important to the agent's explicit reasoning will it be represented, though not necessarily in the same way as at a lower level. For example, knowledge about the physical size and weight of an object might become available at the Knowledge Level through the agent's actively using measuring devices like a ruler or a scale, but this knowledge is not the same as the embodied knowledge about dimensions and weight represented at the Perceptuo-Motor level for the particular object or its object class. As a rule of thumb, representations at this level are limited to objects, events, and states of affairs that the agent needs to be consciously aware of in order to reason and plan, and in order to communicate with other agents at the grain size of natural language. The Knowledge level can be implemented using different KRR and/or planning systems.

Traditional use of the concept of world modeling refers to building models of interactions between the agent and its environment at the conscious level. These models maintain internal states for the agent. The difference in our use of the term "world model" is that we do not intend to have a precise model of all objects in the environment. Instead, we want to model only the entities relevant to the agent's interaction with its world. This requires filtering out some details accessible at the Perceptuo-motor level as the entities are aligned with their counterparts on the Knowledge level. This is known as "perceptual reduction". Physical details of interaction with entities are handled at the Perceptuo-motor level. Representations at the Knowledge level are needed only for explicit reasoning about entities, and contain only the information necessary for doing so. That might include details about physical characteristics in some cases, but it need not. In other cases, it may be limited to a nondescript intensional representation [57] of an object. Conversely, some entities

---

on Friday. Today is Thursday and it is near the end of the business day. Also, the agency won't accept telephone reservations. This example is suggested in [50].

may be represented at the Knowledge level but not at the Perceptuo-Motor level (abstract concepts, for instance). Knowledge level representations are needed for reasoning about entities; Perceptuo-Motor level representations are needed for physically interacting with entities.

The Perceptuo-Motor Level. The *Perceptuo-Motor level* uses a more fine-grained representation of events, objects, and states of affairs. For instance, they specify such things as size, weight, and location of objects on the kinematic side, and shape, texture, color, distance, pitch, loudness, smell, taste, weight, and tactile features on the perceptual side. At this level, enough detail must be provided to enable the precise control of actuators, and sensors or motor memory must be able to provide some or all of this detail for particular objects and situations. The Perceptuo-Motor level is partly *aligned* with the Knowledge level, in that there is a correspondence between some object identifiers at the Knowledge level and some objects at the Perceptuo-motor level.

Kinematic and perceptual representations of particular objects or typical object class instances may be unified or separate, and both kinds of representations may be incomplete. Also at this level are elementary categorial representations; the kinds of representations that function as the grounding for elementary symbols at the Knowledge level, i.e., sensory-invariant representations constructed from sensory data by the perceptual processor [19].

The representations at this level are *embodied* (cf. [39]), meaning that they depend on the body of the agent, its particular dimensions and characteristics. Robots will therefore have different representations at this level than people would, and different robots will have different representations as well. These representations are agent-centered and agent-specific. For instance, they would not be in terms of grams and meters, but in terms of how much torque to apply to an object to lift it,[13] or what percentage of the maximum to open the hand to grasp an object. Weights of things in this kind of representation are relative to the agent's lifting capacity, which is effectively the maximum weight representable. An agent may have a conscious (Knowledge level) understanding and representation of weights far exceeding its own lifting capacity, but that is irrelevant to the Perceptuo-Motor level. When it comes to lifting it, a thousand-pound object is as heavy as a ten-

---

[13] Of course this also depends on how far the object is removed from the body, or how far the arm is stretched out, but that can be taken into account (also in body-specific terms). People's Perceptuo-Motor level idea of how heavy something is is most likely not in terms of grams, either (in fact, a conscious estimate in grams can be far off), but in terms of how much effort to apply to something to lift it. That estimate can be off, too, which results in either throwing the object up in the air or not being able to lift it at the first attempt, something we have all experienced. On the other hand, having a wrong conscious estimate of the weight of an object in grams does not necessarily influence one's manipulation of the object.

thousand-pound one, if the capacity is only a hundred. Similarly, sizes are relative to the agent's own size. Manipulating small things is not the same as manipulating large things, even if they are just scaled versions of each other. A consequence of using embodied representations is that using different "body parts" (actuators or sensors) requires different representations to be programmed or (preferably) learned. While that may be a drawback at first, once the representations are learned they make for faster processing and reactive potential. Representations are direct; there is no need to convert from an object-centered model to agent-centered specifications. This makes the computations at this level more like table lookup than like traditional kinematics computations, which can be quite involved. Learning new representations for new objects is also much simpler; it is almost as easy as trying to grasp or manipulate an object, and merely recording one's efforts in one's own terms. The same holds, mutatis mutandis, for perceptual representations.

There are a number of behaviors that originate at this level: some are performed in service of other levels (particularly deliberative behaviors), some are performed in service of other behaviors at this level, a few are ongoing, and some others yet are in direct response to external stimuli. An agent may consciously decide to perform perceptuo-motor actions such as looking, as in *look for all red objects*, or to perform a motor action, such as *grasp a cup*. These actions originate at the Knowledge level and are propagated to this level for realization [33, 34, 36]. An agent has to perform special perceptual tasks to serve other behaviors, such as to *find the grasp point of a cup* in order to *grasp a cup*. These perceptual tasks may originate at this or another level.

At the Perceptuo-Motor level, an agent has a close coupling between its behaviors, i.e., responses, and stimuli, i.e., significant world states. We observe that, for a typical agent, there are a finite (manageably small) number of primitive ("innate") behaviors available. As the agent interacts with its environment, it may learn sophisticated ways of combining its behaviors and add these to its repertoire of primitive behaviors. We will consider only an agent's primitive abilities for now. We further assume that the agent starts out with a finite number of ways of connecting world states to behaviors, i.e., reflex/reactive rules. Following these observations, we suggest that at this level, the agent's behavior-generating mechanism is much like a finite state automaton. As we noted earlier, learning will change this automaton. The agent starts with an automaton with limited acuity, and uses its conscious level to deal with world states not recognizable at the Perceptuo-Motor level. For instance, the Perceptuo-Motor level of a person beginning to learn how to drive, is not sophisticated enough to respond to driving conditions automatically. As the agent becomes a better driver, the conscious level is freed to attend to other things while driving. This is called *automaticity* in psychology. We discuss an implementation mechanism for these automated behaviors later in this paper.

The Sensori-Actuator Level. The *Sensori-Actuator level* is the level of primitive motor and sensory actions, for instance "move from $\langle x, y, z \rangle$ to $\langle x', y', z' \rangle$" or "look at $\langle x, y, z \rangle$". At this level, there are no object representations as there are at the Knowledge level and the Perceptuo-Motor level. There are no explicit declarative representations of any kind, only procedural representations (on the actuator side) and sensor data (on the sensory side). Primitive motor actions may typically be implemented in a robot control language like VAL, and some elementary data processing routines may be implemented in a sensory sub-system, like dedicated vision hardware. At this level, we also situate *reflexes*, which we consider to be low-level loops from sensors to actuators, controlled by simple thresholding devices, operating independently of higher-level mechanisms, and able to pre-empt the latter. We see reflexes as primitive mechanisms whose main purpose is prevention of damage to the hardware, or to put it in anthropomorphic terms, survival of the organism. As such they take precedence over any other behavior. When reflexes are triggered, the higher levels are made "aware" of this by the propagation of a signal, but they have no control over the reflex's execution, which is brief and simple (like a withdrawal reflex seen in people when they unintentionally stick their hand into a fire).[14] [15] After the completion of a reflex, the higher levels regain control and must decide on how to continue or discontinue the activity that was interrupted by the reflex. Reflex-like processes may also be used to shift the focus of attention of the Knowledge level.

2.3 Symbol Grounding: A Non-Tarskian Semantics

> Tarskian Semantics has nothing to say about how descriptions of objects in plans relate to the objects in the world [47, p. 13].

Let's digress for a moment to some esoteric matters of semantics and reference. One problem an agent has to solve is how to find and maintain a correspondence between a referent in the world and a symbol in an agent's world model. As noted above, the referent in the world is (by necessity) only indirectly considered via its embodied Perceptuo-Motor level representation, hence the problem becomes one of aligning the Knowledge level representations with the Perceptuo-Motor level representations. From the perspective of cognitive science, the problem has been labeled the *symbol grounding problem*

---

[14] An appropriate reflex for a robot (arm) might be to withdraw or stop when it meets too much mechanical resistance to its movement, as evidenced for instance by a sharp rise in motor current draw. Such a reflex could supplant the more primitive fuse protection of motors, and make an appropriate response by the system possible. Needless to say, a robot that can detect and correct problems is much more useful than one that merely blows a fuse and stops working altogether.

[15] The fact that the withdrawal reflex may not be as strong, or not present at all, when doing this intentionally may point to the need for top-down inhibition as well.

[19]. The question is how to make the semantics of a robot's systematically interpretable Knowledge level symbols cohere equally systematically with the robot's interactions with the world, such that the symbols refer to the world on their own, rather than merely because of an external interpretation we place on them. This requires that the robot be able to discriminate, identify, and manipulate the objects, events, and states of affairs that its symbols refer to [20]. Grounding is accomplished in our architecture in part through the alignment of the Knowledge and Perceptuo-Motor levels. If we think of the Perceptuo-Motor level as implementing categorial perception (and perhaps "categorial action"), then the elementary symbols of the Knowledge level are the names attached to the categories. In other words, the alignment of the Knowledge and Perceptuo-Motor level constitutes an *internal referential semantic model* of elementary symbols. Note that, like McDermott, we do not take the Tarskian stance which requires the referents of symbols to be in the world; rather, they are system-internal, similar to what Hausser proposes [21], or what Harnad calls iconic representations: "proximal sensory projections of distal objects, events, and states of affairs in the world" [19]. The Knowledge level is the only level that is accessible for conscious reasoning, and also the only level that is accessible for inter-agent communication. Access to the Perceptuo-Motor level and the Sensori-Actuator level would not be useful for communication, as the representations and processing at these levels are too agent-centered and too agent-specific to be informative to other agents.

Since the Perceptuo-Motor level representations serving as the grounding for symbols of the Knowledge level are embodied (Sect. 2.4), equivalent symbols may have somewhat different semantics for different agents having different bodies. We don't see that as a problem, as long as the differences are not too large.[16] Indeed, we believe that this is quite realistic in human terms as well; no two persons are likely to have *exactly* the same semantics for their concepts, which nevertheless does not prevent them from understanding each other, grosso modo at least (cf. [51]). The problems of translation and communication in general consist at least in part of establishing a correspondence between concepts (and symbols) used by the participants. It is helpful to be able to use referents in the external world as landmarks in the semantic landscape, but one consequence of embodied semantics is that *even* if it is possible to establish these common external referents for symbols, there is still no guarantee that the symbols will actually *mean* exactly the same thing, because in effect the same referent in the world is *not* the same thing to different agents. If we accept this view, it is clear that approaches to semantics based on traditional logical model theory are doomed to fail, because they *presuppose* "identity of referents" and an unambiguous mapping from symbols to referents, the same one for all agents. Another problem is of course the presupposition that all objects are uniquely identifiable. The

---

[16] It is never a problem as long as agents need not communicate with the outside world (other agents), of course, cf. [63].

use of deictic representations does not impose such a condition; as far as our agents are concerned, if it looks and feels the same, it is the same.[17] Nothing hinges on whether or not the objects in the agent's surroundings are *really* extensionally the same as the identical-looking ones that were there a moment ago or will be there a moment later.

In Sect. 3.1 we present an implementation of our architecture that illustrates our ideas on symbol grounding, in the domain of color perception and color naming.

2.4 Embodied Representation

In Sect. 2.2 we already mentioned the use of embodied representations at the Perceptuo-Motor level. We now look at the principle of embodiment from a more abstract point of view.

One of the most general motivations behind our work is the desire to be able to "program" a robotic autonomous agent by requesting it to do something and have it "understand", rather than telling it how to do something in terms of primitive motions with little or no "understanding". For instance, we want to tell it to go find a red pen, pick it up, and bring it to us, and not have to program it at a low level to do these things.[18] One might say that we want to communicate with the robot at the *speech act* level. To do this, the agent needs a set of general-purpose perceptual and motor capabilities along with an "understanding" of these capabilities. The agent also needs a set of concepts which are similar enough to ours to enable easy communication. The best way to accomplish this is to endow the agent with embodied concepts, grounded in perception and action.

We define *embodiment* as the notion that the representation and extension of high level concepts is in part determined by the physiology (the bodily functions) of an agent, and in part by the interaction of the agent with the world. For instance, the extension of color concepts is in part determined by the physiology of our color perception mechanism, and in part by the visual stimuli we look at. The result is the establishment of a mapping between color concepts and certain properties of both the color perception mechanism and objects in the world. Another example is the extension of concepts of action: it is partly determined by the physiology of the agent's motor mechanisms, and partly by the interaction with objects in the world. The result is the establishment of a mapping between concepts of action and certain properties of both the motor mechanisms and objects in the world (what we might call "the shapes of acts").

At an abstract level, the way to provide an autonomous agent with human-like embodied concepts is to intersect the set of human physiological capabilities with the set of the agent's potential physiological capabilities, and endow

---

[17] This is of course the "duck test", made famous by a former US president.

[18] Retrieving "canned" parameterized routines is still a low-level programming style that we want to avoid.

the agent with what is in this intersection. To determine an agent's potential physiological capabilities, we consider it to be made up of a set of primitive actuators and sensors, combined with a general purpose computational mechanism. The physical limitations of the sensors, actuators, and computational mechanism bound the set of potential capabilities. For instance with respect to color perception, if the agent uses a CCD color camera (whose spectral sensitivity is usually wider than that of the human eye), combined with a powerful computational mechanism, we consider its potential capabilities wider than the human ones, and thus restrict the implemented capabilities to the human ones. We endow the agent with a color perception mechanism whose functional properties reflect the physiology of human color perception. That results in color concepts that are similar to human color concepts. With respect to the manipulation of objects, most robot manipulators are inferior to human arms and hands in terms of dexterity, hence we restrict the implemented capabilities to the ones that are allowed by the robot's physiology. The robot's motor mechanism then reflects the properties of its own physiology, rather than those of the human physiology. This results in a set of motor concepts that is a subset of the human one. Embodiment also calls for body-centered and body-measured representations, relative to the agent's own physiology. We provide more details on embodiment in GLAIR in [26].

2.5 Alignment

When a GLAIR-agent notices something in its environment, it registers that it has come to know of an object. Regardless of whether the agent recognizes the type of the object, we want it to explicitly represent the existence of the object in the Knowledge level while processing sensory information about the object at the Perceptuo-Motor level. Similarly, when properties of objects or relationships among objects are sensed by the GLAIR-agent, we want it to explicitly represent these properties and relationships, even if no more is known about them than the fact that they exist. We use unnamed intensional concepts for this purpose [57].

Having sensed an object, an assertion is made about the object being sensed at the GLAIR Knowledge level. Once the object is no longer in the "field of perception", the assertion about its being sensed is removed. This is tantamount to disconnecting the relationship between the symbolic representation and the world. If at the Perceptuo-Motor level a previously sensed object is again being sensed, we reassert the fact that the object, the same one represented before at the Knowledge level, is being sensed. An example of this type of (unconscious) perception is when we look at an object, look away, and then look back at the same object. The unconscious level can provide a short term sensory memory in which memories of objects are stored, and when we see them from time to time, the conscious layer is alerted to that fact. We can think of this phenomenon as a type of *continuity in perception* at the unconscious level. We believe that if we assume this continuity, we should re-use previously constructed representations to represent again-sensed objects. In order for a GLAIR-agent to re-use its previously established representations about objects for again-sensed objects, we either have to assume that the agent has a continuity of perception at the unconscious layer or that a conscious matching of existing representations to sensed objects is performed.

2.6 Engaged and Disengaged Reasoning

Our architecture allows us to elegantly model two different modes of reasoning and planning, which we call *engaged reasoning* and *disengaged reasoning*. Engaged reasoning takes place when all the elementary symbols at the Knowledge level that are involved in the current reasoning activity are immediately aligned with representations at the Perceptuo-Motor level. In practical terms, this means that the agent is reasoning about objects within its field of perception. This may require active perception to keep track of objects, or to shift attention to new objects as the reasoning progresses. For the purpose of object tracking and attention shifting, Knowledge level sensory actions are defined and can be reasoned about like ordinary actions [36]. Engaged reasoning can be done while the actions being reasoned about are actually being carried out, in a kind of plan-as-you-go mode with continuous monitoring of progress being made, or in a more hypothetical mode with no actions being carried out, but potentially affected objects being gauged while the plan is being developed.

Disengaged reasoning occurs when there is no immediate alignment between the Knowledge level symbols involved in the current reasoning activity and Perceptuo-Motor level representations, e.g., when developing a plan for another place and/or another time, in a purely hypothetical fashion. This is the mode that traditional planners used to operate in all the time, by necessity. Intermediate forms of reasoning, between engaged and disengaged, are possible as well.

2.7 Consciousness

As we pointed out above, we identify the Knowledge level with consciously accessible data and processing; the Perceptuo-Motor level with "hard-wired", not consciously accessible processing and data involved with motor control and perceptual processing; and the Sensori-Actuator level with the lowest-level muscular and sensor control, also not consciously accessible. The distinction of conscious (Knowledge) levels vs. unconscious (Perceptuo-Motor and Sensori-Actuator) levels is convenient as an anthropomorphic metaphor, as it allows us to separate explicitly represented and reasoned about knowledge from implicitly represented and processed knowledge. This corresponds *grosso modo* to consciously accessible and not consciously accessible knowledge for people.[19] Although we are aware of the pitfalls of introspection, this provides us with a rule of thumb for assigning knowledge (and skills, behaviors, etc.) to the various levels of the architecture. We believe that our organization is to some extent psychologically relevant, although we have not yet undertaken any experimental investigations in this respect. The real test for our architecture is its usefulness in applications to physical (robotic) autonomous agents (Section 3).

Knowledge in GLAIR can migrate from conscious to unconscious levels. In [24] we show how a video-game playing agent learns how to dynamically "compile" a game playing strategy that is initially formulated as explicit reasoning rules at the Knowledge level into an implicit form of knowledge at the Perceptuo-Motor level, a Perceptuo-Motor Automaton (PMA).

There are also clear computational advantages to our architectural organization. A Knowledge Representation and Reasoning system as used for the conscious Knowledge level is by its very nature slow and requires lots of computational resources.[20] The implementation mechanisms we use for the unconscious levels, such as PMAs, are much faster and require much less resources. Since the three levels of our architecture are semi-independent, they can be implemented in a (coarse-grained) parallel distributed fashion; at least each level may be implemented on distinct hardware, and even separate mechanisms within the levels (such as individual reflex behaviors) may be.

3   Applications

Our architecture as described in Sect. 2 can be populated with components that make up the machinery for mapping sensory inputs to response actions, as does Russell in [54]. We now discuss some applications of GLAIR that we have been developing.

Some important general features of GLAIR-agents are the following:

- Varieties of behaviors are integrated: We distinguish between deliberative, reactive, and reflexive behaviors. At the unconscious level, behavior is generated by mechanisms with the computational power of a finite state machine (or less), whereas, at the conscious level, behavior is generated via reasoning (of Turing Machine capabilities). As we move down the architectural levels, computational and representational power (and generality) is traded off for better response time and simplicity of control. Embodied representations aid in this respect (Sect. 2.4).

---

[19] The term "knowledge" should be taken in a very broad sense here.

[20] As we all know, many reasoning problems are NP-complete, meaning there are no polynomial-time deterministic algorithms known for solving them, or in plain English: they are very hard to solve in a reasonable amount of time (see e.g. [43]). Elephants don't even stand a chance in this respect.

- We assume agents to possess a set of primitive motor capabilities. The motor capabilities are primitive in the sense that (a) they cannot be further decomposed, (b) they are described in terms of the agent's physiology, and (c) no reference is made to external objects. The second property of motor capabilities is so that the success of performing an action should depend only on the agent's bodily functions and proprioceptive sensing. For example, for a robot arm, we might have the following as its motor abilities: calibrate, close-hand, raise-hand, lower-hand, move.
- Our architecture provides a natural framework for modeling four distinct types of behavior, which we call reflexive, reactive, situated, and deliberative. Reflexive and reactive behaviors are predominantly unconscious behaviors, whereas situated and deliberative actions are conscious behaviors.

Reflexive behavior[21] occurs when sensed data produces a response, with little or no processing of the data. A reflex is immediate. The agent has no expectations about the outcome of its reflex. The reflexive response is not generated based on a history of prior events or projections of changing events, e.g., a gradual temperature rise. Instead, reflexive responses are generated based on spontaneous changes in the environment of the agent, e.g. a sudden sharp rise in temperature. In anthropomorphic terms, this is innate behavior that serves directly to protect the organism from damage in situations where there is no time for conscious thought and decision making, e.g., the withdrawal reflex when inadvertently touching something hot. Reflexive behavior does not require conscious reasoning or detailed sensory processing, so our lowest level, the Sensori-Actuator level, is charged with producing these behaviors. Our initial mechanism for modeling reflexive behavior is to design processes of the form $T \mapsto A$, where T is a trigger and A is an action. A trigger can be a simple temporal-thresholding gate. The action A is limited to what can be expressed at the Sensori-Actuator level, and is simple and fast.

Reactive behavior requires some processing of data and results in *situated action* [62]. However, its generation is subconscious. *Situated action* refers to an action that is appropriate in the environment of the agent. In anthropomorphic terms, this is learned behavior. An example would be gripping harder when one feels an object is slipping from one's fingers, or driving a car and tracking the road. We use the term *tracking* to refer to an action that requires continual adjustments, like steering while driving. Examples of this type of reactive behavior are given in [48, 5]. Situated behavior requires assessment of the state the system finds itself in (in some state space) and acting on the basis of that. It might be modeled by the workings of a finite state automaton, for example, the Micronesian behavior described in [62]. Situated action is used in *reactive planning*[2, 16, 55].

---

[21] E.g., visual reflexes in [52]: Here responses are generated to certain visual stimuli that do not require detailed spatial analysis.

Deliberative behavior requires considerable processing of data and reasoning which results in action. In anthropomorphic terms, this is learned behavior that requires reasoning that can be modeled by a Turing Machine (or first order logic), for example explicit planning and action.

We have developed an implementation mechanism for the Perceptuo-Motor-level which we call Perceptuo-Motor-Automata [28]. A PMA is a finite state machine in which each state is associated with an act and arcs are associated with perceptions. In each PMA, a distinguished state is used to correspond to the no-op act. Each state also contains an auxiliary part we call Internal State (IS). An IS is used in arbitrating among competing arcs. Arcs in a PMA are situations that the agent perceives in the environment. When a PMA arc emanating from a state becomes active, it behaves like an asynchronous interrupt to the act in execution in the state. This causes the PMA to stop executing the act in the state and to start executing the act at the next state at the end of the arc connecting the two states. This means that in our model the agent is never idle, and it is always executing an act. The primary mode of acquiring PMAs in GLAIR is by converting plans in the Knowledge level into PMAs through a process described in [28]. A PMA may become active as the result of an intention to execute an action at the Knowledge level [35]. Once a PMA becomes active, sensory perception will be used by the PMA to move along the arcs. The sensory perceptions that form the situations on the arcs as well as subsequent actions on the PMA may be monitored at the Knowledge level. In general, the sensory information is filtered into separate streams for PMAs and for the Knowledge level.

3.1 A Physical Instance: the Color Labeling Agent

We now present an instance of an agent conforming to our architecture, the Color Labeling Agent (CLA). It has a set of *grounded* or *embodied* concepts represented as terms at its Knowledge Level, a "sub-conscious" color space at its Perceptuo-Motor Level, and a color camera at its Sensory-Actuator Level, as well as a color monitor which it uses as a primitive actuator to point to things in its field of view (Fig. 2).

Using a normalized Gaussian function of perceptual color space coordinates as the basic category model, the CLA is able to

1. *Name* real colors in response to real visual stimuli (camera images), and provide a confidence or "goodness of example" or (fuzzy) membership value.
2. *Point out* examples of named colors in real images and provide a confidence rating, and as a derivative of this capability, pick the *best example* of a named color from a set of color samples, or from an image in general.

The model is at present constrained to the eleven so-called *basic color categories* and their corresponding names in English, as defined first in the work

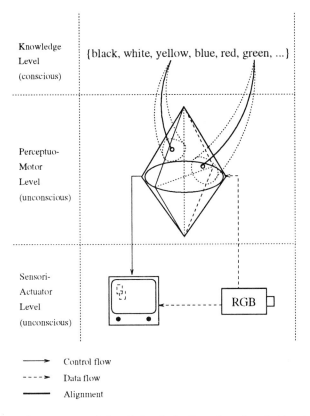

Knowledge
Level
(conscious)

{black, white, yellow, blue, red, green, ...}

Perceptuo-
Motor
Level
(unconscious)

Sensori-
Actuator
Level
(unconscious)

RGB

⟶   Control flow

- - - ➤   Data flow

▬▬▬   Alignment

Fig. 2. The various parts of the Color Labeling Agent, relative to the levels of our architecture

of [8]. The CLA's performance on these tasks has been quantitatively shown to be reasonably consistent with human performance [41]. In particular, this means that

1. The model places the *foci* of the basic color categories in the same regions of the color space as human subjects do.
2. The model places the *boundaries* of the basic color categories in the same regions of the color space as human subjects do.

As such, the color categorization application can be seen as a (partly) *embodied* [39, 32, 26] or *grounded* [19] system, or as an instance of *situated cognition* [62].

The current implementation of the CLA consists of two separate parts, one concerned with selection and display of samples from images, and the other concerned with the actual color perception and categorization model (Fig. 3).

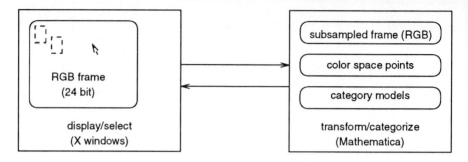

Fig. 3. Outline of the color naming/pointing-out/selecting application, consisting of a display and select part (left), implemented as an X windows client, and a transformation and categorization part (right), implemented in Mathematica code. The two parts communicate asynchronously via a simple file protocol.

The display program runs under X windows, and allows one to display a 24-bit RGB image. It also lets one select samples of a certain pre-defined size (currently 12×16 pixels) from the image using the mouse, which will be passed to the categorization program. It can subsample the entire image using the same blob size, and pass the result to the categorization program. Finally, it can draw boxes around blobs, whose center coordinates it gets from the categorization program.

The categorization program is a collection of Mathematica functions that can (1) *Name* the color of a blob pointed to on the image, and provide a membership value, (2) *Point out* examples of a named color category on the image, and provide a membership value, and (3) *Select* one from a number of samples whose color best fits a named category. The names returned can be simple or complex, and the best n candidates can be returned. The category membership threshold θ may be specified, as well as the underlying color space to use. The names specified for the pointing-out function can be simple or complex as well, and the threshold θ can be specified, as well as the color space to use. The function can either return any n examples (the first exceeding the threshold) or the best n examples. The select function always points to the best example of the specified category within the set of samples provided, using a specifiable underlying color space. It will ask the user to provide it with a set of samples from the image first, to get around the absence of any image segmentation or object recognition algorithms.

For a detailed quantitative and qualitative analysis of the performance of the Color Labeling Agent we refer to [41] and [40]. Suffice it here to say that the performance is reasonably consistent with human performance on the same tasks, which means to us that we have succeeded in grounding a small set of elementary symbols in perception and action – although the active side of the agent is relatively primitive.

3.2 A Simulation Study: Air Battle

We are interested in modeling behavior generation by agents that function in dynamic environments. We make the following assumptions for the agent:

- The environment demands continual and rapid acting, e.g., playing a video-game.
- The impact of the agent's actions depends on the situations under which actions are applied and on other agents' actions.
- Other agents' actions are nondeterministic.
- The agent does not know about the long term consequences (i.e., beyond the current situation) of its actions.
- The agent is computationally resource bounded. We assume that the agent needs time to think about the best action and in general there is not enough time.

To cope in dynamic environments, an agent which is resource bound needs to rely on different types of behaviors, for instance, reflexive, reactive, situated, and deliberative behaviors. Reflexive and reactive behaviors are predominantly "unconscious" behaviors, situated action may be either "unconscious" or "conscious", and deliberative actions are predominantly "conscious" behaviors. We assume that in general "conscious" behavior generation takes more time than "unconscious" behavior generation.

We have written a program, Air Battle Simulation (ABS), that simulates World War I style airplane dog-fights. ABS is an interactive video-game where a human player plays against a computer driven agent. The game starts up by displaying a game window and a control panel window (Fig. 4). The human player's plane is always displayed in the center of the screen. The aerial two-dimensional position of the enemy plane is displayed on the screen with the direction of flight relative to the human player's plane. The human player uses the control panel to choose a move, which is a combination of changing altitude, speed, and direction. When (s)he presses the go button, the computer agent also selects a move. The game simulator then considers both moves to determine the outcome, and updates the screen and the accumulated damage to planes. ABS simulates simultaneous moves this way. If a player's plane is close in altitude and position to the enemy plane, and the enemy is in frontal sight, the latter is fired on automatically (i.e., firing is not a separate action). The levels of damage are recorded in a side panel, and the game ends when one or both of the player's planes are destroyed.

The computer agent has been developed in accordance with the principles of the GLAIR architecture. Figure 5 schematically represents its structure. Initially, the agent does not have any PMAs available, and uses conscious level reasoning to decide what move to make. Once situation transitions are learned and cached in a PMA, the agent uses the PMA for deciding its next move whenever possible. Hence, by adding learning strategies, a PMA can be developed that caches moves decided at the Knowledge level for future use.

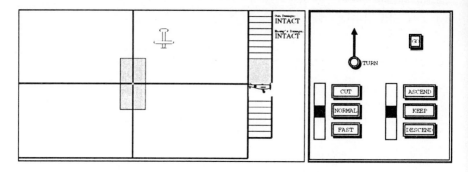

Fig. 4. Air Battle Simulation game window and control panel (see text)

Learning can be used to mark PMA moves that prove unwise and to reinforce moves that turn out to be successful. We are exploring these learning issues. On trial runs, we started ABS with an empty PMA and as the game was played, transitions of the PMA were learned. Also, when similar situations occurred and there was an appropriate PMA response, the PMA executed the corresponding action. As the game was played, we observed that the agent became more reactive since the PMA was increasingly used to generate behaviors instead of the Knowledge level.

Improving "Unconscious" Behaviors. The rules of a PMA are situation/action pairs. As it turns out, a situation can be paired up with multiple actions. The object of learning here is to learn which actions when associated with a situation yield a better result, i.e., the pilot ends up in a more desirable situation.

Some situations in ABS are more desirable for the pilot than others, e.g., being right behind the enemy and within firing range. Let's assume that we can assign a goodness value $G(s)$ to each situation s between -1 and 1. As the pilot makes a move, he finds himself in a new situation. This new situation is not known beforehand to the pilot since it also depends on the other pilot's move. Since the new situation is not uniquely determined by the pilot's move, his view of the game is not Markovian. Let $Q(s,a)$ be the evaluation of how appropriate action a is in situation s, and $R(s,a)$ be the goodness value of the state that the pilot finds himself in after performing a in situation s. The $R(s,a)$ values are determined as the game is played and cannot be determined beforehand. This is called the immediate reward. We let $Q(s,a) = R(s,a) + \gamma \max_k Q(s',k)$ where situation s' results after the pilot performs a in s. γ is a parameter between 0 and 1 that is known as the discount factor in reinforcement based learning. At the start of game, all $Q(s,a)$ in the PMA are set to 1. As the game is played, Q is updated. As of this writing we are experimenting with setting appropriate parameters for Q.

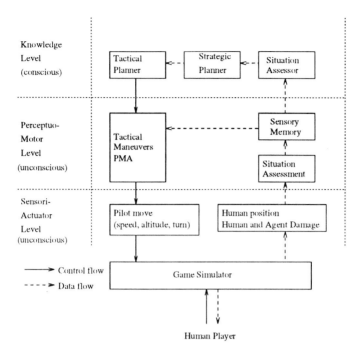

Fig. 5. Schematic representation of the Air Battle Simulation GLAIR-agent

Observing Successful Patterns of Interaction in the World. We assume that the agent does not know about the long term consequences of its actions. Furthermore, the reinforcement based learning we have used assumes a Markovian environment. That is, the agent believes the world changes only due to its own actions. This makes it necessary to observe interactions with the world in order to learn sequences of actions. Over a finite number of actions, when the agent observes a substantially improved situation, chances are he has found a successful *routine*. We record such detected routines and as they reoccur, we increase our confidence in them. When the confidence in a Routine reaches a certain level, a concept is created at the Knowledge level of GLAIR for the routine and from then on, this routine can be treated as an atomic action at that level [23].

We plan to explore other learning techniques such as experimentation as a form of learning [60]. We are also interested in developing experiments that will help in psychological validation of GLAIR and the learning strategies used in ABS. As of the time of writing ABS is fully operational, but several issues are still being investigated, as noted above.

3.3   A Simulation Study: the Mobile Robot Lab

We now describe the Mobile Robot Lab (MRL), a simulation environment
we have developed for mobile robots that function as GLAIR-conformant autonomous agents. The simulation is relatively simple, but nevertheless provides a rich and realistic enough environment to function as a testbed for
the development of physical GLAIR-agents. A complete setup using MRL
consists of a GLAIR-agent, a simulator with an incorporated description of
a physical environment, and a graphical interface (Fig. 6).

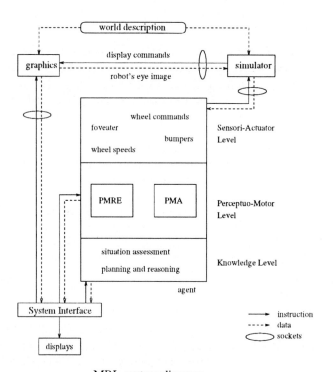

MRL system diagram

Fig. 6. Overview of a complete setup using MLR. It consists of a
GLAIR-agent, a simulator with an incorporated model of a physical environment, and a graphical interface. Arrows represent direction of data flow
among the components.

Emergent Behaviors. A major objective for this project is learning emergent behaviors. Like Agre with his improvised actions [2] and Brooks with
his subsumption architecture [9] we believe complex behaviors emerge from
interaction of the agent with its environment without planning. However, previous work in this area hard-coded a lot of primitive actions. Furthermore,

it did not attempt to learn the improvised behavior. In this simulation, we plan to start with a minimal number of primitive actions and sensations. Our basis for this minimality and the choice of primitive actions is physiological. In other words, we will choose actions that are physically basic for the agent's body as primitive. We then instruct the agent to perform tasks and in the midst of accomplishing this, we expect it to notice some types of behaviors emerge. An example of an emergent behavior we will explore is *moving toward an object*. We expect the agent to be learning to coordinate its wheel motions, starting from nothing more than the primitive sensation of contact with an external object, and the primitive actions of turning its motors independently on or off.

The Physical Environment Description. The simulator uses a description of the physical environment that the simulated robot operates in. This description is easily modifiable (without reprogramming). It includes the physical characteristics of the mobile robot and the space in which it moves. A 2D bird's eye view of a typical room setup with a robot inside is show in Fig. 7.

The room the robot moves in has a polygonal floor plan and vertical walls, and contains a number of solid objects with convex polygonal bases and vertical faces, each with an associated user-defined spectral power distribution (SPD) specifying the surface spectral reflectance (color).

Any number of robots may inhabit the room. They have two independently driven wheels on either side, a bumper bar front and back, with contact and force sensors built in, and a color camera on top, parallel to the direction of the driven wheels. The camera is fixed and mounted horizontally.

The Simulator. The simulator interfaces with the agent and with the graphical interface. It takes care of any I/O with the agent that would otherwise come from the sensors and go to the actuators of a real mobile robot. It also takes care of any I/O with the graphical interface, needed to keep the graphical display of the robot and its physical environment updated.

The simulator incorporates a simplified model of the physics of motion and sensing for the mobile robot. It continually updates the position of the robot depending on the rotation speed and direction of its wheels, and provides the agent with appropriate sensory data about wheel rotation and contact with objects. It also prevents the robot from going "through" walls or objects. It provides simulated camera input to the agent. Camera input can be simplified in different ways, e.g. using a space-variant downsampling to create a "rexelated" image as used in Hierarchical Foveal Machine Vision [7, 45]. This simplified camera view is computed and passed to the simulator by the graphical interface, on the basis of the 3D perspective views.

The simulator incorporates a simplified lighting model to determine the appearance (color) of objects in the room. Light sources can either be point sources or homogeneous diffuse sources. Each light source has its own SPD. Each object has its own spectral reflectance function. All objects are assumed

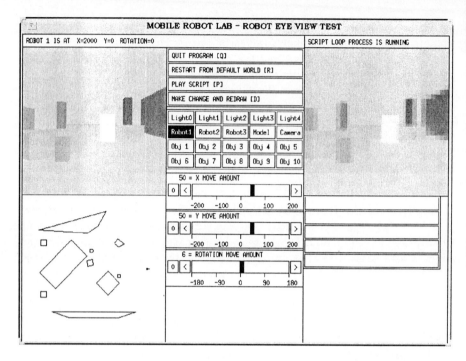

Fig. 7. The graphical interface of the mobile robot simulator. Upper left: color perspective view of the environment, from the robot's point of view. Upper right: same, but in a "rexelated" view from the robot's rectangular foveal camera (see text). Lower left: floor plan of the environment. The robot's current position and orientation is indicated by the small triangle near the right edge, halfway from the top. Middle: interface and simulator control panel, including camera and lighting model, and movement controls. Courtesy of Amherst Systems Inc, Buffalo NY.

to be Lambertian reflectors.

The Mobile Robot Lab has been used to study GLAIR-agent based gaze control in the context of space exploration, in a man-machine cooperative scenario [7], in addition to some other explorative work.

4 Concluding Remarks

We have presented a general architecture for autonomous embodied agents that integrates behavior-based architectures with traditional architectures for symbolic systems. The architecture specifies how an agent establishes and maintains a conscious connection with its environment while mostly unconsciously processing sensory data, and filtering information for conscious processing as well as for reflexive and reactive acting. We ended our paper by instantiating the architecture with a few (physical and simulated) agents embedded in their environment, in various stages of implementation. Some additional aspects of our work are discussed in [25], [27], and [29].

We believe our work can contribute towards integrating traditional ungrounded symbol systems with the newer physically grounded systems. Combining an elephant's body with a man's[22] mind makes for an awesome combination.

5 Acknowledgements

We appreciate comments made on an earlier draft of this paper by Phil Agre, Chris Brown, Stevan Harnad, Donald Nute, John Pollock, Beth Preston, William Rapaport, and Tim Smithers. Goofs are entirely attributable to the authors, of course. Thanks to Cesar Bandera and Robert J. Makar of Amherst Systems Inc. for kindly providing the Mobile Robot Simulator interface snapshot. Apologies to John Steinbeck, who carries no blame for this paper.

References

1. P. Agre. The dynamic structure of everyday life. Technical Report 1085, MIT Artificial Intelligence Laboratory, MIT, 1988

2. P. E. Agre and D. Chapman. Pengi: An implementation of a theory of activity. In Proceedings of AAAI-87, Seattle WA, pp. 268–272, July 1987

3. J. Albus, A. Barbera, and R. Nagel. Theory and practice of hierarchical control. In 23rd International IEEE Computer Society Conference, pp. 18–38, 1981

4. J. R. Anderson. The Architecture of Cognition. Cambridge, MA: Harvard University Press, 1983

5. S. Anderson, D. Hart, and P. Cohen. Two ways to act. In ACM SIGART Bulletin, pp. 20–24. ACM publications, 1991

6. D. H. Ballard and C. M. Brown. Computer Vision. Prentice-Hall, Englewood Cliffs, NJ, 1982

7. C. Bandera, H. Hexmoor, and S. C. Shapiro. Foveal machine vision for robots using agent based gaze control – final report. Technical Report SBIR-NAS 9-19004, Amherst Systems Inc., Buffalo, NY, September 1994

8. B. Berlin and P. Kay. Basic Color Terms: Their Universality and Evolution. University of California Press, Berkeley, CA, 1991 edition, 1969

9. R. Brooks. A robust layered control system for a mobile robot. Technical Report 864, MIT AI Labs, MIT, 1985

10. R. Brooks. Planning is just a way of avoiding figuring out what to do next. Technical Report 303, MIT AI Labs, 1987

11. R. A. Brooks. Elephants don't play chess. Robotics and Autonomous Systems, 6: 3–15, 1990

---

[22] He-man or She-man.

12. S. Card, T. Moran, and A. Newell. The Psychology of Human-Computer Interaction. Erlbaum, Hillsdale, NJ, 1983

13. D. Chapman. Vision, instruction, and action. Technical Report 1204, MIT Artificial Intelligence Laboratory, MIT, 1990

14. J. Connell. SSS: A hybrid architecture applied to robot navigation. In IEEE Conference on Robotics and Automation, pp. 2719–2724, 1992

15. J. Culbertson. The Minds of Robots. U. of Illinois Press, 1963

16. R. J. Firby. An investigation into reactive planning in complex domains. In Proceedings of AAAI-87, pp. 202–206, 1987

17. J. Fodor. The Modularity of Mind. MIT Press, 1983

18. E. Gat. Reliable goal-directed reactive control of autonomous mobile robots. Technical report, Dept. of Computer Science, Virginia Polytechnic Institute and State University, 1991

19. S. Harnad. The symbol grounding problem. Physica D, 42(1-3): 335–346, 1990

20. S. Harnad. Electronic symposium on computation, cognition and the symbol grounding problem. Ftp archive at princeton.edu:/pub/harnad/sg.comp.arch*, 1992

21. R. Hausser. Computation of Language: An Essay on Syntax, Semantics and Pragmatics in Natural Man-Machine Communication. Springer-Verlag, New York, NY, 1989

22. H. Hexmoor. An architecture for reactive sensor-based robots. In NASA Goddard conference on AI, Greenbelt, MD, 1989

23. H. Hexmoor. Representing and learning successful routine activities. Unpublished PhD Proposal, 1992

24. H. Hexmoor, G. Caicedo, F. Bidwell, and S. Shapiro. Air battle simulation: An agent with conscious and unconscious layers. In University at Buffalo Graduate Conference on Computer Science 93 (TR-93-14). Dept. of Computer Science, SUNY at Buffalo, New York, 1993

25. H. Hexmoor, J. Lammens, G. Caicedo, and S. C. Shapiro. Behavior based AI, cognitive processes, and emergent behaviors in autonomous agents. In G. Rzevski, J. Pastor, and R. Adey (eds.), Applications of AI in Engineering VIII, Vol. 2, Applications and Techniques, pp. 447–461. CMI/Elsevier, 1993. Reprint available TR-93-15, CS dept., SUNY/Buffalo

26. H. Hexmoor, J. Lammens, and S. Shapiro. Embodiment in GLAIR: A grounded layered architecture with integrated reasoning for autonomous agents. In D. D. Dankel (ed.), Proceedings of the 6th Florida AI Research Symposium, pp. 325–329. Florida AI Research Society, 1993

27. H. Hexmoor, J. Lammens, and S. C. Shapiro. An autonomous agent architecture for integrating "unconscious" and "conscious", reasoned behaviors. In Proceedings of Computer Architectures for Machine Perception, New Orleans, LA, 1993. Preprint available as TR-93-36, CS dept., SUNY/Buffalo

28. H. Hexmoor and D. Nute. Methods for deciding what to do next and learning. Technical Report AI-1992-01, AI Programs, The University of Georgia, Athens, Georgia, 1992. Also available as TR-92-23, CS dept., SUNY/Buffalo

29. H. Hexmoor and S. C. Shapiro. Examining the expert reveals expertise. In Proceedings of the Third International Workshop on Human and Machine Cognition: Expertise in Context (Seaside, FL), 1993

30. L. Kaelbling. Goals as parallel program specifications. In Proceedings of AAAI-88. Morgan Kaufmann, 1988

31. L. Kaelbling and S. Rosenschein. Action and planning in embedded agents. In P. Maes (ed.), Designing Autonomous Agents, pp. 35–48. MIT Press, 1990

32. P. Kay and C. K. McDaniel. The linguistic significance of the meaning of Basic Color Terms. Language, 54(3): 610–646, 1978

33. D. Kumar. An integrated model of acting and inference. In D. Kumar (ed.), Current Trends in SNePS–Semantic Network Processing System: Proceedings of the First Annual SNePS Workshop, pp. 55–65, Buffalo, NY, 1990. Springer-Verlag

34. D. Kumar. From Beliefs and Goals to Intentions and Actions: An Amalgamated Model of Inference and Acting. PhD thesis, Technical Report 94-04, Department of Computer Science, State University of New York at Buffalo, Buffalo, NY, 1994

35. D. Kumar. The SNePS BDI architecture. Journal of Decision Support Systems—Special Issue on Logic Modeling, 1994. Forthcoming

36. D. Kumar and S. C. Shapiro. Acting in service of inference (and vice versa). In D. D. Dankel II (ed.), Proceedings of the Seventh Florida Artificial Intelligence Research Symposium, pp. 207–211. the Florida AI Research Society, St. Petersburg, FL, May 1994

37. J. Laird, M. Huka, E. Yager, and C. Tucker. Robo-SOAR: An integration of external interaction, planning, and learning, using SOAR. In Robotics and Autonomous Systems, 1991

38. J. E. Laird, A. Newell, and P. S. Rosenbloom. SOAR: An architecture for general intelligence. Artificial Intelligence, 33: 1–64, 1987

39. G. Lakoff. Women, Fire, and Dangerous Things: What Categories Reveal about the Mind. University of Chicago Press, Chicago, IL, 1987

40. J. M. Lammens. A somewhat fuzzy color categorization model. Submitted to ICCV-95

41. J. M. Lammens. A Computational Model of Color Perception and Color Naming. PhD thesis, State University of New York at Buffalo, 1994. Also available as TR-94-26, CS dept., SUNY/Buffalo

42. P. Langley, K. McKusick, and J. Allen. A design for the ICARUS architecture. In ACM SIGART Bulletin, pp. 104–109. ACM publications, 1991

43. H. J. Levesque. Logic and the complexity of reasoning. Journal of Philosophical Logic, 17: 355–389, 1988

44. P. Maes. Action selection. In Proceedings of the Cognitive Science Society Conference, 1991

45. R. J. Makar. GLAIR: Mobile robot lab camera simulation. Master's project dept. of Computer Science, SUNY at Buffalo, NY, 1994

46. J. L. McClelland, D. E. Rumelhart, and G. E. Hinton. The appeal of parallel distributed processing. In D. Rumelhart, J. McClelland, and the PDP Research Group (eds.), Parallel Distributed Processing, chapter 1, pp. 3–44. MIT Press, Cambridge, MA MA, 1986

47. D. McDermott. Robot planning. Technical Report CS-861, Yale University, 1991

48. D. Payton. An architecture for reflexive autonomous vehicle control. In Proceedings of Robotics Automation, pp. 1838–1845. IEEE, 1986

49. J. Pollock. How to Build a Person. MIT Press, 1989
50. J. Pollock. New foundations for practical reasoning. In Minds and Machines, 1992
51. W. J. Rapaport. Syntactic semantics: Foundations of computational natural-language understanding. In J. H. Fetzer (ed.), Aspects of Artificial Intelligence, pp. 81–131. Kluwer Academic, New York, 1988
52. D. Regan and K. Beverly. Looming detectors in the human visual pathways. In Vision Research 18, pp. 209–212. 1978
53. J. K. Rosenblatt and D. Payton. A fine-grained alternative to the subsumption architecture for mobile robot control. In Proceedings of the International Joint Conference on Neural Networks, 1989
54. S. Russell. An architecture for bounded rationality. In ACM SIGART Bulletin, pp. 146–150. ACM publications, 1991
55. M. J. Schoppers. Universal plans for unpredictable environments. In Proceedings 10th IJCAI, pp. 1039–1046, 1987
56. S. C. Shapiro. Cables, paths, and 'subconscious' reasoning in propositional semantic networks. In Principles of Semantic Networks. Morgan Kaufmann, 1990
57. S. C. Shapiro and W. J. Rapaport. SNePS considered as a fully intensional propositional semantic network. In N. Cercone and G. McCalla (eds.), The knowledge frontier: essays in the representation of knowledge, pp. 262–315. Springer Verlag, New York, 1987
58. S. C. Shapiro and W. J. Rapaport. The SNePS family. Computers Math. Applic., 23(2–5): 243–275, 1992
59. S. C. Shapiro and the SNePS Implementation Group. SNePS 2.1 User's Manual. Department of Computer Science, SUNY at Buffalo, 1994
60. W.-M. Shen. Learning from the Environment Based on Actions and Percepts. PhD thesis, Carnegie Mellon University, 1989
61. R. Simmons. An architecture for coordinating planning, sensing, and action. In Proceedings of the DARPA planning workshop, pp. 292–297, 1990
62. L. A. Suchman. Plans and Situated Actions: The Problem of Human Machine Communication. Cambridge University Press, 1988
63. P. H. Winston. Learning structural descriptions from examples. In R. J. Brachman and H. J. Levesque (eds.), Readings in Knowledge Representation, pp. 141–168. Morgan Kaufmann, San Mateo, CA, 1985 (orig. 1975)

How Do You Choose Your Agents?
How Do You Distribute Your Processes?

Miles Pebody[1]

Department of Computer Science, University College London, Gower St, London
WC1E 6BT UK. E-Mail: M.Pebody@cs.ucl.ac.uk

Abstract. This paper sets out some ideas and thoughts that resulted from an attendance at the NATO Advanced Study Institute "The Biology and Technology of Intelligent Autonomous Agents" that took place at the Castle Ivano, Italy in March 1993. The discussion is presented in the form of a project proposal for work that aims to examine the nature of applying some of the ideas and techniques covered in the institute's talks and discussions to an industrial problem domain (active sensing of material faults and defects). First some working characterisations are covered that provide some flavour of the main directions of the proposed work. These are followed by a description of the industrial domain and a test-bed that has been built to support experimentation. The conclusion provides discussion of interesting research directions.

Keywords. Industrial application, distributed control, multi-agent architectures.

1. Introduction to a Study of Distributed Control

Over the last few years there has been an increasing interest within the systems engineering, computer science and Artificial Intelligence (AI) communities towards the implementation of systems as groups of independent co-operating processes or agents i.e. distributed systems. These groups are approaching the topic from different angles as each has its own goals and reasons for its interest. The point to note is that although many people talk about distributed systems they are not necessarily talking about the same thing and there is consequently much confusion. The result is a fuzzy region which overlaps the subjects of AI, computer science and systems engineering. This can be seen by comparing the contents of the distributed industrial control articles of [Jones91] with for example the distributed artificial intelligence paper [Hayes-Roth88] and the paper published by [Steels89] which is indicative of another area of AI. This multi-directional, multi-disciplined approach is perhaps an inevitable result of the work of many communities into more complex systems and is not necessarily a

---

[1]Funded by the EPSRC and Sira Ltd.

problem, except when a project has to be defined as being a project in distributed control. Consequently one of the main aims of this paper is to clarify the subject that is to be studied.

The common idea behind all these directions of study is that a distributed system is viewed as a collection of interacting subsystems or processes. Each process is modular and designed to realise a particular task or sub-problem. Other details are taken care of by other member units and the overall task is achieved as a result of an apparent global functionality of the system, a result of the interactions between many of the separate processes. The constituent properties of a system, network or colony may vary immensely. Some may contain hundreds or thousands (or even more) of small simple processes whilst others may consist of a few complex ones. The assumption as to what constitutes a distributed system and the meaning of many descriptive terms varies depending on the interests of different research and engineering groups. The next part of this section covers the range in more detail.

1.1 Fields of Distributed Systems Work

In many cases the work on distributed systems has come about as a result of the ever increasing complexity of systems being designed and built. Some are inherently distributed such as the variety of machines found in a manufacturing environment whilst others are explicitly designed to be distributed, for example the real-time computer image generation equipment that is used on aircraft simulators. Between both extremes there have been countless developments in computer architectures and communication techniques to allow the many separate parts of the system to co-operate in order to more efficiently achieve the desired outcome. Many of these technologies have appeared as the direct result of the needs of a customer and so have consequently been engineered and developed from existing equipment. In contrast to this industrial evolution, the academic community studies new concepts in subject areas such as computer science and these can result in the introduction of brand new ideas, many of which involve the use of distributed systems. Examples can be found in the development of the object oriented programming paradigm and the extensive research into parallel computer architectures. These find their way into everyday life as spin-offs from scientific research rather than from continued engineering development

Parallel computing is a major area of research that falls under the collective title of distributed systems. However this has largely concentrated on the development of mechanisms and architectures that are able to execute some form of program. The main goal has been to produce systems that are general purpose machines, their functionality is defined by the software programs that they run. Consequently the nature of the research is abstracted away from the real world. The development of programming languages, multitasking operating systems and computer hardware provides tools for other people to use, people who have specific real world requirements.

The field of AI is somewhat different from that of systems engineering and computer science. In its purest form it is the study of the phenomenon of intelligence and it uses computer technology as a tool to experiment with mechanisms of intelligence in the artificial [Smithers91]. This area also has its commercial spin-offs, most noticeably in the form of expert systems and knowledge engineering. However, like computer science, AI has many internal groups and areas of interest two important ones in respect to this document being distributed AI (DAI) and behaviour based AI (BBAI). These two though, are not (or should not be) entirely separable as many of the mechanisms and ideas explored by the behaviour based approach naturally incorporate distributed processes. BBAI takes a bottom-up approach to the exploration of highly self sufficient systems and looks at them in the context of the environment and the task that the system has to manage and achieve, see [Brooks86b, Brooks91].

1.2 About This Document

In general terms the project being described in this document will set out to explore the characteristics of design frameworks, architectures and mechanisms that can be used to control agents that operate in the real world. Many of these have emerged from the behaviour based branch of Artificial Intelligence but there are also other techniques such as neural networks and genetic algorithms which may prove to be interesting and useful. These will be dealt with in the future. An important and interesting aspect of the project is that the mechanisms will be examined within a structure other than that of mobile robots which have been used as the basis for most of the experimental work done in the field to date. It is hoped that this will provide an insight as to whether the solutions to the many problems of controlling mobile robots have any relevance when transferred to other real-world situations. The project will particularly look at the nature of systems with differing distribution of processes and agents and differing levels of communication. Also of interest is the effect resulting from different agent/process assignments for example a comparison of functionally distributed agents versus physically distributed ones.

It will be apparent from the previous paragraphs that the ideas and meanings behind the term *distributed system* are diverse and that some form of sub-grouping would be useful, for example DAI, distributed control and distributed computing. However this discussion would probably be extensive and also inconclusive and so in order to put the work of the project into context the next section of this document will introduce some working characterisations which will provide a basis for the work and discussions in the rest of this document. The third section continues the focusing vein by describing the experimental test-bed that is to be used: The Image Automation industrial inspection equipment developed by Sira Ltd. Having set the scene for the project the next section (4) will introduce aspects of control mechanisms that have been developed as a result of AI research and an example which will be used to build a first system.

2. Some Working Characterisations

A distributed system consists of a group or colony of processes or agents which interact and co-operate to achieve an emergent or global functionality of the system. This may be observed of the collective behaviour of an ant colony or the result of the component parts of a television set operating together to receive and present pictures and sound. The working characterisations that are outlined in this section are not intended to be definitive and will certainly be subject to change as the project develops. These terms are explained and set out in order to clarify descriptions that will be used in the project and to introduce some of the ideas and interests that will be a major part of the work. It will also provide an introduction to some of the ideas and philosophy behind the design frameworks of Behaviour Based AI. A useful discussion along these lines can also be found in [Smithers91].

Behaviour Based AI tends towards the study of autonomous agents acting within a real world environment, for example mobile robot vehicles that must move around a room and search for objects without becoming trapped or stuck. The autonomous agent paradigm, especially as practised by the BBAI groups, encompasses a need to understand not only the inner mechanisms of the agent but also the mechanisms of the agent's world, i.e. the processes of the agent's environment that effect the operation of the agent. In the case of BBAI the emphasis is on the resulting behaviour of an agent rather than on the characteristics of its internal mechanism. It is generally accepted that as long as an agent performs reliably and robustly it does not matter what mechanism is generating that behaviour (although for the purposes of design and implementation of mechanisms there is clearly a significant need for appropriate tools and techniques). The important point is that an agent has a suitable behavioural repertoire for its environment.

The fact that BBAI concentrates on the complete system, agent and environment, is that which separates it from the other areas of AI. Traditionally the search for the mechanisms of intelligence has concentrated on the processing of information in the form of symbols, accepting that a complete abstraction from the real world is a necessary characteristic of the logical language frameworks that have been developed. However BBAI practitioners argue that this abstraction from the real world is a fundamental problem and that any system based on this form of symbolic reasoning alone cannot function adequately, let alone reliably and robustly, see [Brooks86a, Brooks86b, Smithers91, Steels91, Pfeifer and Verschure 91]. The search has turned to finding robust and reliable mechanisms rather than mechanisms that are apparently "intelligent" in some way. It is one of the aims of this project to continue along these lines and to examine the use of such mechanisms for engineering equipment that must operate in an industrial environment.

2.1 Agents and Processes

In the AI literature there is not always a clear distinction between an *agent* and a *process*, both are used to indicate some form of mechanism running and interacting concurrently with others. In DAI terms an agent may be an abstract mechanism existing within a software environment (for example [Dean90]) with some form of complexity or self sufficiency measurement differentiating it from a process. However, it may be that a more useful distinction is that an agent is an embodied entity that operates within the real world (for example a mobile robot) which may then consist of a number of internally distributed processes along with various sensors and actuators. The term agent is taken to have definite physical connotations[Smithers91] and [Brooks90]. Smithers refers to an agent as being a coherent organisation or architecture of processes and their dynamic interaction with the world. It may be argued that a process is simply a more abstract term and can quite adequately cover the above description of an agent.

An agent or process can be seen as a self contained unit with a set of inputs, a mechanism and a set of actuatory outputs. These parts are necessary for the elements of the system to function individually. Although the boundaries between these parts may not be clear-cut they may be identified in the members of any distributed system. The following list provides more detail:

Sensors: An agent's input originates from discrete physical sensors or perhaps from its own internal state. What ever its form, the input can be regarded as the state of the agent's world and as such is the complete set of the agent's sensory information. All of the agent's actions are based on this information. Even if some form of memory based knowledge is used it must still be based on the agent's own perception of its world.

Actuators: The agent's output can be regarded as its actuation or its method of effecting its world. Again as for sensors, the actuators may either be physical devices such as motors in a robot, or they may be more abstract data streams connected to the inputs of other agent's which may or may not result in communication. In some circumstances the state of an agent's actuators also provide a useful input to the same agent.

Mechanism: The internal mechanism of an agent translates the sensory input to an appropriate actuatory output. The purpose of the mechanism is to enable the agent to reliably and robustly work towards achieving its current task. The process may incorporate varying degrees of memory which can be used to augment the sensory input with the agent's previous experiences. However, the type of mechanism is not an important issue, it may be clockwork, hardwired electronics, a simple computer program, a symbol based inference system, a neural network or indeed, a lower level set of distributed co-operating processes. The only criterion is that the correct behaviour is realised within the constraints of the agent's task including any possible time limits.

It could be claimed that there is a problem with a 3 way input/mechanism/output split of an agent in that it does not allow for the agent's response to any unexpected external effects due to other agents or erratic events in the environment. Surely, if the agent is suddenly picked up and moved to an other location it should be considered as a valid input influence? However, whilst the influence is certainly a valid one if the agent in question has no ability to sense these effects then there can be no way that it can control its reactions to those events.

2.2 The Agent's World: Situatedness

The agent's world encompasses all external objects and processes that can effect its behaviour. These effects may either be detected by sensors or simply operate on the agent without its knowledge (for example human beings are unaware of radiation exposure until it is too late). The world will consist of other autonomous agent's and all manner of physical processes and static objects. The ecological scenario of a biological agent occupying a niche in its environment is useful here as it emphasises the intricate relationship between an agent and its world. An animal has evolved to suit a particular situation and role in its environment. Both the animal and its environment interact intricately with one another and if separated both will function differently and possibly unsuccessfully. This is usefully encompassed with the concept of situatedness.

An agent is designed (or has evolved) to function within its niche and this includes all aspects from its physiology to its behavioural repertoire. As a result there must be limits to the agent's ability to deal with unusual situations, that is to perform its task robustly and reliably. For example a spider can cope admirably in a hedgerow, spinning webs despite all manner of weather conditions but if confronted with the totally alien environment of a bath tub is as good as useless [Smithers91a]. Human beings are successful by virtue of their extensive physiological and behavioural recourses and their ability to learn from experience but despite this are still fallible. An agent then, is effective only so long as its current situation is within the scope of its resources and as designer's of artificial autonomous agents we try to ensure that the agent is as well equipped as possible.

It is important that the designers of artificial agents are aware of the necessity to *situate* the agent in its world and to ensure that the agent's processes lead to actions that are based on the agent's sensory input and not on the designers own view of the situation which may make unrealistic assumptions on the data coming from the agent's sensors. Therefore the amount of 'pre-programmed' or 'a priori' knowledge used by the agent must be carefully considered. The reasons for giving the agent that information in the first place must be fully understood.

2.3 Multiple Agent and Multiple Process Groups

This section covers a number of characteristics that apply to groups of more than one agent or process and this will be a particularly important aspect of the project being proposed. Much of the work will be covering the nature of the control and communication structures of multiple process/agent groups. In the following discussion the nature of the group element is not important; that is a member of a group may be an agent or a process or any other descriptive term that is useful, the important point is that there is more than one and that they are able to interact.

The behaviour observed of a group is usually the result of the interactions between the group members. It may be said that the group behaviour is an emergent or global behaviour as it is not usually possible to find to find any mechanism or representation in an individual member of the group that directly relates to the global phenomenon. For example the behaviour of a swarm of bees cannot be explained by examining a single bee. Likewise the intricacies of a banks computer system cannot be ascertained by simply looking at the mechanism of an automatic teller machine[2]. One of the main working characterisations of BBAI is that all agents that exist in the real world consist of such collections of interacting processes and mechanisms. This suggests that at whatever level you look at a system it can always i) be seen to consists of a number of interacting processes and ii) may be part of a higher order of interacting processes. Indeed, this characterisation is often used by biologists when describing organisms that range from simple single cell animals to complex mammals.

A distributed system of agents or processes can vary massively. Some colonies may consist of members that are extremely diverse in appearance and highly specialised whilst others may have members that are all identical. Members of colonies may interact in many ways. At one extreme is a group of formally communicating elements, for example [Durfee87], whilst at the other is a group who's members loosely interact without any observable co-operation, see [Steels89] in which the self organisation of agent cooperation is discussed. [Mataric92] also adds light to the discussion by describing research into the behaviour of and resulting from agent interactions.

The following subsections present some characteristics and descriptive terms of group behaviour which cannot be applied to individual elements but which are important aspects that must be taken into account when designing distributed systems.

2.3.1 System Granularity

Granularity is a term used to describe the size and density of atomic elements in an interacting group or distributed system at a certain level of abstraction. A

---

[2]Example due to John Gilby.

colony that consists of a few large agents would have a course granularity whilst and system with a fine granularity would consist of many smaller, perhaps thousands or more, agents. Granularity is useful as a descriptive term when comparing distributed architectures.

2.3.2 Inter Agent Cooperation

Groups or colonies of agents may co-operate (for example termite colonies or the many processor units in a parallel computer system) to perform a task or simply in order to survive. These groups may be made up of identical agents or of many different types of agent which are specialised to perform a particular sub function of a global task. Within a group of agents the overall functionality is not explicitly encoded or designed into each agent (this would probably be prohibitively expensive and is in any case not necessary), rather it is an emergent characteristic resulting from the interactions between the agents in the colony. Therefore each agent is only concerned with its own world, i.e. its sensed input state. The benefits of this type of operation are many, for example there is an increased robustness (of the colony) as the loss of a few individuals is unlikely to be significant and also the design of individual agents is kept relatively simple.

The nature of inter-agent co-operation can be roughly categorised into two main areas, passive and active, and these are closely related to the areas outlined in the following discussion on communication. Active co-operation occurs when the participating agents communicate, possibly involving complex negotiations. In the case of passive co-operation the agents function individually, they have no knowledge of the operation of other agents except that which can be sensed locally from the environment, stored and recalled for future use. Any such co-operation is a side effect resulting from the dynamics of the colony. [Parker93] discusses inter-agent co-operation and the nature of local and global information.

An example of passive co-operation could be a "light switching off" agent that waits for lights to be switched on before acting. The fact that another agent switches the lights on is immaterial, the "light switching off" agent can only sense that a light has appeared in its world and that the state is correct for it to act. There is no explicit communication between the two agents but the effects of their actions serve the same purpose. See [Mein91] for a real world implementation of this and for another example see [Pebody91] in which a single mobile robot is controlled by a set of co-operating processes without any explicit communication.

2.3.3 Inter Agent Communication

Communication is seen as a symbolic interchange of information between agents which have a common understanding of the symbols or tokens being exchanged. For example Morse code is only a useful form of communication when the operators at transmitter and receiver can both understand the dots and dashes, they may in turn translate the symbols into English (or any other language for that matter) and communicate the contents of the message to another person.

Communication can either be unidirectional or bi-directional. For example a radio station only transmits information whilst a telephone provides two way (or more) conversation. These two types of communication may also be described as active or passive. The radio transmission is passive in that a message is sent with out the assurance that there is anyone listening, this is also like putting a message in a bottle and throwing it in the sea. Active communication occurs when both agents are exhibiting communication behaviours and some form of rigid protocol.

Communication must not be confused with co-operation which may or may not involve the direct transfer of information. Co-operation can also be active (communicating) or passive (non communicating). However, both of these phenomena are similar in that they are an emergent behaviour only observable as a result of the interactions of more than one element of a group.

3. A Real World Experimental Test Bed: The Image Automation Inspection Equipment

As suggested in the introduction, in contrast to the work of the academic community there has been a steady evolution of the use of distributed control techniques within the process and manufacturing industries. A useful overview of this appears in articles by [Jones91] who describe commercially available distributed control systems. A useful comparison to academic material is provided. These examples present an interesting real-world environment for research into the interaction of different agents. The systems often consist of separate machines or cells that have been interconnected by a communications link and can thus either communicate between themselves or, as is more often the case at the moment, be sent commands from a central control computer or operator. It is the ability of these machines to take care of their own individual control requirements locally that leads to the reason for calling such systems distributed. The writers indicate that most of these systems are conservative in design approach and it would seem that this is due not to a lack of real world resources (e.g. sensing, processing and actuator hardware) but to a shortage of techniques for controlling large numbers and types of sensor and actuator mechanisms. Perhaps then, this is an area in which the control mechanisms and techniques that are emerging from the field of AI and Robotics may be usefully applied.

Unfortunately the enormous cost of industrial machinery prevents its use for experimentation and so other less expensive systems have to be found. Many AI researchers (especially those involved in behaviour based AI) have turned to the use of small mobile vehicles for an experimental base. These less expensive devices are also less complex but they do retain the real world operation environment that is important to AI and autonomous systems research. Even so, at some stage a transfer from these experimental robots to a more industrial environment would be desirable as this will provide an opportunity for both the industrial and academic worlds to exchange and test new ideas and should lead to a more fertile research and development environment for both groups. This will be a major aspect of the project which will involve working with a real world test bed other than that of mobile robots. The rest of this section overviews the inspection system. A description of its operation environment is given along with an outline and brief functional description of the major constituent parts based on Sira technical documentation [Sira92a] and [Sira92b].

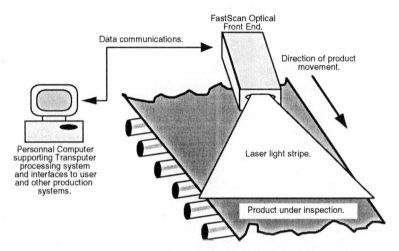

Figure 3.1 The laser scanning inspection system

3.1 The Inspection System and its Operation Environment

Materials such as film, glass and acetate are used in numerous modern industries and tiny flaws occurring during production can be critical and even cause the costly wastage. Many manufacturers currently rely on the human eye to spot bubbles, distortions and scratches. Results can be unreliable, defects missed and product thrown away when it could have been salvaged if the precise nature and location of the fault had been identified. The system presented here is specifically designed to search out and identify faults automatically while production is in progress. This results in a better quality of material and reduced rejection.

The inspection system currently operates by producing and storing a high resolution map of a complete roll or surface of the product as it is being produced, allowing any flaws to be assessed accurately and appropriate action to be taken. Production can then be rapidly corrected to eliminate defects and spoiled sections of finished rolls can be located and removed. If defects are not critical then sub-standard rolls can be downgraded. The system works by focusing and sweeping a laser beam across the continuously moving product surface. Each of the different types of fault (for example scratches bubbles and holes) has a characteristic effect on the laser beam which allows them to be located, identified and classified.

Information from the inspection system can be incorporated into statistical process control mechanisms and kept to satisfy quality control procedures.

There are two main components to the system: i) the optical front-end (which contains the laser light sources and accompanying optics, light detector devices and control electronics) and ii) a personal computer hosted data processing hardware modules. It is the optical front-end that is the target of this research project. The rest of this introduction overviews the PC based part of the inspection system and is followed by more detail on the optical front-end and its constituent parts in section 3.2. Figure 3.1 details the interconnection of these items.

The data processing and user interface part of the system is hosted by an IBM personal computer. The main task of processing the laser scan video and formatting it for presentation to the user is performed in three stages of electronics hardware, the first being specific for each input video signal (dependant on the number of receivers in the optical front-end). The first two stages are implemented on special purpose hardware that has been installed on the PCs input/output expansion bus. The final information formatting stage is performed by a number of Transputers running on an Inmos B008 transputer mother board. The PC provides disc storage facilities and a windows operating environment for the user.

3.2 The Optical Front-End

The optical front-end of the inspection system consists of a laser, a multi-faceted spinning polygon, an optical system, and light sensitive detectors. These items and communication with the PC based part of the system are controlled by various interlinked Transputer microprocessor circuit cards which may in future experiments be considered as agents in their own right as they can be seen to be embodied, albeit within the framework of the optical front-end subsystem.

The optical system focuses the laser beam onto a spinning polygon. The result is a striping laser beam that scans across the material under inspection. The detection of the laser light depends on the material. For example in the case of painted surfaces the laser stripe is angled so as to reflect onto a strip of retro-reflective material, the light then retraces its route back into the system, back through the rotating polygon to where a beam splitting device redirects it to the light sensitive detectors.

Currently the system uses photomultiplier tubes (PMT) to detect the incoming laser beam. These are used as they provide a high order of sensitivity and resolution, however the control of these devices is reasonably complex due to the non-linear response of the PMT to the EHT gain controlling voltage. It is for this reason that a programmable computer is used to provide the control of this device.

3.2.1 Optics Controller

The optics controllers task can be broken down into three main groups:

- Host PC - Front-end communications.
- Laser switching and status.
- Miscellaneous control and monitoring.

The intercommunication between the controllers within the optical front-end and the host PC is via the transputer serial communication link protocol. The optics controller is situated at the head of a chain that consists of it and the photomultiplier tube (PMT) controllers. The transputer link to the PC uses an RS232 serial driver in order to transfer messages over the greater distance, however the protocol remains the same although the data rate is slower. The optics controller routes relevant messages from the host on down the chain to the PMT controllers.

The optics controller can, as long as the relevant switching hardware is installed, control the length of a section of a single scan by switching the laser off. This can be used to reduce the area of product to be scanned and is currently controlled by commands sent from the host.

In addition to the above, the optics controller is also responsible for various miscellaneous functions. In fact this covers all the tasks that are not taken care of by the PMT controllers and includes the monitoring and subsequent reaction to a number of safety interlocks. Other system status such as laser power output is also monitored.

3.2.2 Photomultiplier/EHT Controllers

The photomultiplier tube (PMT) controller is responsible for ensuring that the gain of the PMT is always sufficient to allow the correct resolution of product defects to be detected, or in other words to maintain the required amplitude of the output video signal. To do this an analogue to digital converter samples the video signal emitted from the PMT (this being the optical front-ends primary sensory input) the output of which represents an analogue voltage between 0 and 10 volts. This is used to control the gain of an EHT amplifier which in turn controls the gain of the PMT. However the PMTs gain as a function of the applied EHT is highly non-linear and this is why a programmable controller is necessary to provide the correct feedback for stable control.

The PMT controller operates by sampling and digitising a complete laser scan stripe into 1024 eight bit code words, the resolution is thus a 10 mV ($2^8 = 256$) per bit resolution, a sample is acquired at a rate of 500 KHz. The PMT gain is set to provide a video level at a prespecified percentile of the sampled signal which corresponds to an input digital word of 32h, the input data is used to calculate the variation of the actual signal with the desired state and then generate a 16bit output word that is converted into an analogue 0-10V EHT control signal. This is currently done in two ways depending on the size of the error. In the case of a large error the control output is varied by a count of +/- 4, the next sample will then be used to reassess the situation. During normal operation the error will be minimal and so a function of the log of the calculated input video level is used to adjust the error such that the set point of 50 ADC counts is maintained.

Other tasks of the PMT controller are to monitor the output EHT control level and to ensure that it does not exceed the maximum EHT voltage level allowed for the PMT. This is done by comparing a pre-set value with the current output from the PMT control process. The PMT control function can also be halted (in the current implementation) when an EHTHold signal is sent from the host PC, this can be used to prevent instability during brakes in the material being inspected. Finally the EHT amplifier is disabled and relevant status registered should a number of safety interlocks be triggered.

3.3 An Experimental Test-Bed for Distributed Control

The optical front-end of the inspection equipment is a virtually self contained system, in fact it can be likened to a single agent. It has a number of controllers which are responsible for maintaining the laser scanning mechanism and the light detectors and these are based around a number of interconnected Inmos Transputer microprocessors. The agent aspect of the optical head is evident in the nature of its operation in fact, as previously mentioned, individual parts of the system such as the Optics and PMT controllers may also be regarded as agents as they too have sensors and actuators embedded in the real world. The system has numerous internal and external sensory inputs varying from simple safety cover interlocks to laser power values and the input of digitised video from the light detectors. As the material under inspection moves past the scanner this information is used to maintain the desired characteristics of the light detectors, which have a highly non-linear gain function. This is to ensure that the system outputs a clean analogue video signal to the host computer. This task becomes increasingly difficult as the product under examination becomes more complex.

The inspection system is currently operational with its controllers programmed in a more traditional manner than that proposed for the project. This will prove to be useful baseline from which to compare the results of developments using the paradigms of BBAI. It is not intended that the work of the proposed project necessarily produce an alternative control structure with increased functionality (although this might be a useful side effect of the work) but rather to simply use the existing hardware technology as a real world test-bed for agent based control architectures. Experiments will be undertaken in two distinct stages. The first will be a re-implementation of the current capabilities using the Subsumption Architecture [Brooks86b] taking into account aspects of system granularity and agent intercommunication. After this is complete and evaluated a number of extensions to the behaviour and robustness of the system will be explored. Further experiments using other mechanisms, for example the Process Description Language [Steels92], are also planned.

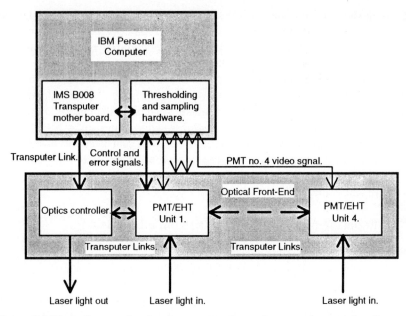

Figure 3.2 Block diagram showing the complete inspection system processing elements

4. Design Experimentation Frameworks and Test-Bed Control

This project will research into the control of systems that operate in the real world with particular emphasis on the mechanisms and methods that are being developed in AI. The fact that such mechanisms must be embedded or situated in the environment within which they are to operate has already been discussed and for these reasons it has been decided to use a test-bed based on the Image Automation industrial inspection system as an alternative to the more standard mobile robots that are used for most of the academic work. It is hoped that this will enable both a transfer of useful ideas to industry and be of benefit to research into the nature of interactions between distributed autonomous agents. The test-bed will consist of a number of standard components mounted on a special open frame which will allow access to the working parts, as the project progresses it may become desirable to mount additional sensors and other hardware. To provide a platform for the development of the transputer software and to control the optical front-end an IBM compatible PC will be used in conjunction with additional transputer based hardware and software.

Various design frameworks, control strategies, methodologies, architectures and other structures that come under many other collective names have been proposed as solutions to the problem of controlling an autonomous agent in a real world environment. Part of the scope of this project will be to compare and contrast the nature and applicability of these mechanisms for use on the test-bed inspection system. This can be taken regarded as an agent with a different environment and task set than the normal AI mobile robot but which has a similar sensor/actuator control problem. The other aspect of the project, an investigation into the nature of distributed process/agent interaction and communication, will be carried out concurrently with this work.

In the literature different researchers provide different levels and qualities of detail about their pet programming frameworks or design paradigms, examples of which are the Subsumption Architecture [Brooks86b] and the Process Description Language [Steels91]. It may well turn out that the systems are interchangeable and that the basic principal of bottom-up design is the most significant aspect of a systems success. It is certainly the case that several of these frameworks are interchangeable. For example the Subsumption architecture and its accompanying Behaviour Language has recently been extended to include an implementation of a spreading activation network similar to that of [Maes89] which had previously been programmed in LISP and other frameworks [Pebody91]. Other techniques such as neural nets and genetic algorithms are also of interest as they may provide useful solutions to some problems, for example the interpretation of large arrays of sensory data.

The use of a design framework or methodology for the project is particularly important as it will allow a more effective comparison with the results of work on mobile robots. The use of a design framework will also contribute towards the comparison of different individual versions of the inspection system software. These advantages were both particularly evident in the results of the work described in [Pebody91]. The Subsumption Architecture has been chosen as the first design framework for the development of control programs for two main reasons: i) Use of the Subsumption Architecture and its accompanying Behaviour Language has been reported on for a number of years and there have been many working demonstrations of successful implementations [Brooks90]. ii) The Subsumption Architecture and the Behaviour Language are one of the most fully documented of the design frameworks originating from BBAI. The next section overviews the main characteristics of the Subsumption Architecture and is followed by discussion about the inspection system and its potential for experimentation.

4.1 The Subsumption Architecture

The Subsumption Architecture was amongst the first frameworks to emerge from the AI world that advocated the bottom-up design methodology for intelligent robotics. It has since been continuously under development and a programming language called the Behaviour Language has been designed to support it at a higher level. This work was started by Rodney Brooks of the Massachusetts Institute Of Technology and has lead the way in the up and coming field of BBAI. The work so far has concentrated on the building of autonomous mobile robot agents and so the use of these methods for building the control mechanisms of another type of agent, the inspection system optical head, will be interesting.

The Subsumption Architecture provides a mechanism which enables the real-time control mechanisms of mobile robots to be built up in layers starting with very basic, reactive response and reflexes. Following layers of the system can be added without the need to modify already operational lower ones. One of the main characteristics of the system, one that typifies that of all BBAI research, is the nature of the interaction of the layers with sensors and actuators and thus the real-world. Rather than using a functional approach which leads to a sequential pipe-line mechanism of sensors-sensor processing-map building-action planning-action realisation-actuators (or some such) the Subsumption Architecture results in a massively parallel mechanism with modules which have access to virtually any sensory or internal information. The redundancy that results enhances the systems performance by making it more robust and more able to respond quickly (as a reflex) to dangerous situations.

The mechanism consists of a number of finite state machines that have been augmented with real-time clock information, hence: Augmented Finite State Machines: AFSM. The AFSMs are interconnected via single element buffers and "wires". Information is not stored in the buffer, it simply contains the most recently arrived message, it is up to the AFSM to respond to this in some suitable way as messages can be lost if a new one arrives before the old has been used. The inputs and outputs of a particular AFSM may be inhibited or suppressed by a wire output of another AFSM. Inhibition is applied to a modules output and prevents the modules output from reaching its destination. It is pre-programmed to last for a set time period and must be continuously updated by the causing AFSM if it is to continue its effect. Suppression effects the input to an AFSM. Its effect is similar to that of an inhibition but additionally the suppressing AFSM forces a new value onto the wire. Each AFSM also has a reset default input wire. Figure 4.1 below presents a single AFSM.

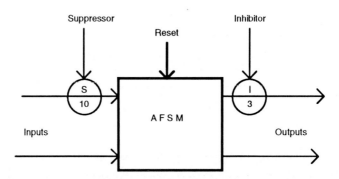

Figure 4.1 An augmented finite state machine and its interconnections

A single layer of a system will consists of several AFSMs and as higher layers are added the lower layers become totally embedded in the system. The higher layers influence the lower layers by inhibition and suppression and also by their effect on the actuators and hence the vehicles situation in the world. More detail on programming agents using the Subsumption Architecture and details of robots that have been built can be found in (amongst others) [Brooks86b], [Brooks90] and [Brooks91].

It is envisaged that the layered AFSM approach of the Subsumption Architecture will adapt well to the distributed processing hardware of the experimental test-bed system. In the same way that each AFSM is an isolated processing machine only connected by its input and output registers, each Transputer process only shares data with other processes via channels. The communication between discrete Transputers and multiple processes running on them would thus appear to be a useful characteristic which can be made use of when implementing the wires of a Subsumption network. Another aspect that supports the potential suitability of the Subsumption Architecture is the fact that the characteristics of enhancing a system by adding further behavioural layers is similar in principal to adding extra Transputer based hardware in order to increase system resources.

4.2 Test-Bed System Control

Section 3 outlined the basic control requirements of the inspection system as they are currently implemented and introduced the system as a stationary agent with its world passing by in front of its sensors. This section will show how these relate to the lower level control layers of a system designed using the Subsumption Architecture and then suggest possible higher levels of implementation that will increase the systems reliability and robustness. The assignment of tasks to the various hardware controller elements of the system is not yet determined, indeed this is one variable that this project sets out to explore.

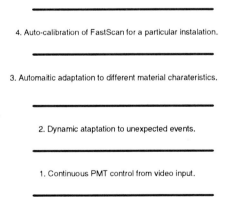

4. Auto-calibration of FastScan for a particular instalation.

3. Automaitic adaptation to different material charateristics.

2. Dynamic ataptation to unexpected events.

1. Continuous PMT control from video input.

Figure 4.2 Suggested control layers for the test-bed system

The operation of the inspection system can be categorised into four main levels based on a number of criterion including the amount of time that the system spends on certain operations, the amount of pre-programming or fixed knowledge that the system must be provided with to usefully function (conversely the degree of autonomy of the system) and thirdly the frequency of the phenomenon that stimulates the characteristic behavioural mode. See figure 4.2.

Given the levels above it is possible to envisage each being implemented with differing degrees of automation dependant on the frequency of the events that characterise the level. Whilst all would agree that it is vital for level 1 to be automatic it becomes increasingly less vital for the higher levels. Indeed, the current system has very little ability to cope with changing circumstances in levels 2, 3, and 4 as these are mostly controlled by fixed, non-dynamic aspects of the system. Ultimately it would be desirable for the complete system to be automated. The following points outline some possible features that would contribute to the automation of these levels.

Levels 2,3: A process could be introduced into the optics controller that would monitor the laser power and provide indications as to reasons for the laser status. For example the laser may be off due to certain safety interlocks being open or it may be outputting low power because it is nearing the end of its working life. Certain aspects may be indicated by the status of other control processes within the system head.

Levels 2,3: A process could be introduced to monitor the life and status of the PMTs in a similar way to that outlined for the lasers above. Automatic PMT shutdown is essential if the EHT voltage becomes excessive.

Levels 1,2: The system should be able to dynamically and automatically cope with changing speeds of inspected material whether in discrete panels or in continuous feed format (this could include changing rolls of material).

Levels 1,2: The system should be able to recognise and adapt to materials that have different characteristics.

All levels: Self test and diagnostics.

All levels: Automatic calibration. In fact this may apply more to levels 3 and 4 as the lower levels could be described as continuously performing this task.

These additions can be made to a system that has been built up using a framework such as the Subsumption Architecture. Figure 4.2 presents a number of control layers which are similar to those used to describe the control mechanism of a mobile robot in [Brooks86b] and which were developed using the bottom-up approach that is a main characteristic of this methodology. There is also scope for further experimentation resulting from the distributed nature of the inspection system. In order to implement extra layers of competence various parts of the system must be able to co-operate. For example there are two PMT controllers each sampling the received laser light and it is most probable that an interaction involving some form of result verification would be necessary in order to facilitate the ability to adapt to different materials without disturbing the scanning process. This introduces both an aspect of robustness and a question of communication as the two PMT controllers should also be able to cope with the possibility that the other may be damaged and sending erroneous information. The other additional competencies of the system that are outlined above would require similar increases in ability for each of the distributed controllers of the inspection system as would the addition of further hardware sensing modules.

This subsection has outlined a few features that could be added to the industrial inspection system and the project, in part, aims to look at the nature of designing systems that can be incrementally enhanced in this way. It will be interesting to see how the methodologies of the Subsumption architecture can be used along with the modularity and distributed nature of the test-bed inspection system for these implementations.

5. Conclusions

This paper has presented an outline for a program of research and experimentation into the control of a multi-agent, multi-process real-world system. It started by attempting to locate the project into the area of distributed systems research by focusing on distributed control and DAI. A number of working characterisations based on the ideas of BBAI were then given followed by an overview of an experimental test-bed that is planned. Finally an example design methodology that is to be used and ideas for extensions to increase the robustness and reliability of the current system were outlined.

Acknowledgements

I would like to thank both John Campbell of UCL and John Gilby of Image Automation Ltd. for the many hours of discussion and help they have given in order to help me define this project. Finally this project is made possible by the cooperation between the Sira Postgraduate Centre and the University College London.

References

Brooks86a: Achieving artificial intelligence through building robots. Rodney A. Brooks. AI Memo 899, Massachusetts Institute Of Technology Artificial Intelligence Laboratory, May 1986

Brooks86b: A robust layered control system for a mobile robot, Rodney A. Brooks, IEEE journal of robotics and automation. Vol. RA-2, No 1, March 1986

Brooks90: Elephants don't play chess. Rodney A. Brooks, Robotics and Autonomous Systems 6 (1990).

Brooks91: Intelligence without reason. AI memo 1293, Massachusetts Institute of Technology AI laboratory

Dean90: Cooperating agents - A database perspective. S. M. Dean. Cooperating Knowledge Based Systems 1990. Edited by S.M. Dean. Springer-Verlag

Durfee87: Cooperation through communication in a distributed problem solving network. Edmund H. Durfee, Victor R. Lesser and Daniel D. Corkill. Research Notes in Artificial Intelligence: Distributed Artificial Intelligence. Ed. Michael N. Huhns. 1987

Hayes-Roth88: A blackboard architecture for control. Barbara Hayes-Roth, Readings In Distributed Artificial Intelligence, 1988. Edited by Alan H. Bond and Les Gasser

Maes89: How to do the right thing. Pattie Maes, Connection Science, Vol. 1, No 3, 1989

Mataric92: Designing emergent behaviours: from local interactions to collective intelligence. Maja J. Mataric. Proceedings From Animals To Animats, Second International Conference On Simulation of Adaptive Behaviour. (SAB92) MIT Press

Mein91: Co-operative behaviour in uniformly and differentially programmed Lego vehicles, Richard Mein, MSc Thesis, Department of Artificial Intelligence, University of Edinburgh

Parker92: Adaptive action selection for cooperative agent teams. Lynne E. Parker. Proceeding of the second International Conference on Simulation of Adaptive Behaviour (SAB-92)

Pfifer and Verschure91: Distributed adaptive control: a paradigm for designing autonomous agents. Vrije Universitiet Brussel AI Memo 91-7

Pebody91: How to make a Lego vehicle do the right thing, MSc Thesis, Department of Artificial Intelligence, University of Edinburgh

Jones91: Distributed process control system survey. Journal: Control & Instrumentation Vol.: 23, 5, 30-71. May 1991 Country of Publication: UK

Sira92a: FastScan head hardware specification. DS00133 000 05. 12 March 1992. Sira R&D

Sira92b: Optical front-end software requirements specification. DS00133 000 03. S.R. Hattersley. 3 September 1992 Sira R&D

Smithers91: Taking eliminative materialism seriously: a methodology for autonomous systems research. Tim Smithers, Proceedings For The First European Conference On Artificial Life 1991

Smithers91a: What is artificial intelligence? A personal view. Tim Smithers. Presented as lecture material for the knowledge representation and inference course, University of Edinburgh, Dept of AI. 1991

Steels89: Cooperation between distributed agents through self-organisation. Luc Steels in Workshop On Multi-Agent Cooperation, Cambridge UK. Pub: North Holland

Seels92: The PDL reference manual. Luc Steels. Vrije Universitiet Brussel AI Memo 92-5. March 1992

Steels93: Building agents out of autonomous behaviours. Luc Steels in: The artificial life route to artificial intelligence: Building Situated Embodied Agents. L Steels and R. Brooks (eds) 1993. New Haven: Lawrence Erlbaum Associates

The Reactive Accompanist: Adaptation and Behavior Decomposition in a Music System

Joanna Bryson

The Artificial Intelligence Laboratory, MIT NE43-825, Cambridge, MA 02139, USA

joanna@ai.mit.edu

Abstract This paper describes a reactive, behavior-based system that mimics a human musical skill — chord accompaniment of unfamiliar melodies. The system was constructed under the subsumption architecture methodology. This paper discusses the design task of behavior decomposition for such a system, and recommends a strategy of modularizing the minimal adaptive requirements for the desired competence.

Keywords Music, subsumption architecture, behavior decomposition, adaptation, learning, neural networks

1 Introduction

Reactive, behavior-based systems have become widely used and accepted within some AI communities. A behavior-based system is one where control has been decomposed into a number of simpler and relatively autonomous *actions* that run concurrently. The intended overall behavior emerges from the continuous interaction of these components. In contrast, the traditional approach for software design is to decompose control into general *functions* and then construct the overall by stringing sequences of these behavior-segments together. If a behavior-based system is reactive, then each component behavior's output is a reflexive consequence of its inputs — there is no hidden complex computation or shared global state.

Although this new paradigm has its origins in general studies of mind and communication (see (Minsky 1986) or (Agre & Chapman 1988)) it has been most widely used and popularized as a method for controlling mobile robots. Where robots operating under the traditional control sequences had found the problem of moving across a dynamic free space almost intractable, robots using the new strategy can operate in such environments at speeds comparable to animals, using minimal computation (Brooks 1991 a). This success has been duplicated in the area of assembly robotics, where behavior-based control has been used to address the problems of sensor calibration and motion control, greatly simplifying the task of central assembly planning (Malcolm & Smithers 1990).

So far, however, the range of tasks done by fully reactive, behavior-based systems has been fairly limited, mostly to types of navigation and exploration. While the followers of the new practice are claiming to be establishing a new approach to artificial intelligence, skeptics have doubted whether "truly human" cognition can be achieved (Kirsh 1991). This question actually has at least two parts:

1. Can a reactive, behavior-based system perform tasks we consider to be acts of "higher cognition"?
2. If so, how do we know what constituent behaviors to create in order for an overall competence in these tasks to "emerge?"

As a beginning of an answer to both of these questions, this paper describes a recent research project, the Reactive Accompanist. The accompanist was developed under an approach called *subsumption architecture* as outlined in (Brooks 1991 b). Previously, subsumption architecture had been used exclusively for small, autonomous mobile robots — that is, robots with all their computation and power requirements carried on board. The RA carries the principles of this approach into a new domain — real-time musical expertise.

2   The Reactive Accompanist

The goal of the Reactive Accompanist is to derive chord structure from a melody in real time. This mimics the ability of skilled musicians to accompany unfamiliar melodies on chord instruments, such as the piano or the guitar. The input melodies are played live on traditional instruments, and the processing is totally reactive, with no transcription of the melody nor rules of harmonization applied. This research is most thoroughly described in (Bryson 1992). It was conducted as part of a Masters project for the Department of Artificial Intelligence at the University of Edinburgh in the summer of 1992. The Accompanist was written primarily in GNU C++ and implemented on a SUN SPARCstation SLC, using a the standard SUN sound equipment for audio input and output.

2.1   High Level Structure

The behaviors of a subsumption architecture program are organized into groups or *layers*. The layers each represent a level of competency. They are designed and implemented sequentially, beginning with the most basic. In the RA, we start with Pitch, which interprets microphone input as a pitch in a predetermined scale.[1] The next layer is Chord recognition, which transforms the output from pitches into chords. The following level is Time recognition.

---

[1]The RA was designed for use in any harmonic system, with the number of potential pitch classes set as a compile-time constant. However, it was only tested on the standard Western 12-tone chromatic scale.

Although the basic competence of this level is to recognize and anticipate the beats of its input melody, the output of the program at this level is not just rhythmic information, but the modified output of the chord layer. Now instead of a single chord, it outputs chords in a timed sequence.

SONG
STRUCTURE
TIME
CHORD
PITCH

Figure 1: Basic layers of a complete machine. Bold indicates implementation.

A proposed extension to the Accompanist includes a next layer that tries to recognize Structure in the piece. That is, it would bias chord choice towards recurring patterns[2]. The final level of competence would be Song recognition, whereby the program would learn to recognize song types. This would subsume Structure in its ability to anticipate the number of times a pattern would repeat, and what the next pattern is likely to be. Although these levels have not yet been broached, they are included here as further examples of high level decomposition.

2.2 Pitch

Pitch really consists of two behaviors, but one of them is currently external to the rest of the Accompanist software. That behavior is to translate microphone input into a set of frequency/gain pairs. Currently this is done via fast Fourier transformation using csound (Vercoe 1991). Unfortunately this relatively slow process moves the current implementation of the Accompanist out of real time, and necessitates the use of prerecorded SUN sound files.

The first behavior of the Accompanist proper is NOTE. This module is an artificial neural network. More specifically, it is a simple, fully connected, two-dimensional grid competitive net. The input lines correspond to ranges of frequencies, and the output nodes corresponding to the pitch classes. In order to bias the network to interpreting sound similarly to people, the ranges of the input lines are scaled logarithmically so that there are the same number of input lines between any two notes humans perceive as being octaves (that is, where the lower frequency is half the higher one). We chose on trial and error to have 96 frequency input lines per octave — we chose a low number for computational efficiency, but too few input lines lower the accuracy.

For our trials, this net was trained using supervision (a training file with known "correct" results). The training file consisted of slightly over four

---

[2]Rhythmic pattern bias is already implemented in the Time layer by the TIMED module.

minutes of a mandolin being plucked at the first twelve fret positions of each pair of strings. Which of these forty-eight intervals would be used was chosen at random, then the precise frame was chosen randomly from that interval. Note the frame could then consist of plucking noise, string noise, attack or any part of the note decay. Peak recognition tended to be about 86% correct, a figure that could partly reflect the noise of the file. The net trained up to 66% accuracy within 80 samples, and peaked by 200. After training, errors made often reflected dominant harmonics of a note.

If the NOTE net is trained without supervision, that is, the winning output node is reinforced regardless of accuracy, it will sometimes learn to distinguish multiple "pitches". These reflect actual pitch classes, but usually clustered unpredictably (neither necessarily harmonically nor sequentially) on just a few of the outputs. The fact that unsupervised training can reflect pitch rather than some other factor (such as plucking noise) probably indicates the utility of the initial frequency clustering bias.

The network trained on the mandolin file generalized well to penny whistle, Scottish small pipes, and four strings of an electric guitar. It failed on male and female human voices, a wood recorder, and the two (non-consecutive) other strings of the guitar.[3] The output of the network was taken to be the weights of *all* the output units, though in the training session only the winner was attended to. Using the full output array resulted in the Accompanist being able to exploit simple polyphonic (more than one note simultaneously) input for better chord choice.

2.3 Chord

The second layer is composed of only a single behavior, CHORD which is also a neural network. This is also a two-dimensional grid network, its design a hybrid of conventional associative and competitive networks. The input nodes correspond to each pitch class, with the input typically being an output or a summation of outputs from the NOTE module. The output nodes correspond to absolute chords, such as F minor or C major. The weights for each output node are referred to as a "chord template," and are algorithmically determined based on the notes actually present in the conventional definition of that chord. The output of any chord template that has received a single instance of a perfect match is 1, its output if it receives all pitch classes weighted evenly is 0. As with NOTE an array of the output values is preserved.

As an interesting incidental behavior, if an entire song is presented to the CHORD module as a single summed input, it will generally report the key of the song. This is a natural consequence of the dominance of the notes of the chord corresponding to that key in the song overall.

---

[3] We failed to find a pattern to what instruments this net could classify. All inputs were checked with conventional tuning devices. Microphone response may have been a factor.

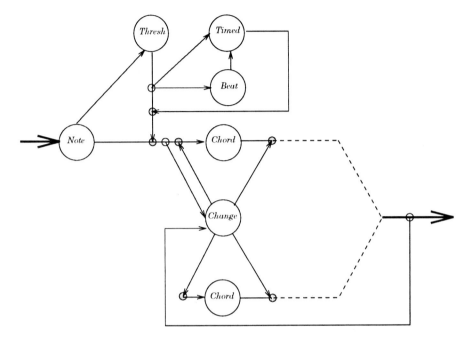

Figure 2: The interactions of the components of the Reactive Accompanist—circles represent modules, lines the flow of information

2.4 Time

The modules of the time layer all operate at the rate of the frame sampling of the Fourier transformation.[4] Note that were the Accompanist running in real time, it might be necessary to drop some frames to save processing time. All of the behaviors are by design robust with respect to this loss, providing the frame drops are distributed evenly (not clustered over harmonically significant events) and the rhythm modules are either aware of the *actual* frame number or else the frequency of frame droppings are consistent throughout a piece.

The time layer consists of four modules. The first three are responsible for determining the location of the beats within the melody. THRESH looks for the threshold of a note by noticing if a single pitch classification persists after a change. It consists of just a few register variables — the pitch and

---

[4] The experiments sampled with a frame size of 256 milliseconds and a frequency of 128 milliseconds.

the frame count. BEAT recognizes beats and pulses in the music using a strategy inspired by reported incidental results in (Desain & Honing 1992). It consists primarily of a one-dimensional neural network (just a vector array) with each (indexed) weight associated with a number of frames. It notes the intervals between positive results from THRESH and sums them (in a normal curve which includes neighboring weights) onto its weight array at that interval. Previous weights are decayed by the same amount as the new interval introduces; the overall sum of the weights is constant. Consequently over time the peak weight/index values can shift if the piece is changing tempo. Notice peaks will emerge at each common length for a note, thus pulses, beats and even measures may be found in the same array. TIMED outputs a true value at what it expects to be the threshold of a beat. It reacts to the peaks of BEAT in determining its interval and THRESH in determining its timing. Like THRESH it consists just of a few counters and register variables.

The other module of Time, CHANGE, is what introduces the possibility of chord change into the overall output of the accompanist. CHANGE maintains two competing copies of the CHORD behavior — feeding them both the input from NOTE. The output of whichever copy seems to be dominating is reported as the output of the whole module. The challenging CHORD could become dominant if the relative weight of its winning chord to the number of inputs seen is increasing while the dominant chord's is decreasing.[5] On a signal from TIMED of a new beat, CHANGE resets the value of the challenging chord. This makes a new chord more likely to emerge quickly, since it didn't need to compete with existing weights. (Chord changes are most likely to occur on the beat.) The correct new chord was found more quickly and accurately if the challenging chord was "primed" with weights that reflected seeing the current dominant (best guess) chord for a fraction of a time interval. This utility of context is due to the nature of traditional music — a following chord is likely to contain several of the notes of the previous one.

2.5   Results

On an input song on one of its preferred instruments, the Reactive Accompanist makes acceptable chord choices, on average, about 60% of the time. For simple songs or polyphonic input, performance can go up over 75%. Testing was somewhat arbitrary and qualitative, but was performed by several trained musicians. On examining the transcripts, the RA was found to make some of the same errors as beginning harmony students. It also had the entertaining if pathological behavior of refusing to change chords when accompanying the smallpipes, even when the drones [the continuous, single-tone pipes] were removed. This is probably a consequence of acoustic properties of the chanter. The accompanist did pick the correct drone chord.

---

[5] An alternate, simpler strategy of straight relative chord strength was found equivalent 80% of the time, but generally less accurate where there were discrepancies.

There are few comparable systems to the Reactive Accompanist — most intelligent systems for musical accompaniment follow scores. Those that improvise tend to take rhythmic and harmonic information as input and produce melodic lines. Nearly all systems work with MIDI (Musical Instrument Digital Interface) input rather than microphones. The RA was developed in just three and a half months, has unbiased access to 24 chords (major and minor of all 12 pitch classes), and produces output that people perceive as intelligent and interesting.

3 Adaptation and Behavior Decomposition

Having presented an already functional decomposition, the question becomes "how did you know to do it that way?". The answer for this project is largely the same as it would be for any of the traditional systems we construct — thought, intuition, experience in programming [6], and a certain amount of trial and error. However, we can trace some of the fundamental decisions involved.

The first step was to divide the task into Brooks' "levels of competence" (Brooks 1991 b). It is widely accepted that pitch and time are the fundamental components of music, but the decision to concentrate first on pitch was more arbitrary. The primary motivation there was that the ability to recognize pitch and harmony seemed more fundamental and easily separable from the overall task than the ability to recognize rhythmic events.

What is probably more interesting is the nature and number of the behavior modules within the levels. In the RA, why are some of the modules neural networks, while others are not? Why are there several different neural networks, for example the chord and note competences, instead of one large general network? Again, the original reason can be put down to intuition. But we suspect the reason this approach proved successful is because it divides the problem along the lines of when and how behavior needed to be adapted. Some information, such as what nodes compose a chord, is completely external to the project, and thus behaviors dependent on that information need never adapt. Other information, such as what notes are being played *right now*, is completely dynamic and must be learned in real time. Before a more complete discussion of these divisions in the Accompanist, we first define these different forms of learning.

---

[6] Although this was my first experience in subsumption architecture, my knowledge of traditional programming algorithms certainly gave me some guidance as to what could and could not be expected to be achieved in translating a set of inputs into a desired set of outputs.

3.1   Three Types of Learning

First, it must be acknowledged that the use of the term "learning" is dangerous, since many researchers associate the word with specific kinds of processes. Within the context of this paper, *learning* means any process which equips an agent with a new behavior which it can perform autonomously. This definition is deliberately broad enough to include such things as hardware design.

Under this definition, forms of learning can be broken into three classes:

- provided learning,
- required learning, and
- open learning.

Provided learning is the behaviors determined without any action by the intelligent agent itself. For animals, this includes the bodies they are born with, their reflexes, and some extent of what might be called "instinctive behavior". For robots, this is their physical hardware, and any software behaviors that have no variable storage content, that is, that are not influenced by the agent's own previous experience but only by its direct current environment. In the RA, an example is the chord templates.

Required learning is domain-specific, preordained learning performed by the individual agent that is required for its normal "adult" behavior. Examples from biology include things like the learning of stellar maps and songs in birds, or bees learning the ephemeris of the sun for navigation. In robotics, examples include self-calibration of sensors or preliminary training of recognition networks, like the Accompanist's NOTE net. This kind of learning can often be thought of in terms of parameter adjustment. Often it occurs in animals during specific periods of their early development, while in robots it might occur before the the robot is considered fully functional. Some "recalibration" sorts of required learning persist through the whole lifetime of the agent.

Open learning is the closest equivalent to the traditional definition of learning — that of a general associative mechanism that allows us to classify and respond to all our surroundings. It is freer and more general than required learning, and is likely to occur over the entire lifetime of the agent. Examples of this are basic conditioning in animals and the ability to "compile" action sequences into expert skills in humans.

Open learning is still not necessarily the kind of complete, general, logical learning long proposed as the whole story by traditional psychology. Not only does open learning coexist with, and in most cases play a subordinate role to, provided or required learning, but it may well be constrained to specific domains. For an example from classical conditioning, pigeons can learn to peck for food, but cannot learn to peck to avoid a shock. They can, however, learn to flap their wings to avoid a shock (Hineline & Rachlin 1969). This research is cited in (Gallistel, Brown, Carey, Gelman & Keil 1991) as evidence that most dynamic learning in animals tends to be specialized and constrained, *not* generalized. Gallistel et al. (1991) in fact suggests that nature only uses individual adaptation as a last resort, for behaviors that cannot be provided on an evolutionary time scale. For example, the stellar maps that some migratory birds imprint on in the nest change too fast for natural selection to keep up.

We introduce these classifications of learning largely to understand and to reference the distinctions in animal learning brought out in (Gallistel et al. 1991)'s survey. Particularly in biology, the border-lines between classifications may be unclear in some cases; for example, the physical growth of the organism. In an artifact, however, these distinctions are clear because one can easily distinguish the mechanisms that underlie them. Provided learning will never change over the artifact's lifetime. Required learning adjusts specific parameters in narrow, well defined ways, using special purpose mechanisms. Open learning uses general techniques to add new skills, new reactions. Unfortunately, the organic ambiguity is reintroduced in the engineers' decision of which behaviors should fall into which class of learning. Our recommendation is to follow the hypothesis of Gallistel et al. (1991), that learning should be no more general than necessary.

3.2 The Reactive Accompanist in Terms of Modularized Adaptivity

The Reactive Accompanist needs to adapt to environmental factors along many axes. For example:

- The number and intervals of *pitch classes* in a scale.
- The number and description of *chord templates* in a harmonic system.
- The *frequency signature* a particular instrument makes in creating a tone.
- The *note* that is currently being played.
- The occurrence of a *pitch change*.
- The *tempo* at which the piece is currently being played.
- The number of different notes that might be played within a given beat (*pulses*).

Nothing in the Accompanist actually reacts *to* a chord or a chord change in the melody. Rather, the *apparent interpretation* of a melody into a sequence of chords emerges from the behavior of the constituent parts.

Pitch classes and chord templates don't vary at all within the scope of an execution for the Accompanist, even if that execution were to encompass several songs. Thus they are provided learning. The frequency signature might also be, though adapting to different instruments would be useful since there is not complete generalization. The signature learning is required instead of provided more because of its complexity – it is not practical to provide it in a reasonable time frame.

The remaining variables must be adapted to ("learned" as defined by (Gallistel et al. 1991)) in real time. They are each handled in a highly specialized provided learning mechanism segregated into a single behavior. Note recognition takes place in NOTE, Pitch change in THRESH, tempo and pulses in BEAT.

Other behaviors that are not dealing directly with adaptation to external events are dealing with integrating the results of other modules into useful form. CHORD connects note and the provided chord templates, TIMED integrates beat and thresh for CHANGE, and CHANGE integrates the rhythm information with CHORD.

Notice the Reactive Accompanist has no open learning mechanisms at all. Few subsumption-architecture projects do, but this is not out of necessity. Mataric (1990) uses a highly specialized open learning module for map learning in navigation.[7] Open learning could have obvious uses in higher levels of control programs designed to learn experientially. Probably the reason it has been avoided thus far is that open learning is normally conceived of only in terms of standard "general learning" with its requisite symbolic-logic structure. Implementing such a system would violate one of the basic tenets of subsumption architecture — no representation (Brooks 1991 b). The positive result of this restriction has been very fast, robust behavior in the reactive, behavior-based systems. The negative side has been complaints that the systems don't do enough "interesting" things. This complaint is not only addressed in the RA, but also in other current research (see below). Regardless, the subsumption architecture approach to control has proven naturally conducive to following the strategy of specializing learning as much as possible.

4 Current Research

There is much more work to be done on the Reactive Accompanist. As it stands, the RA has little sense of context on the level of musical phrase, let alone song. Consequently, it can make what to humans seem like obvious mistakes, such as failing to chose the appropriate final chord to match the single held tone that typically ends a phrase of folk music. Also, from a technological standpoint, there are several new strategies available for conducting the frequency analysis necessary for the input to the initial network that could move the project truly into real time. Hopefully these expansions will be made within the next few years.

Currently we are involved in a more ambitious project of human-level cognition being conducted at MIT. This research, referred to as "the humanoid project" is a major large-team effort to replicate many aspects of human behavior. In concentrating on human infants as a model, the humanoid project

---

[7](Bryson 1992) suggests using a similar system to enable higher levels of the Reactive Accompanist to "map" songs and musical phrases.

is provided with a well-researched natural decomposition and ordering of levels of competence. We will be testing many models of learning and adaptation on this project, but acquisition of musical skills is still some time off in the future.

5 Conclusions

Both reactive and behavior-based models of cognition are still very young, even for a field like artificial intelligence. Nevertheless, they are proving fertile, not only in providing engineering solutions, but also to research in ethology and philosophy in mind. Projects like the Reactive Accompanist can help us explore both the new techniques of engineering and the potentials and limitations of the new cognitive models.

In the Reactive Accompanist, we have demonstrated an implementation of a human competence well outside the conventional domain of strict physical reflexive behavior some have suggested is the limit of the utility of subsumption-architecture styled systems. The Accompanist shows a musical competence human listeners consider to sound intelligent, if not particularly gifted.

The Reactive Accompanist also serves as a new example of behavior decomposition under subsumption architecture. This paper has introduced a dichotomy of behavior acquisition — into provided, required, and open learning — as a framework for analyzing tasks. We support the strategy of limiting and specializing adaptability as much as possible. We suggest compartmentalizing the different learning competencies as both a development methodology for and one of the strengths of behavior-based systems.

Acknowledgments

Several people have urged me to write a paper on the decomposition problem, most notably my supervisors on the Reactive Accompanist project, Alan Smaill and Geraint Wiggins, and also Morten Christiansen who suggested some of the direction of this paper. The learning and ethology research was conducted at the suggestion of Lynn Stein and Rodney Brooks. Thanks to Bill Smart for his contributions and proofreading. Besides myself, musicians for the Reactive Accompanist were Manuel Trucco, Sheila Tuli, Colin Macnee and Helen Lowe.

References

Agre, P. E. & Chapman, D. (1988), What are plans for?, AI memo 1050, MIT, Cambridge, MA

Brooks, R. A. (1991 a), Intelligence without reason, in Proceedings of the 1991 International Joint Conference on Artificial Intelligence, pp. 569–595

Brooks, R. A. (1991 b), Intelligence without representation, Artificial Intelligence 47(1–3), 139–160

Bryson, J. (1992), The subsumption strategy development of a music modelling system, Master's thesis, Department of Artificial Intelligence, University of Edinburgh

Desain, P. & Honing, H. (1992), Music, Mind and Machine: Studies in Computer Music, Music Cognition, and Artificial Intelligence, Thesis Publishers, Amsterdam

Gallistel, C., Brown, A. L., Carey, S., Gelman, R. & Keil, F. C. (1991), Lessons from animal learning for the study of cognitive development, in S. Carey & R. Gelman (eds.) The Epigenesis of Mind, Lawrence Erlbaum, Hillsdale, NJ

Hineline, P. & Rachlin, H. (1969), Escape and avoidance of shock by pigeons pecking a key, Journal of Experimental Analysis of Behavior 12, 533–538

Kirsh, D. (1991), Today the earwig, tomorrow man?, Artificial Intelligence 47, 161–184

Malcolm, C. & Smithers, T. (1990), Symbol grounding via a hybrid architecture in an autonomous assembly system, in P. Maes (ed.) Designing Autonomous Agents: Theory and Practice from Biology to Engineering and Back, MIT Press, Cambridge, MA, pp. 123–144

Mataric, M. J. (1990), A distributed model for mobile robot environment-learning and navigation, AI memo 1228, MIT, Cambridge, MA

Minsky, M. (1986), The Society of Mind, Simon and Schuster, New York, NY

Vercoe, B. (1991), The Csound Reference Manual, Cambridge, MA

Behavior-Based Architecture with Distributed Selection

Luís Correia[*] and A. Steiger-Garção[**]

[*] Computer Science Department; [**] Electrotechnical Engineering Department
Universidade Nova de Lisboa - FCT, Quinta da Torre, 2825 Monte de Caparica, Portugal

Abstract. A behavior based architecture is presented in which behaviors are sparsely interconnected, using an arbitration structure which solves the problem of behavior selection in a distributed way. The arbitration structure is constituted by sets of blockers, each blocker resolving conflicts between two behaviors. There will be a blocker set for each actuator in order to allow independent actions from non interfering behaviors. The behavior model is also described comprising two sub-modules, one for the action and another for the priority output of the behavior. The user when designing a particular vehicle must also define, for each behavior, a fixed priority input plus a fatigue and a recovery times that complete the behavior characterization. A case study is depicted with analysis of the vehicle global behavior for different parameter values. The implications of the architecture model for the control of autonomous vehicles are discussed and future developments are advanced.

1   Introduction

Autonomous Vehicles (AVs) can be considered as artificial beings in the sense that they should be adapted to evolve in a certain type of environment and there they must be able to handle "any" particular problems encountered during their lives (of course occasionally they will fail to do so, as it happens also with natural beings). Furthermore the control architecture of AVs can be composed of a set of behaviors which compete among them to be selected, depending on the external environment perceived by the vehicle and on its internal status. This parallel between artificial and natural beings has been considered in several research works ([1-4]) and is also followed in this paper.

In addition to the behavior composition of the architecture, our approach is oriented towards a completely distributed behavior selection, thus avoiding any centralized module that would be essential for the operation of the vehicle. Also, our proposal allows for easy reconfiguration of an architecture, by a mere change of a few behaviors' parameters.

Behavior hierarchies are believed to exist in animals - allowing coordination, for problem solving, among lower behaviors - although it supposedly does not end in a unique top layer. Moreover behaviors at different levels have always some degree of interaction and deal with most situations among themselves ([5-7]) which is also the case in our model.

The model of the architecture, in particular the constitution of arbitration structures, is described in the next section, with the remark that we are restricting it here to a single layer (an hierarchical extension may be found in [8]). Section 4 depicts the behavior model that underlies the whole architecture. Section 4 presents a simple case study using these concepts and in section 5 some conclusions are drawn and a few aspects of future work are introduced.

2   Architecture

Taking into account the goals and restrictions desired for the control architecture of an AV - namely an emergent behavior solution, based on the interaction of independent behavior modules, with distributed decision regarding behavior selection and flexibility to endure easy design changes in the implementation of specific vehicles - we propose a general model in which modules direct interaction (inhibition, stimulation, or communication in general) is reduced to a minimum as an orientation guide-line. In fact, presently, in our model their only interaction is indirect through the world, using a common structure *(arbitration structure)* responsible for the resolution of the behavior selection problem. Our aim is to minimize direct connections between behaviors - without hampering possibilities of overall performance. We feel that by unrestricting direct behavior interconnections the resulting architecture tends to present a very intricate structure which will condition the design of a vehicle.

These are only orientation principles. It is possible that there will be need for one behavior to feed another with some kind of input (see for example [9]). Although these will be considered exceptions and limited to the strictly necessary, the model should also be able to support them. On the other hand the use of internal sensors, for example, can be of interest in several cases as an alternative to direct behavior connections (see section 4).

With an arbitration structure and a convenient model of the behaviors (presented in the next section) it is possible to debug and adjust each behavior separately and then to integrate the behaviors needed for a particular vehicle and tune the global architecture - by varying the relative priorities and the other parameters of the different modules - hopefully with no need to change the configuration previously found for every particular behavior.

A behavior based architecture for the control of an Autonomous Vehicle produces a global behavior as a result of the interaction of its constituents. Thus it is important for each one to act no more nor less than it needs to accomplish its specific design. It must be able to control whatever it needs, but it must not over-exert its influence since it may hinder other behaviors which would otherwise be able to express themselves. In this way we propose a decomposition of the arbiter into multiple structures (blocker structures), one for each independent actuator of the AV (see fig.1) - e.g. in a car-like AV there will be two arbitration structures, one for propulsion and another for steering[1].

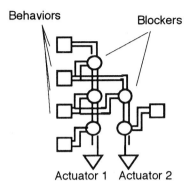

Fig. 1. Example of a control architecture with two blocker structures

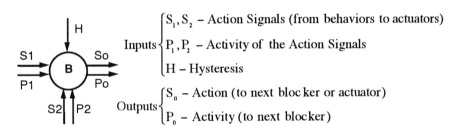

S_1, S_2 – Action Signals (from behaviors to actuators)

Inputs $\{$ P_1, P_2 – Activity of the Action Signals

H – Hysteresis

S_0 – Action (to next blocker or actuator)

Outputs $\{$ P_0 – Activity (to next blocker)

Fig. 2. Detailed representation of a blocker

The idea is to have each behavior using only the actuators needed for its demands freeing the other actuators which may then be controlled by *orthogonal* behaviors. By this we mean behaviors that produce independent activities, thus not interfering with one another. For instance the depth control in an Autonomous Underwater Vehicle (AUV) can usually be performed independently of the motion in the horizontal plane in many situations, and a manipulator can be moved while a vehicle is in motion (in this case, much like in a stopped situation, the free workspace must be considered).

---

[1] This approach can also be followed in more complex actuation structures such as manipulators with several degrees of freedom if we consider that behaviors interact with a dynamic controller (or executor [9]) which decomposes movements along some referential producing multiple independent virtual actuators (e.g. x, y, z and θ, φ, ψ).

Also, from our point of view, a behavior module when controlling some actuator should not have its action changed by the action of other behavior, unless the later is more prioritary and in that case it will impose its own action. This means that the solution for the arbitration structure should not perform combination of behavior actions.

The constitution of the arbitration structure is completely based on blockers, represented (in a simplified form - without the hysteresis input) as circles in fig.1. Each blocker (see fig.2) solves competition among two behaviors with the following output functions transitions:

$$S_0 = \begin{cases} \to 0 & \text{if} & |P_1 - P_2| \le H & \text{and} & \text{sgn}(S_1 S_2) \ne 1 \\ * \to S_2 & \text{if} & P_2 > P_1 + H \\ * \to S_1 & \text{if} & P_1 > P_2 + H & * \equiv \text{any output} \end{cases} \tag{2.1}$$

$$P_0 = \begin{cases} P_1 \to P_2 & \text{if} & P_2 > P_1 + H \\ P_2 \to P_1 & \text{if} & P_1 > P_2 + H \end{cases} \tag{2.2}$$

The function sgn(x) in equation (2.1) is the signal function [10]:

$$\text{sgn}(x) = \begin{cases} 1 & \text{if} & x > 0 \\ 0 & \text{if} & x = 0 \\ -1 & \text{if} & x < 0 \end{cases} \tag{2.3}$$

Due to the hysteresis in the blockers the output functions' transitions are a more understandable representation than the output functions themselves - alternatively the blockers can also be characterized as simple state machines with the above output transitions.

As it was already stated a blocker has the ability of choosing, between two behaviors, the one with higher priority[2]. In doing this it passes on to the next blocker both the priority and the activity signal of the "wining" behavior. These inputs have to stand another comparison with the inputs from a third behavior and this process goes on till the last blocker of the corresponding blocker structure, which outputs the action signal directly to the actuator

Hence behavior modules will have each different actuation output connected to a distinct blocker structure (see fig.1). There will be n_i-1 connections in each blocker structure, n_i being the number of behaviors using the corresponding actuator. This may amount to a large number of connections. However these connections can be computed automatically from the list of behaviors used in a particular architecture. Since the former are all of the same type the problem is

---

[2] This is not strictly true since the effect of hysteresis has to be considered. But for this reasoning we can allow the simplification.

rather simplified. And in spite of the sequential character of the arbitration structure this is a completely distributed way of behavior selection[3].

The hysteresis input serves to prevent the repeated commutation between two behaviors presenting a similar priority, and thus it contributes to stabilize the vehicle's global behavior. Without this input in the blockers the vehicle could often get into situations of repeated behavior selection, due to small fluctuations in the sensors readings[4].

Another variation of this same problem prone to occur in autonomous vehicles based on the interaction of modular behaviors is the oscillation of the AV between two conflicting behaviors ([11, 12]). In animals this problem is solved mostly with the mutual cancellation of the two behaviors ([13, 14]), giving rise to the display of what ethologists call *displacement behaviors* (in [11] a solution for an animat which displays this type of behavior is also presented). In our model of the blocker (fig.2 and equations 2.1 and 2.2) the activity signal output is set to zero when two behaviors having a similar priority try to control the same actuator in opposite ways - here similar meaning that the two priorities differ less than the value of hysteresis[5]. Thus blockers force a cancellation of equal priority conflicting behaviors (see footnote 2) solving the oscillation problem.

We discussed this as if the two conflicting behaviors were physically connected to the same blocker but the reasoning is general since the blockers let go through the most prioritary of the inputs (see also footnote 2). Therefore whenever the two most prioritary behaviors in a blocker structure are conflicting their outputs will eventually be resolved at one unique blocker.

We are assuming that the activity outputs of the behaviors have symmetric signals in case they are conflicting (positive or negative activation of the actuator). Such an assumption is acceptable if the use of one arbitration structure for each actuator is kept. Notice that the priority output in this case is not set to zero in order to still block other behaviors less prioritary than these two. Conversely, when behaviors with similar priorities but not opposing in their actions compete at a blocker the action output is kept as well as the priority since the behaviors are competing but they are not in conflict regarding the motion of that actuator.

Another consequence of both this solution and the assignment of a blocker structure for each actuator is that, in case of behavior conflict in one of the blocker

---

[3] It is very much like a competition access method of a ring local network, though the topology is not exactly that of a ring (rather the one of an open loop).

[4] From a psychology point of view we can say that the hysteresis calms down the vehicle which otherwise would be too nervous! And from a control perspective this can be considered merely as a low-pass filter!

[5] The existence of hysteresis (H) is not enough. Suppose two conflicting behaviors, one trying to approach an object and the other trying to move away from it. Without the annulment of the activity, one of them would act until the priority of the second would rise H above the priority of the first. Then the second one would undo the action of the first behavior until the priority of the first got higher again. And this would repeat indefinitely in case the external stimulus were maintained.

structures, the vehicle may still be enabled to move actuators under other blocker structures. Sometimes this is as much as necessary to remove the conflict situation (another solution for this problem is presented in next section).

The structure thus defined combined with a suitable behavior model allows for a high flexibility in the definition of the architecture for a particular AV. The fact that priorities are separated from the actions of the behaviors plus the continuous character of those same priorities (see next section), that can be dealt with gracefully by the blockers, leaves room to a less rigid control structure. By allowing continuous varying priorities the arbitration structure complies with behaviors whose instantaneous priorities change accordingly to their input stimulus. In the next section we will see how the user can also introduce offsets in the behaviors' priorities to reflect different importance of distinct behaviors.

3 Behavior Model

The main components of this architecture are the behavior modules, responsible for limited aspects of the performance of the Autonomous Vehicle, and from whose interaction emerges the global behavior of the AV. Seen that these modules will interact much in a mutual competition form, the behavior model must cope with the need to resolve competition situations among different modules. At each moment the control of the AV will have to consider external and internal stimulus and, in a behavior based architecture such as this one, select the behavior (or behaviors) that will take over the control of the vehicle. Here we must also take into account the considerations made in the previous sections regarding our purpose of arriving at a distributed form of the decision process and at avoiding direct connections among behaviors.

Then a behavior should produce as outputs an activity value (which is effectively its priority) and an action value, both of them being used as inputs to the blocker structures described above.

One way to accomplish this goal is to have a behavior divided into two sub-modules. Here we named them Action and Activity sub-modules. A similar type of division has already been presented in [2], though in that case the Activity sub-module (there named Applicability Predicate) serves only as an internal modulator (by inhibition) of the Action (there called Transfer Function). Here we have the whole behavior module producing two independent Action and Activity outputs instead of only the Action (fig.3).

The Action sub-module is responsible for producing the commands, to be delivered to the actuators, which will enable the VA to exhibit the specific aspects of behavior that concern its module. The Activity sub-module will output a value that represents the urgency (or priority) the behavior has in expressing itself - taking into account the stimulus inputs it is receiving and its internal state. We can say that Action is the fundamental component of the behavior module, since it is the responsible for its actuation output. And in case there was no other behavior competing there would be no need for the Activity sub-module.

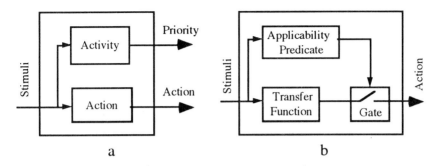

Fig. 3. Behavior Models: a) Proposed; b) Connell's

In this way we separate the instantaneous action produced by each behavior module from the instantaneous priority it presents. This allows for a flexible selection of behaviors depending on their output priorities unlike solutions in which this selection is predefined by hardwiring at the moment of the construction of the vehicle. Also this type of solution - with two separate activity and action outputs - is necessary if variable priorities are to be considered. In fact a model with only the action output underlies fixed priorities of the behaviors, with at the most a commutation between an all active and an all inactive state for each behavior, step type of function in fig.4.

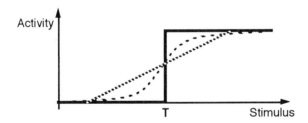

Fig. 4. Behavior activity

In fig.4, other forms of the Activity function are also depicted, with shapes of continuous functions. All of them are monotonic increasing functions of the stimulus[6] - we consider a domain (e.g.[0,1]) equal for all activities. This means that behaviors will be selected accordingly to the value of their input stimulus -

---

[6] Notice that there is no need for all the activity functions of the behaviors to have the same shape. In fact they may well reflect peculiarities of the behaviors (e.g.. rise heavily to the maximum with low stimulus, or only with high stimulus, etc.)

the stronger the stimulus the higher the activity and therefore the behavior has more chances of being selected. Naturally a vehicle limited to such a solution would have a completely stimulus driven character, meaning that the behavior with the stronger stimulus response would always be selected, disregarding any considerations on the relative importance of the behaviors.

Thus in order to allow the user to assign a relative emphasis to the different behaviors each of these modules has a *fixed priority* input parameter. It is combined with the variable priority, consisting of the activity value, by a product function, in order to produce a final priority output of the behavior[7]. So, the behaviors will start of with uneven priorities according to their importance for the global behavior of the AV (it is possible that a behavior with a strong stimulus response is beaten by another which, in spite of having a lower stimulus response, is more vital for the vehicle).

In a given behavior architecture, the set of fixed priorities defines the character of the vehicle, or in other words it outlines a certain type of AV giving it the necessary basis for the kind of task it is supposed to perform (see case study in section 4). On the other hand the variable priority, generated internally by a behavior, defines the situation in which the vehicle is at that moment, as "seen" by the behavior.

A further feature exists in our behavior model. It draws on the fundamental properties of animal behavior that apply even to reflex behaviors. There are four of these properties worth considering [13]. A brief description follows:

Latency - Delay from "stimulus on" till action display. This means that the response of the behavior is not instantaneous.

After-discharge - Delay from "stimulus off" till end of action. Here also the response is not immediate.

Warm-up - Time from action displayed till maximum action intensity. The response does not jump from zero to maximum, but rather increases gradually for a while.

Fatigue - This feature is the decrease of action intensity down to complete lack of response in presence of a continued or frequently repeated stimulus. The behavior needs a further recovery time before the action can be performed again. Fatigue can be defined by three parameters: time from beginning of action till beginning of decrease (*fatigue time*), *decrease time*, and *recovery time* (taken from stimulus ceasing, on).

In our behavior model we feel necessary only to consider explicitly the fatigue effect and here only fatigue and recovery times. The other properties will still be there as a consequence of the particular implementation of the behavior - namely as a result of the technology used, of the type of sensors necessary for the

---

[7] Functions other than the product are admissible for this combination. It is only required that their domains comply with the priority domain (e.g..[0,1]). The product has also the advantage of producing a zero result when any of its inputs is zero.

behavior input, and of the behavior selection process. The complete behavior model is as represented in fig.5.

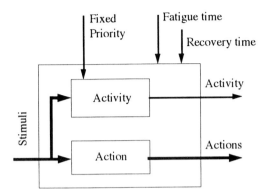

Fig. 5. Behavior model

Why then keep the fatigue and recovery times? This option helps in solving a problem characteristic of decentralized architectures which is the deadlock. It usually happens as a consequence of a conflict between two behaviors, whose myopic views, of the vehicle in the environment, are unable to detect the situation - other solution, based on watch-dog behaviors, has also been proposed ([9, 15]) in previous versions of the architecture, however such an approach is more situation dependent. By using only the above results inspired on ethology research (fatigue and recovery times) we can obtain a general solution already quite efficient, being able to get around many deadlock situations. When one of the two conflicting behaviors eventually reaches a fatigue situation the other will gain control.

Fixed priority, fatigue and recovery times are all defined by the user as part of the behavior architecture of a particular AV. While the first of those is only relevant to the Activity module, the fatigue parameters affect both sub-modules.

4   Simple Case Study

We present as an example of the use of this architecture the TO, an auto-like vehicle with three infra-red sensors - two pointing ahead and one backwards - and three behaviors: Avoid Obstacles, Wander and Follow Objects.

The vehicle has two independent actuators: propulsion (with three possible commands, forward, back and stop) and direction (also with three commands, left, right and ahead). The direction actions are not independent of the propulsion actions due to the non-holonomic character of the vehicle.

This example is based on a work described in [16] where a detailed definition of each behavior's workings can be found. A short description follows. The Avoid

Obstacles only gets active when the sensors detect close objects, while the vehicle is either moving or stopped. The Wander behavior is active for some periods moving ahead and turning to either side in a pseudo-random way. It has as input a timer used to control the periods of action and inaction of the behavior (when inactive its activity output will be zero). This timer input is a clear example of an internal stimulus (which is not exactly a sensor). The Follow Objects detects an approaching object (only when the vehicle is stopped) and tries to follow it until it looses "interest". That happens when the object stops for a time long enough (fatigue).

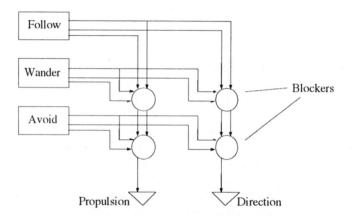

Fig. 6. Control Architecture of TO

In this case all the composing behaviors need to control both the vehicle actuators (fig.6). Therefore in this example there is no case of orthogonal behaviors. Such a situation (of orthogonal behaviors) happens for instance in the interaction of a wander behavior with a phototaxis behavior. The wander behavior controls only the vehicle speed (in the simplest case, stopped or forward) and the phototaxis controls only the vehicle turning motion (left or right). In such case we can see the vehicle wander but tending to stay around zones where the light is brighter[8].

It is interesting to analyze the effect of the fixed priorities in the global behavior of the vehicle. The Avoid Obstacles is considered to have the highest fixed priority. If the Follow Objects has a higher fixed priority than the Wander, the AV will be more eager to follow things, thus it will be useful for surveillance purposes - the wander behavior only gets active when the vehicle loses interest in the object being followed, and it may help to find another object to follow. On the

---

[8] Such a vehicle was actually developed in the Lego lab sessions of the Trento ASI. It had also three other behaviors, follow objects, avoid collisions and resolve collisions.

other hand if the wander behavior has a higher fixed priority the vehicle will have a higher tendency to move about, which may be useful in (vacuum) cleaning tasks - in this case the follow behavior can be used to lead the AV to other areas.

Another interesting result happens if we loosen the above restriction of having the Avoid behavior with the highest fixed priority and we set the fixed priorities such as: $P_{follow} > P_{avoid} > P_{wander}$. In this case the vehicle will approach the followed object until contact (or very near if the activity of the Avoid is high enough to compensate for the difference in the fixed priorities). Such a vehicle may be viewed as aggressive or friendly, depending on the observers evaluation of the consequences of colliding.

5   Conclusion

The proposed architecture, here presented in its simpler one-level form, shows good capabilities compared to other behavior based solutions plus a very good modularity. The behavior parameters are efficient for modifying the vehicle's global behavior.

We do not claim to have produced a solution that tackles all the problems without failures. In fact we believe that any vehicle sufficiently exposed will eventually find a situation in which it will fail. In designing a control for an autonomous vehicle one only needs to anticipate major difficulties that may turn up, configure the architecture to be able to overcome them and let the vehicle handle the minor difficulties, hopefully in a correct way. Our architecture, being modular (behavior modules and decentralized selection) and flexible (with configuration parameters: fixed priorities, fatigue and recovery times, and hysteresis of the blockers) provides a basis for a fast test and development of AVs.

It has been clearly shown the importance of the fixed priorities' relative values to radically change the aptitude of the vehicle to perform one or other type of task. Also the support of variable priorities in the behavior modules gives additional flexibility since the vehicle thus combines the offset (fixed) priorities defined by the user with situation dependent values.

The completely decentralized method of behavior selection, in which there is no entity whatsoever having a global picture of the behavior priorities, is also an important asset more in accordance with the spirit of a behavior based architecture than a central decision entity.

More recent work has focused on the extension of the architecture to multiple hierarchical levels, improving results regarding problems of behavior locality and sequencing, and global variables support [8]. These can be seen as a kind of hormone, which may affect many behaviors. A vehicle controller with 14 behaviors based on this extended model has also been developed with good results [17].

As a future development we plan to work on adaptation mechanisms for the model. These will enhance the chances of a good performance of the vehicle after an adaptation phase in which the vehicle produces, in runtime, alterations to the configuration parameters, either supervised or not by a user.

The application of these results to other fields distinct from autonomous vehicles - such as manipulator control and active camera surveillance in a building - is also being considered.

Acknowledgements

This work was partially supported by Junta Nacional de Investigação Científica e Tecnológica with a fellowship (Proc° 29702/INIC) for L.Correia, and by the EC through SNECOW project of MAST program.

References

1. Brooks, R.A., A robust layered control system for a mobile robot. IEEE Journal of Robotics and Automation, 1986. vol. RA-2(No.1, March)
2. Connell, J.H., Minimalist mobile robotics. 1990, Academic Press
3. Maes, P., (ed.), Designing autonomous agents. 1990, MIT Press
4. Meyer, J.-A. and A. Guillot, Simulation of adaptive behavior in animats: review and prospect. in First International Conference on Simulation of Adaptive Behavior. 1991. Paris, September 24-28, 1990: MIT Press
5. McFarland, D., Problems of animal behaviour. 1989, Longman
6. Bolles, R.C., Species-typical response predispositions, in Workshop on the Biology of Learning. 1983. Berlin: Springer-Verlag
7. von Holst, E. and U. von Saint Paul, Comportamento electricamente controlado, in Psicobiologia, as bases biológicas do comportamento - textos do Scientific American (Mar 62), ed. 1977. 1962, Livros Técnicos e Científicos, S.A
8. Correia, L. and A. Steiger-Garção, A model of hierarchical behavior control for an autonomous vehicle, in PerAc'94 From Perception to Action. 1994. Lausanne, Switzerland, 7-9 Sep.: IEEE Computer Society Press
9. Correia, L. and A. Steiger-Garção, An AUV architecture and world model, in Fifth International Conference on Advanced Robotics - Robots in Unstructured Environments. 1991. Pisa - Italy, June 19-22: IEEE
10. Courant, R. and F. John, Introduction to calculus and analysis, vol.1. 1965, Wiley-Interscience
11. Maes, P., A bottom-up mechanism for behaviour selection in an artificial creature, in First International Conference on Simulation of Adaptive Behavior. 1991. MIT Press
12. Bellingham, J.G., et al., Keeping layered control simple, in Symposium on Autonomous Underwater Vehicle Technology. 1990. Washington, DC, USA, 5-6 June

13. Slater, P.J.B., An introduction to ethology. 1985, Cambridge University Press

14. McFarland, D., Animal behaviour. 1985, Pitman

15. Correia, L. and A. Steiger-Garção, A reactive architecture and a world model for an autonomous underwater vehicle, in IEEE/RSJ International Workshop on Intelligent Robots and Systems '91. 1991. Osaka-Japan, November 3-5: IEEE

16. Correia, L. and L. Gomes, O três-olhos (TÓ para os amigos). 1990, internal report, UNL-FCT/DI

17. Correia, L., Veículos Autónomos Baseados em comportamentos - um modelo de controlo de decisão. 1994, PhD dissertation (to be published) UNL

Reporting Experiments on Integration of Learning Algorithms and Reactive Behaviour-Oriented Control Systems on a Real Mobile Robot

Filip Vertommen

Artificial Intelligence Laboratory, Vrije Universiteit Brussel
Pleinlaan 2, 1050 Brussels, Belgium. E-mail: filipv@arti.vub.ac.be

Abstract. This article reports on experiments that integrate reinforcement connectionist learning and self-organising neural networks with reactive behaviour-oriented mechanisms. The application concerns navigation towards a goal location with which a light source is associated. The experiments are executed on a real mobile robot. We compare the results of the experiments and look at the capability of the mechanisms to control a real mobile robot. We also propose changes to the navigation mechanisms that can improve their performance.

1 Introduction

This article is concerned with the integration of multiple navigation mechanisms for controlling a real mobile robot. Two navigation mechanisms controlled by different learning systems are combined with a reactive and behaviour-oriented navigation mechanism. Navigation is in this context the process of reaching a point B starting from a point A in the environment of a robot. To accomplish a navigation task, the robot may need additional skills e.g. the capability to avoid obstacles in the environment. Navigation mechanisms differ in the way these different skills are implemented and combined.

1.1 Navigation Mechanisms

Three navigation mechanisms were investigated. The first two control the robot by means of situation-action rules, but each has a different approach to construct these rules. The first navigation mechanism uses reinforcement connectionist learning, the second exploits self-organisation in neural networks. The third navigation mechanism is based on photo-taxis and is implemented in a reactive behaviour-oriented framework.

The first navigation mechanism has been designed by Jose del R. Millan and Carme Torras [5]. It uses connectionist reinforcement learning to construct situation-action rules that control the robot. In the view of Millan and Torras a connectionist reinforcement learning framework in which the robot learns by doing is the best choice to overcome typical restrictions of other situation-action rules based control systems, such as the need for a teacher or the need to represent all possible situation-action rules. Their reinforcement learning mechanism derives situation-action rules by trying to associate with each situation X an action Y that maximises a performance feedback signal or reinforcement signal z. If the action Y is followed by a high reinforcement signal, the connections between situation X and action Y are strengthened. Otherwise the connections are weakened. The reinforcement learning mechanism must also occasionally consider to execute non-optimal actions to determine whether these actions may lead to a higher global performance. The navigation mechanism that Millan and Torras propose has been tested in simulation and yielded good generalisations, so that not all possible situation-action rules need to be represented. It was also able to adapt the set of situation-action rules to take care of new situations.

The second navigation mechanism has been designed by Jukka Heikkonen, Pasi Koikkalainen and Erkki Oja and is based on self-organising neural networks or self-organising maps (SOM) [1]. This navigation mechanism also uses situation-action rules to control the robot, but represents them in a SOM and updates the connections between situations and actions by means of the Kohonen rule [2]. The updating of the connections will only happen when a collision situation occurs. A collision situation indicates that the current set of situation-action rules is not appropriate to handle the current situation and needs to be changed. As long as no collision occurs, the SOM remains unchanged. This approach increases the reactivity of the robot if no collisions occur, since no intermediate computations need to be processed. But when the robot moves into a different environment, more nodes need to be updated, which can slow down reactivity.

The third navigation mechanism is based on photo-taxis and is reactive and behaviour-oriented. No learning is involved in this mechanism. The goal location is indicated by a light source that can be detected by the robot. The robot uses the light source to navigate towards the goal. The navigation mechanism is reactive in that no model of the world is used to direct the actions of the robot and in that the navigation mechanism reacts instantaneously without considering any internal states. The navigation mechanism is behaviour-oriented in that it is constructed out of miscellaneous behaviour modules that co-exist. The global behaviour is due to interactions between these behaviour modules. Although no learning is involved in this scheme, it turns out that it can handle almost every collision situation that it is submitted to. A disadvantage of not having a learning mechanism is, of course, the inability to improve efficiency of the navigation mechanism based on experiences.

This article reports on two experiments. The first experiment integrates the navigation mechanism based on reinforcement learning and the navigation mechanism based on photo-taxis into the same behaviour-oriented framework. The algorithm is then used to control a mobile robot in a navigation task. The second experiment is similar but integrates the navigation mechanism based on SOM and the navigation mechanism based on photo-taxis.

1.2 Objectives of the Experiments

The objectives of these experiments were threefold. First, we wanted to investigate whether the first two navigation mechanisms were indeed capable to control a real mobile robot. Or in other words that they could react appropriately to all kinds of events happening in a real-world environment. Prior to our experiments, these navigation mechanisms were only tested on simulated robots. A second objective involved an investigation of the behaviour-oriented framework and its capability to host computatiohally more complex algorithms that, in addition, were not originally designed for this framework. The third objective was to investigate whether two different navigation mechanisms could co-operate in the same framework to improve the performance of the robot and this without being explicitly tuned for each other. It was not a main objective of these experiments to compare the two learning mechanisms for their learning performances. Nor did we want to challenge any fundamental ideas behind the two learning mechanisms. We only wanted to test the usefulness of specific implementations of the learning mechanisms in a specific control framework for a specific robot.

To reach the objectives, we had to re-implement the learning mechanisms to fit into the behaviour-oriented framework. We had to find out a scheme that allowed the learning mechanism to work together with the reactive photo-taxis mechanism. The forced collaboration with the photo-taxis mechanism was due to two reasons. First, for safety reasons. The photo-taxis mechanism and the accompanying obstacle avoidance mechanism was purely reactive and would prevent the robot from damage. Although the two learning navigation mechanisms also claimed to be reactive, their reactivity was not proven yet on a real mobile robot. Therefore we wanted to be able to fall back, if necessary, on a mechanism that was already tested for this matter. The second reason for the forced collaboration concerned the speed of the learning phase of the first two navigation mechanisms. If the learning mechanisms would need to choose an action randomly out of the complete action space every time a situation-action rule turned out to be wrong, it could take a long time before the optimal action for that situation would be found. Instead, the learning mechanisms can look at the action proposed by the reactive photo-taxis mechanism and this way speed up the search for an appropriate action.

1.3 Major Conclusions

We experienced that it was fairly straightforward to re-implement the learning algorithms to fit into the behaviour-oriented framework. Co-operation between the learning algorithm and the photo-taxis behaviour modules also turned out to be easy. The main problems of the integration project showed up in the achieved response time of the total control system when executing on the real mobile robot. This response time became rapidly unacceptably slow in both experiments and inhibited the robot to continue its navigation task quite early in the execution of the experiments.

The main conclusion of the experiments was that a direct integration of the different navigation mechanisms led to a control system that was unable to safely guide the real mobile robot towards its goal. More research towards more fine-tuned mechanisms should be done. The encountered problems mainly concern the large amount of parameters involved with the learning mechanisms. A lot of parameters need to be tuned by hand in both learning mechanisms and many times a small change involves a important difference in behaviour. A mechanism should be thought of that reduces the amount of parameters or that allows the adaptation of these parameters, regulated by the robot's experiences. An equally important cause of problems lies in the relation between the degree of computational complexity of the first two navigation mechanisms and the available processing power of the robot's on-board computer. The learning mechanisms require a huge amount of statements to be processed at each time step and quickly exhaust the on-board computer's resources. Another design of the navigation mechanisms that makes better use of the robot's resources should be examined.

In the first two sections of this article we briefly describe the implementation of the two learning navigation mechanisms. In the third section we describe the navigation mechanism based on photo-taxis. The following section focuses on the real mobile robot Lola, that we used to run the experiments on. In section five and six we will explain the details of the two integration experiments and also report the results of each integration project. The next section provides a discussion of those results and in the final section we draw conclusions from these first trials to integrate different control mechanisms and propose possible improvements that can be elaborated in further research.

2 Navigation by Reinforcement Connectionist Learning

This section summarises the most important issues of the navigation mechanism that is based on reinforcement connectionist learning and that is delineated in [5]. Later in this text we will show the results of our trial to integrate this learning mechanism within a behaviour-oriented robot control system.

This implementation of reinforcement connectionist learning was originally already constructed with the intention to control a real mobile robot in a real world environment. For that purpose its designers had to solve a couple of aspects that in their view are due to the real world. They name three issues of real applications that their proposed algorithm solve. The first issue they mention is determination of the search space of the suitable action for each situation. The second concerns the speeding up of the learning phase. The third issue is the construction of a system with powerful generalisation abilities, so that it can survive in environments where it is not trained for.

The goal of the authors of [5] was to develop a reactive system for navigation that, amongst other features, could cope with continuous-valued inputs and outputs, a limited perception range of the agent and strong performance demands in terms of shortest paths. The architecture that has been built to address these demands consists of three elements: a reactive system, an evaluator and a planner. The reactive system processes the sensory inputs and computes actions. Sensory inputs are matched with already existing exemplars of the different categories of situations that are created during the execution of the algorithm. Linked with the categories are prototypical actions and an expected reinforcement signal. If there is a match, an action is generated that depends on the prototypical action of the current category plus a variation due to the situation the robot is in at that moment.

After reacting the real reinforcement signal is computed by the evaluator and used for learning. Learning in this system is done in three ways. One way concerns the topology of the network, another involves weight modification and the last is the adaptation of the expected reinforcement signal of the category into which the current sensor pattern is classified. The first way of learning takes care of constructing the network itself which is preliminary non-existent, but expands during execution. The level of expansion will depend on the level of similarities in the test environment. The less different sensor patterns are obtained, the less nodes in the network will be constructed. The second way of learning is concerned with fine tuning the weights of the connections from the exemplars to the action units. This fine tuning is controlled by the reinforcement learning rule:

$$w_{pi}^{j}(t+1) = w_{pi}^{j}(t) + \alpha * \left[z(t+1) - b_{j}(t)\right] * e_{pi}^{j}(t)$$

where w_{pi}^{j} is the weight associated to the connection between the ith exemplar and the pth action unit at time step t, α is the learning rate, z is the real reinforcement signal, b is the expected reinforcement signal and e_{pi}^{j} is the eligibility factor of w_{pi}^{j} or a measure of the amount of influence of that weight in choosing the action.

A third way of learning updates the expected reinforcement signal parameter b that is connected with every category. Therefore, the designers use the simplest form of the temporal difference method: TD(0). For more information on this matter, we refer to [5] and [9]. The previous learning rules show that the control system is highly conducted by the robot's experiences. Because of this, very little information has to be known about the environment before hand. The network is completely constructed by using information it receives from the environment itself.

As soon as the reactive system produces an unsuitable action or can not produce an action at all, the planner will be invoked. The planner will perform a one-step planning process on a discretised action space. The output of the planning process is a new category defined by the prototypical action and by the expected reinforcement signal. If the category did not exist before, it will be brought into the system. Otherwise a new exemplar will be added to the category.

In simulation the algorithm works fine and confirms all assumptions stated before. It exhibits strong generalisation abilities, large noise tolerance and a short learning period when compared with other connectionist systems. For more details on these results we refer to [5].

3 Navigation by Self-Organisation in Neural Networks

The second navigation mechanism that we considered was based on a self-organising artificial neural network. The results of the research on navigation with self-organising artificial neural networks that is relevant to this experiment are reported in [1]. This section recapitulates that article's most important topics.

As already stated before is this navigation mechanism based on trial-and-error and self-organising principles. This means that a self-organising map (SOM) is trained by using the robot's own experiences. Each time the SOM fails to give the right action to a specific situation, the weights of the SOM are updated. An action is regarded to be wrong when a collision with an obstacle occurred after the execution of that action.

Weights between nodes in the SOM are updated by making use of the Kohonen rule stated in [1]:

$$W_r^{new} = W_r^{old} + \Psi_{rb_t}(x_t - W_r^{old}), \quad r \in Ne(b_t) \cup b_t, \quad 0 \leq \Psi \leq 1$$

the neighbourhood $Ne(b_t)$ is defined by the Gauss formula :

$$Gauss(x, \sigma) = e^{(\frac{x}{\sigma})^2}$$

The learning mechanism also keeps track of its actions taken in the near past and of the sensor patterns that triggered them. The moment that the weights in the SOM need to be updated, because a collision happened, the mechanism generates training examples. A training vector is a transformation of a vector that is constructed out of the values of the sensors patterns plus the value that indicates the relative orientation of the robot to the goal plus the value representing the action that is executed by the robot.

It is stated in [1] that best performance is obtained when the collision situation is mirrored and when this new data is used again to learn the SOM. This will lead to $2 * M$ training examples if M is the number of actions in the past that have been stored.

The mechanism was tested in three different simulated environments. In each environment the mechanism is tested with 2 to 4 rounds of 500 missions and for each mission different start and goal locations were selected. The SOM that was tested contained 100 nodes and the M factor, that is explained above, was 15. The three test environments contained 3, 6 and 14 obstacles respectively. The agent in these environments was a simulated robot that had 8 distance sensors at its front side, covering a front view of 180 degrees up to a certain maximum distance R. The robot was also able to measure its turning angle between consequent time instances and held two tactile sensors for detecting collisions.

It turned out that the performance of the mechanism improved relative to time of execution. In the last environment, though, which contained 14 obstacles, it took the agent 3 rounds of 500 missions and 124 collisions to deliver a system that did not collide anymore during another round of 500 missions.

The generalisation ability of the implemented SOM was proofed by training a SOM in the third test environment that is just described and to use that trained SOM to control an agent in an environment that contained different shaped obstacles. This resulted in another round of 500 missions without any collisions. The agent could navigate in the new environment without collision although it was trained in another environment. The authors of [1] state that this ability of generalisation is vital to agents living in real world environments and therefore promote the power of self-organisation in learning.

4 Navigation by Photo-Taxis

The third navigation mechanism that is used in the experiments uses photo-taxis. The photo-taxis mechanism is constructed by creating multiple parallel-running behaviour modules. The PDL framework provides tools to create such behaviour modules.

4.1 PDL, a Framework to Implement Behaviour Systems

At the AI-Lab of the Free University of Brussels we have developed a programming abstraction that provides us with the tools to implement dynamical behaviour systems. We call this abstraction "PDL" or "Process Description Language" [6]. At the moment of the writing of this report the abstraction is implemented as a super-set of the C computer language, although versions in LISP or Scheme have also been developed.

The primitive structures of PDL are quantities and processes. Quantities are numerical values that resemble the dynamics of the external environment in which the robot operates and the internal dynamics of the robot itself. Also is every actuator on the robot resembled by a quantity.

Processes are the means by which a behaviour system can propose changes in values of specific quantities. All processes and therefore the management of all proposals for changes in the dynamics, are executed at the same time or in other words run in parallel. There is no explicit priority system involved within PDL that would allow particular processes to be executed faster or more times per time interval.

To make parallelism possible on a sequential processor, we have constructed a small dedicated scheduler. The scheduler discretises time and implements a loop construction. In every step of the time loop that is implemented by the PDL kernel, the following things are taken care of: at the beginning of each time step all quantities that resemble sensor dynamics are updated i.e. the values of the sensor readings are put into the memory locations of the representative quantities. Next step in the scheduler is the execution of all processes that are included in the PDL system. Mostly a process involves the following steps: the investigation of sensor quantities and internal quantities that are of interest for the concerning process and dependent to their values a proposal to change the values of other quantities. This way of implementing processes is, however, not the sole possible way. Sometimes proposals for changes are immediately presented without considering other quantities. Typically, behaviour systems are designed by using mixed types of implementations of processes. Special care has been taken that within the time period of one cycle in the PDL loop no value of any quantity changes. This is to assure that in the period of one cycle all processes will deal with the same data or in other words that integrity of data is guaranteed. This is necessary when parallelism is to be simulated. In this scheme it does not matter in what sequence the processes are executed between two time steps, for they all deal with the same data. The next step of the scheduler begins as soon as all processes are executed. All proposed changes and the current value of each quantity are added and the sum becomes the new value of the quantity. The last step of the scheduler takes care of the actions that need to be taken by the agent by means of its actuators. The nature of the actions will follow from the values of the quantities that resemble the actuators.

The four basic steps in the PDL scheduler just described, are repeated for as many cycles as is indicated by a dedicated variable. This variable can be defined by the PDL programmer. When the execution of these four steps happen fast enough, a sense of parallelism can be created. As soon as the time to execute one cycle takes too long, it will have a direct influence on the reaction time of the agent. The processes that take care of reacting to sudden changes in the environment will, in that case, not be executed in time and fail their objectives.

Many reasons for the fact that executing one cycle can take too long, involve the hardware of the agent. When too many processes need to be executed, the computational power of the hardware might get exhausted. The same holds for too complex processes e.g. if a lot of floating point arithmetic is involved. In extreme cases this can inhibit the agent to function properly in its environment. The PDL programmer has to be aware of these possible deficiencies.

For more information on PDL and its approach of programming robots, we refer to [7] and [8].

4.2 Photo-taxis by using PDL

Navigating towards a light source, a behaviour that is also called photo-taxis, can very easily be implemented using the PDL framework.

The photo-taxis behaviour or any other behaviour in PDL is mostly not implemented as a stand-alone application. Most of the time some quantities and processes will already exist and will normally take care of the most fundamental life saving behaviours or even more abstract behaviours. An example of a life saving behaviour in a mobile robot involves retracting when an obstacle hits the agent at the front. Another is moving forward when the agent is hit at the back or turning away to the right when the agent is hit on the left or turning to the left when a hit is detected on the right.

The power, though, of the PDL way of controlling robots is that the programmer does not have to take the effect of all these other processes into account. He or she only has to focus on the specific task that needs to be taken care of, which is in this case photo-taxis. A well designed PDL program will adapt any possible parameters in the system, so that the global system will operate optimal even if new behaviours are added.

To solve the problem of photo-taxis the following strategy is used in PDL. The mobile robot Lola has two horizontal mounted detectors for visible light. Exploiting this configuration, two additional processes are added to the PDL system. One process inspects the value of the most right mounted light detector and continuously proposes a negative value to the quantity connected to the rotation motor. The magnitude of this value is inversely proportional to the amount of light detected by the right mounted light detector. A negative value of the rotation motor quantity will result in rotation action to the left. A second process has similar functionality but concerns the left most mounted light detector and proposes a positive value to the rotation motor quantity. As already can be expected does a positive value cause the rotation motor to turn to the right.

Some consequences of this strategy show up when the control system is executed on the robot. A first consequence is that the actual amount of turning will depend on the difference between the incoming signal of both the light sensors. In other words, the bigger the difference in the amount of light that one light sensor receives relative to the other one, the higher the influence of its process on the rotation motor quantity. A second consequence is that the agent goes straight forward when the light source is right in front of the agent because the two new processes will propose an influence value of equal magnitude but opposite sign to the rotation motor quantity. The two opposite signed influences will therefore cancel each other and none will have effect. As soon as no other processes influence the rotation motor quantity, the robot will go straight forward. A third consequence of this strategy for photo-taxis reveals when an obstacle shows up in front of the robot and when because of this the light source is only partially visible. Since this situation is interesting also for the rest of this report, we will elaborate it in more detail.

If an obstacle turns up in front of the robot a couple of things can happen depending on the specific context. For example, the obstacle covers the light source only for a part and one light detector can still detect it. This will result in a bigger difference between the proposed influence values to the rotation motor quantity. The process connected to the light sensor that is receiving the least amount of light will propose a bigger value to the rotation motor quantity in its favourite direction. The rest of the processes act the same way as before, because the situation has not changed in their view. The global behaviour that is observed by an external spectator will be a turn of the robot around the obstacle but biased toward the light source. The explanation of the bias towards the light source when avoiding the obstacle is quite simple. Due to the rotation action caused by the process from the light sensor that lost contact with the light source, the contact will smoothly be regained and will result in a less high influence value to the rotation quantity. After some time the influence will be back to its initial value. Overshooting of the rotation quantity can happen, but the inertia of the agent and of the internal dynamics will cause a steady decrement of overshooting and the robot will end up aligned with the light source again. If, on the other hand, this overshooting is cause of a collision with the obstacle, the processes for touch based obstacle avoidance come into play. The influences on the motor quantities proposed by these processes will increase and will dominate for some time the motor actions instead of the processes that implement the photo-taxis behaviour. Note that we say "dominate" and not "inhibit the photo-taxis processes". The influence values from the photo-taxis processes are still proposed but have less "weight in the balance" in the time period that the processes that implement the life-saving actions, are triggered. Because the photo-taxis processes are still active, all actions, also the life saving actions, will be biased towards the light source.

The navigation strategy described here is purely reactive. Every process is a direct reaction to a pre-defined stimulus. But the power of this way of programming is that, although every process is developed for a very specific and simple task, the combination of multiple processes that are implemented for multiple functions, end up in a behaviour that is not explicitly programmed, though beneficial to the agent.

This approach shows that navigation is not necessary only subject to "learning", but can also be obtained solely by reactive behaviour.

5 The Real Mobile Robot Lola

To test the mechanisms that are constructed out of the integration experiments, we use the real mobile robot Lola. Lola was designed and developed at the AI-Lab of the Free University of Brussels.

5.1 General Description of the Lola Robot

The Lola robot is a modular, general purpose real mobile robot. It is modular in the sense that it is very easy to attach new devices to it without affecting the already existing hardware and software. New sensors or actuators just have to be mounted onto the robot and connected to the "spinal cord"-databus which is available throughout the whole robot. Programs just have to address the new sensor by its unique address and the sensor is ready to be used in the programs.

Lola consists of multiple layers of circular aluminium plates which are mounted on top of a commercial available motor base: the B12 mobile robot base manufactured by Real World Interface, Inc.. The base gives shelter to the motors, the batteries and some electronics to control all this. There are two independent motors. One for translation motion and one to steer the orientation of the robot.

The main processor on Lola is an Intel 80386. It is mounted on one of the aluminium plates which we call the "brain brick". There is a possibility to connect a CGA monitor to the main processor, which can be useful for debugging purposes. There is also a 3 1/2" disk drive mounted on the robot at the brain brick layer and is connected to the main processor. Typically, programs are downloaded to the main processor by making use of this disk drive. The disk drive can also be used to store information on disk during or after execution of programs.

5.2 Sensory devices on the Lola Robot

For the experiments described in this article, the following sensors are available on Lola: 12 bumper sensors, 4 photo-transistors which are used as detectors for visible light and 12 active IR sensors. Fig. 1 shows a picture of the Lola robot.

Fig. 1. The Lola robot.

Especially for this experiment we developed and mounted an IR based beacon detector on top of the robot. It consists of 16 IR detectors mounted in a circular way to cover a visual field of 360 degrees. Each IR detector, therefore, needs to cover an area of at least 22.5 degrees. The specification of the IR detectors mention a covering area of 100 degrees before the sensitivity of the detectors goes below half the full sensitivity. This means that there are overlapping covering areas between two IR sensors of the beacon detector, or "beacon eye" as we call it, but the electronics on the beacon eye will choose the IR sensor that receives the most IR as the one that indicates the orientation of the beacon relative to the front of the robot.

The IR detectors on the beacon eye can only detect IR frequencies that are modulated at a frequency of 36 KHz. The reason of the frequency modulation is to eliminate the influence of IR sources other than the beacon.

A small device with 2 IR emitters, emitting IR signals that are modulated at the right frequency, serves as beacon.

6 Integrating Two Learning Navigation Mechanisms and a Photo-Taxis based Reactive Control System

In this section we will delineate the experiments to check the capability of the two learning systems to serve as control system for a real mobile robot. In this first phase of the experiments we only wanted to find out whether it would be possible to do the experiments without changing much to the original implementations of the learning mechanisms. In this phase it was never the aim to build optimal systems. One important prerequisite for the control systems was that the robots should react in real time.

There are two reasons to integrate the implementations into a PDL frame-work. First, we wanted to investigate the capability of PDL to serve as a more general programming tool for implementations of all sorts of mechanisms. Second, we implemented some reflexes in PDL that were intended to protect the robot from damage as soon as a collision occurred. Since the learning navigation mechanisms were only tested in simulation, a more secure solution needed to be available in case of failure to tackle collision situations. The reflexes would run in parallel with the learning mechanism. Besides these life saving reflexes, the two learning algorithms were allowed to take advantage of PDL as much as they would need.

The aim of the control system that would be constructed, is to navigate towards a light source, which we will call the goal location. The goal location is said to be reached as soon as a third light sensor, that is mounted horizontally on the front of the robot, received a certain amount of visible light. Because the environment also contains obstacles, the program should be able to handle collision situations. This is the place where the two learning algorithms come into play. They should allow the system to improve its capability to handle these collisions. This means that the system should learn how to spend as less energy as possible in coping with collision situations. To accomplish this the programs may use the bumper sensors at the bottom layer of the robot as well as the IR sensors on the lowest but one layer of the robot. The perception range of the IR sensors is not fixed. This can depend, for instance, on the texture of the reflecting obstacle. Nevertheless, the sensing range is in most cases higher than the range of contact sensors. Therefore, learning how to use these IR sensors for obstacle avoidance would be beneficial for the system.

6.1 Integrating Reinforcement Connectionist Learning and Photo-Taxis in the PDL Framework

This subsection deals with the details of the first integration experiment. The reinforcement learning mechanism should enhance the navigation ability of the robot over time and will use the photo-taxis behaviour to help to reduce the learning period. Enhancing the navigation ability means that over time less energy should be spent for this task.

6.1.1 Set-up of the Experiment

This integration experiment has two building blocks: an implementation of a navigation mechanism based on reinforcement connectionist learning and the implementation of photo-taxis in PDL.

The implementation of the navigation mechanism based on reinforcement connectionist learning, to which we will refer as the reinforcement navigation algorithm in the rest of this article, was originally implemented in LISP. The implementation was translated into C for a couple of reasons. One reason was that Lola, the mobile robot used in this experiment, was controlled by a Intel 80386 processor and at the moment of this experiment no appropriate LISP environment was directly available for it. A second and more important reason involved the implementation of PDL, which is completely written in C. To avoid that we had to rewrite the kernel of PDL or that we would need special function-call structures, the entire experiment was written in C.

The first step in the development of the total system involved the implementation of the PDL processes that would respond to bumper triggerings. Earlier in this text we have also called these the life-saving processes. As soon as a bumper is triggered this means that an obstacle is very near the robot. If no appropriate response by the robot is activated, this could lead to damage of the robot. Appropriate response is in this case translated into PDL processes that will let the robot turn away from the direction in which the collision is detected. This can be accomplished by four processes that are already explained in a previous section.

Besides the four life-saving processes, some other processes were added to the total system. These were the processes that would lead to a photo-taxis behaviour, which are already describe in detail before. These two additional processes also run in parallel with the ones already mentioned. The reason that these processes where added to this particular system was to guide the learning system in its search through the action space. If the planner in the reinforcement learning algorithm would construct a new category, it would use the action generated by these two processes as the proto-typical action for the new category.

The second main part in the program involves the C version of the reinforcement navigation algorithm. The whole reinforcement navigation algorithm was reduced to two parameterless C functions. One function took care of the specific execution steps that needed to be taken when the goal of the robot was reached it and also suspended the execution of all the processes for some time to give the experimenter the opportunity to move Lola to another start position. The second parameterless function involved the reinforcement navigation algorithm during a run and covered the following steps:

- collect the sensor patterns;
- try to categorise these patterns;
- react with the learned action when the categorisation was successful;
- add a new exemplar to an existent category if possible or construct a new category when categorisation of the sensor patterns failed;
- compute the reinforcement signal for the situation the robot is in at that moment;
- adapt the parameters in the previous time step conform the newly computed reinforcement signal by means of the reinforcement learning rule.

It is a rather simple procedure to hook parameterless C functions, in this case the two main functions of the reinforcement algorithm, into PDL processes. What needs to be done is defining two new processes in PDL and referring to the functions implementing the reinforcement algorithm. This a straightforward operation in PDL. When this is done, the functions, now transformed into processes, will be executed each cycle of the PDL scheduler.

The test environment for the experiments had a very simple physical construction. The whole environment was surrounded by wooden plates to avoid the robot to end up in places that were not very safe for it and that would unnecessary complicate the experiment. The goal location in the environment was indicated by a halogen lamp of 30 Watts. Obstacles consisted of cardboard boxes that were high enough to be detected by the IR sensors but that did not obscure the light source. The halogen lamp was therefore visible from every spot in the environment. Of course, it did not have the same intensity everywhere. It was necessary to make the light source visible from everywhere in the environment because the intensity of the light source is one factor that is important when computing the reinforcement signal. The higher the intensity of the light, the higher the reinforcement signal.

A typical experimental round consisted for the experimenter of the two following simple steps: switch on the halogen lamp and activate the robot. As soon as the robot reached the goal, it informed the experimenter and suspended the execution of the program for some time. During this period the experimenter was able to move the robot to another position in the environment and the next mission of that experimental round could start. After multiple missions the system should have learned from its previous missions how to use its IR sensors and collide less with the obstacles.

6.1.2 Experimental Results

During the experiments the values of interesting quantities were stored into memory. When the experiment had finished, this data was stored on file. This way the data could be used afterwards for analysis purposes with appropriate tools. We used spreadsheet programs for this.

The experiments were executed each time for 800 cycles of the PDL scheduler. This number is quite low compared with other PDL programs, but it will become clear in the rest of this text that in this case this number is justified. From the beginning of the experiments it became clear that the longer the experiment was running, the less the system was able to react to perturbations in the environment within a proper time period. It turned out that to execute 800 cycles the program needed on the average 837 seconds. In that time period, the goal was reached 10 times after which the program was suspended for more or less 30 seconds. The time that the program therefore really needed to execute 800 cycles was about 537 seconds. This means that per second 1.5 cycles were processed or that one cycle took on the average 0.67 seconds to execute. This is far too long when real time reaction is required.

It also turned out that the more exemplars were brought into the system, the longer it took to execute one cycle. This is understandable, because during each cycle all exemplars had to be checked to find out whether there already existed an exemplar that matched the sensor pattern of that cycle. To check for matching exemplars the Euclidean distance between the vectors constructed out of the current sensor pattern and the sensor pattern associated with the exemplar to be checked, was calculated. A sensor pattern in this context means the values of the IR sensors, the values of the photo-transistors and also the value of rotation motor quantity. If the Euclidean distance was below a certain threshold, then the sensor patterns were said to match.

Fig. 2 shows the relation between amount of cycles on the X axis and time and amount of exemplars on the Y axis.

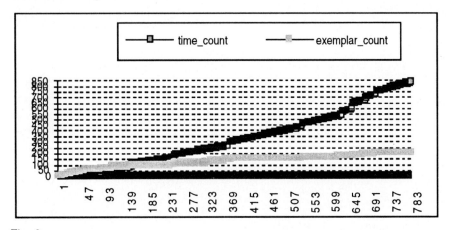

Fig. 2

As can be seen on Fig. 2 is the time of execution of one cycle almost linear to the amount of exemplars in the system. The more cycles elapsed, the more exemplars were added that in the next cycle also needed to be checked and the more the reaction time of the robot increased. In the region between cycle 650 and 730 the robot reached the goal relative quickly because the starting position of that mission was near the goal position. This resulted in more periods in which the program was suspended for 30 seconds. This caused the incontinuities in the graph.

We tried some variations of a typical round to investigate the effects of this on the algorithm. A first variation was a trial to reduce the amount of newly constructed exemplars. We put the robot rather near the light source every time a new mission started. This should reduce the amount of different observed situations and therefore the amount of newly constructed exemplars, because the number of different sensor patterns is reduced. But putting the robot closer to the goal also means that the reinforcement signal stays more or less identical, since the reinforcement signal is dependent on the amount of light that the robot receives from the goal. In the neighbourhood of the goal the intensity of the light does not change a lot and this will therefore result in a reinforcement signal that in each time step is almost identical as in the previous step. And as long as the reinforcement signal is not significant different to the previous situation, the system feels no need to change the weights of the nodes in the network. It would not "learn" a lot. A second variation on a typical mission tried to force a negative reinforcement and to investigate the consequences of this on the learning system. Therefore, we did sometimes put an obstacle in front of the robot. However, the value that the reinforcement algorithm proposed to the rotation motor quantity did not change a lot, although the reinforcement signal in a collision situation is certainly different than in any other case. This has two reasons. A first reason concerns the matching criterion, which turns out to be too severe. Situations that for an observer seemed close to another, were not regarded to be matching. The second reason involves the sparse moments of negative reinforcement signal compared to situations of positive reinforcement signal, since the robot was getting closer to the light source which caused a increasing reinforcement signal. The learning from these situations is minimal.

In Fig. 2 the value that was proposed by the reinforcement algorithm to the rotation motor quantity of the PDL system, which we called infl_value, and the incoming signal of the front light sensor, called PT_front, are plotted on the Y axis against the number of cycles on the X axis. The incoming signal from the front light sensor indicates when the goal was reached. If that signal is high enough (in our case this meant having a value higher than 190) the goal is reached and a new mission could start.

On Fig. 3 we can see that the value that the learning system proposes to the rotation motor quantity is not very high relative to the maximum value of this quantity, which is 100. The learning system is therefore not able to dominate the direction of the rotation motor because the influence of the processes for the phototaxis behaviour is very much higher, as is shown on Fig. 4. On the other hand we see a slight increasing evolution in infl_value, but still too weak in the time period of 800 cycles to be of any interest.

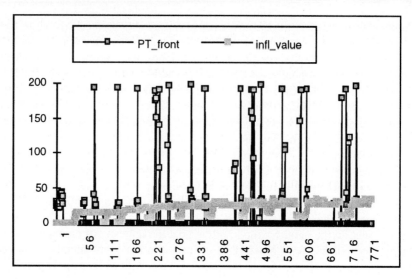

Fig. 3.

In Fig. 4 the values of three quantities are plotted on the Y axis against the number of cycles on the X axis. The three quantities are the left light sensor, called PT_left, the right light sensor, called PT_right and the rotation motor quantity, called turn_direction.

From this chart it is obvious that the processes which implement the photo-taxis behaviour are dominant in these experiments. The reason is obvious when we know that the experiments are done near the goal. On Fig. 4 we can see that the direction of the rotation motor is dependent on the direction of the light source relative to the robot.

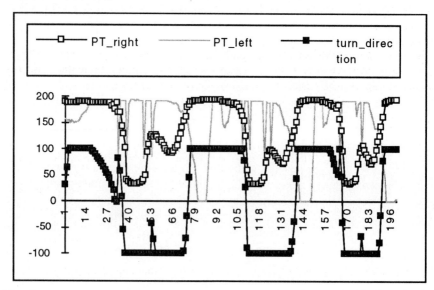

Fig. 4.

A third variation on an experimental mission consisted of experiments starting from positions in the environment that were located further away from the goal. Unfortunately, this always resulted in situations in which the control system of the robot needed too much time to process all the computations that were involved. The reaction time decreased dramatically within a period of a couple of seconds. Mainly due to the larger amount of different sensor patterns, which caused the system to construct more categories. These new categories caused an additional delay in computation time and thus in reaction time, for their examples all needed to be checked on their matching level with the current sensor pattern. It even turned out to be impossible to collect data from any of these runs, because dead-lock situations inhibited to save the interesting data on file. This means a lost of this data, since there is no other means to collect data from the robot at the moment of the writing of this report.

6.1.3 Intermediate Conclusions

The main conclusion that we can derive from this experiment is that the implementation of the reinforcement learning mechanism that we tried to integrate into the PDL framework, is computationally too expensive for the Lola robot. This is especially the case when we want the robot to run around in the world without interruption so that it is able to react continuously to changes in its environment. Due to this, it became very hard to achieve experimental data that could be analysed properly and used to compare with other implementations of learning mechanisms. On the other hand, the few data that was gathered gave a clue what changes to the implementation could improve performance. Later in this text, we will propose changes of the current implementation. These proposed changes can be topic of further research.

6.2 Integrating Self-Organisation and Photo-Taxis in the PDL Framework

This subsection covers the implementation of the second integration experiment. This time, a navigation mechanism based on self-organisation in neural networks is used to improve the navigation skills of the robot over time.

6.2.1 Set-up of the Experiment

Similar to the previous experiment does this one exist of two main parts: an implementation that covers the self-organising learning mechanism and the implementation of photo-taxis in PDL.

The original implementation of the SOM included all the necessary function definitions to implement the simulation environment. The original code, therefore, needed to be filtered first. What remained was one main C function that was the core of the implementation. This function was integrated in the PDL system by constructing a new process of which the body was defined by this core function.

The steps that are taken by the core function that implemented the SOM were:

- collect the sensor patterns;
- find the best matching unit (bmu) in the SOM;
- execute the action that is associated with the bmu;
- if a collision occurred, update the weights of the bmu and of the units in its topological neighbourhood;
- if the goal is reached, suspend the program for some time to let the experimenter move the robot to another start position.

The updating of the weights was directed by the following algorithm:
For all the previous M steps in the execution of the algorithm that are stored, do:
update the weights conform the Kohonen rule that is already stated above;
The format of the training vectors used in the Kohonen rule is the following:

$$x_k = (d_{k1},\ldots,d_{k5},\beta_k,\alpha_k \pm A(k))$$
$$\text{with } A(k) = A_{\max} / (t - k + 1) \text{ and } k = t,\ldots,t - M - 1 \qquad \text{(Eq.1)}$$

- the d_k are the sensor patterns of the node that is currently used, out of the set of M old used nodes of the SOM
- the β_k is the direction of the goal
- the last element in the training vector is the action that in the future needs to be taken if this sensor pattern is encountered again. This parameter is obtained by applying a function on the previous action parameter. The function is described in (Eq.1). $A_{\max}$ gives the maximum value for $A(k)$ at k=t.

- mirror the collision situation and train the network with training vectors that are constructed by this new data. This will speed up the learning phase. The training examples of the mirrored situation will have the following format:

$$x_k = (d_{k8},\ldots,d_{k1},1-\beta_k,1-\alpha_k \pm A(k))$$
$$A(k) \text{ has the same format as in (Eq.1)}$$

The part of the PDL program that was not directly involved with learning, was similar to its counterpart in the previous experiment. It consisted of PDL processes that implemented the life-saving processes and a photo-taxis behaviour. The photo-taxis behaviour served also in this experiment as a guiding parameter when the self-organising system needed to update the actions associated with a particular sensor pattern.

There were a couple of differences between the implementations of the two learning algorithms. One difference is the fact that the implementation of the SOM will only update the weights that define the network when a collision occurs. A collision means that the associated action in that situation is wrong. As long as no collisions take place, the weights are regarded as being satisfactory. Another difference in the implementation of the SOM in contrast with the implementation of reinforcement connectionist learning, is that the amount of nodes in the SOM is fixed from the beginning of the experiment. No new nodes are constructed during execution. The pattern in the SOM that differs the least from the current sensor pattern is regarded as the one to work with in that time step. The level of matching is also in this learning algorithm computed by the Euclidean distance formula.

In contrast with the first experiment, we started to test this implementation in a little different configuration. Instead of immediately executing the program on the on-board computer of the Lola robot, we first tried the program on a desktop computer that had the same specifications as the on-board computer of Lola. The desktop computer was also connected to the "spinal-cord" databus and had similar control over the sensors and actuators of the robot as the on-board computer. Another change in the test configuration was the fact that the wheels of the Lola robot were put off-ground by means of wooden blocks. The benefit of this change in configuration was that experiments could be done without having fear to damage to the robot and that we had immediate control over the robot.

When all tests in this situation would be satisfactory, the robot could be placed in the real test environment.

6.2.2 Experimental Results

First, we did some tests focused on the reactive part of the control system only. This involves all the processes in the PDL system related to the life-saving actions and the photo-taxis behaviour. Remember that this part of the experiment is identical in the previous experiment.

The tests mainly involved measurements of cycles times, for in the previous experiment this turned out to be critical when more complex computation is involved. The tests showed the following results: for a test round of 1000 PDL cycles the system needed 14 seconds. In other words about 71 cycles were executed every second. This is far enough when real time reaction is required. No significant difference was experienced when sensors were triggered. The number of cycles per second stayed equal.

The next step in the testing phase was the integration of the learning system based on SOM in this scheme. The results of this step on the time measurement of PDL cycles was all but encouraging. It took the control system 256 seconds to process all 1000 cycles. This is 18 times as long as in the previous step. This means that per second more or less 4 cycles could be processed. This is too few when real time reaction is required.

On Fig. 5, time is plotted on the Y axis against amount of cycles on the X axis. The two different test configurations are shown. Two graphs represent the elapsed time for a certain amount of cycles respectively when the learning part of the experiment is not included (thin graph) and when it is included (thick graph).

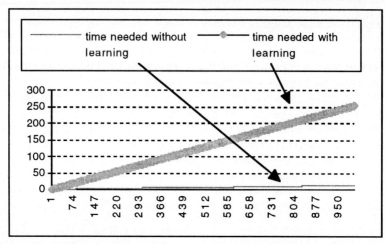

Fig. 5.

The results of the tests turned out to be even worse when a bumper was triggered manually which simulates a collision situation. The SOM interpreted these situations as indicators that the SOM had associated a wrong action with a particular sensor pattern and started updating the weights of the nodes in the SOM. Every updating operation took about 30 seconds. This time period is extremely long and makes the control system useless on the Lola robot. Fig. 6 is a plot of elapsed time on the Y axis against the amount of cycles on the X axis in the case that three collision situations were simulated. These three moments are visible as a sudden increase in time between the execution of two successive PDL cycles.

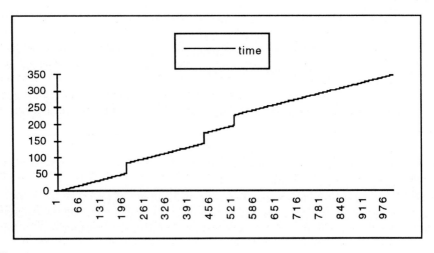

Fig. 6

6.2.3 Intermediate Conclusions

Due to the results that were obtained from the previous experiment, we concluded that it was unrealistic to let the robot run around in the real test environment before the timing problems that are explained before, were solved.

One obvious observation was that the execution time of the cycles stayed equal during all steps. Only when a collision situation had to be resolved by the learning system the execution time of a PDL cycle took longer. This is obvious because the statements that need to be executed are equal during the complete time period of the experiment as long as no collision happens. No new nodes in the SOM are constructed, nor additional datastructures are built during the lifetime of the experiment. This means that, in contrast to the previous experiment, the amount of computation exceeds the computational power of the on-board computer of Lola already from the start. Since a complete re-design of the learning navigation mechanism was out off the scope of this research, we ended the experiment at this point.

7 Discussion

The results of both integration experiments turned out to be unsatisfactory for real world navigation. In both cases the implementations were too computational complex for the processor unit of the test robot. This affected the real time aspects of the control system of the robot. Again, we have to remark that in this phase of the experiments the main goal was to try to fit different approaches of control systems for real mobile robots into the same framework without challenging fundamental issues in either implementation. It would demonstrate the flexibility of the different mechanisms. It would also investigate the ability of the existing implementations of the learning algorithms to be used within a reactive control system on a real robot.

In the first experiment, where an implementation of a navigation mechanism based on reinforcement connectionist learning algorithm was used in collaboration with a reactive behaviour-oriented system, the robot was indeed able to run around in the environment for some short time period. It was, however, in that time period mainly controlled by the reactive system. After a couple of seconds and before the learning system had been able to proof its usefulness, the processor unit of the robot was unable to process the incoming information in real time and ended up in a dead-lock situation.

The experiment of integrating the implementation of a navigation mechanism guided by SOM and the reactive behaviour-oriented control system, resulted in an immediate inability of the processor unit to process all statements within an acceptable time period.

However, it took on the average less time to execute one PDL cycle by using the implementation of the SOM learning system than a cycle of the implementation of the reinforcement connectionist learning system, respectively 0.26 seconds and 0.67 seconds.

Because we are only interested in systems that can exist in continuous evolving environments, we reject a solution that is based on interruption of execution of the algorithm every time a collision situation occurs. This would, if indeed applied, give the learning part of the control system the opportunity to adapt its parameters, or in other words to learn. But real robots do not live in environments that allow them to suspend reacting for a while and continue when they are ready. All kinds of things that are of interest to the robot can happen in the periods of "thinking" to which the robot would not be able to react. Extreme cases of these situations could even lead to physical damage of the robot. However, existence is one of the main concerns of many agents in real world environments and because damage can be regarded as a degradation in existence, this should be avoided as much as possible. Another argument against interrupted execution is that it is possible that environmental information that is necessary for some other possible subpart of the control system, is not processed because the information was only accessible in the periods that the agent suspended a lot of its activities to be able to "learn". Important information would not be observed, though it is available in the environment.

More research should be done on the effects of changing all kinds of parameters that are used in the implementations of both experiments. In the first experiment, for example, there are many parameters that are concerned with the level of matching between two sensor patterns. A first set of parameters in this context concerns the number of sensors that make up one sensor pattern and the value of their possible range. The more sensors that are regarded and the bigger the possible ranges of these sensors, the bigger the state-space and the higher the chance that the patterns will not match if the matching criteria are too severe. It should be thoroughly investigated what the optimal is for these parameters and what the effects are on the performance of the program. Besides these, there are lots of other parameters and thresholds that govern the rest of the control system. No full examination has been carried out yet to investigate the level of influence of these on the performance of the program. A different initialisation of those parameters can result, for instance in the first experiment, in a network construction with less exemplars and categories and therefore in a faster classification. This can improve the reactivity of the total program. But in that case should the performance of the learning mechanism itself be investigated again. Possibly, this gain in reactivity performance will lead to a learning algorithm that is too slow to be useful, relative to the life-time of the experiment, because to few details are covered. Further research is required on this topic.

Another minor remark on the implementation of the first experiment is that the code is written for readability, not for speed considerations. This results in lots of function calls that all need to store and remove their most important data on the stack, which slows down the program in some degree. Also here can improvements be made in further research.

In the implementation of the second experiment, parameters are also concerned with the amount of sensors that are regarded in the vectors to calculate the topological closest nodes in the SOM. In the first experiment the amount of sensors has indirectly influence on the amount of categories to are constructed, what is already explained above. In this experiment, however, will the major improvement in computational performance not be focused on the amount of constructed nodes, because this amount is fixed from start, but be only in the computational effort of the topological distance measurement itself, which will take less time.

Concerning the implementation of the SOM, one remark should be made. At the same time that the second experiment has been constructed, a new version of SOM has been developed. This version is called tree-structured self-organising maps (TS-SOM). It accelerates the search algorithm for the best matching unit drastically and even reduces it to $O(1)$ if a lookup table would be used. This, of course, would accelerate the whole system considerably and can lead to important performance bursts. This topic should certainly be elaborated more. More information on TS-SOM can be found in [3,4].

One of the main parameters in the implementation of the second experiment is the number of previous steps that are stored in memory, or the M parameter as it is called before. The learning system uses training vectors that are constructed out of the sensor patterns achieved in these stored steps, plus the orientation of the goal at that time plus the action. Those examples are then slightly transformed. The more steps are stored, the more training examples are available, but also the longer the execution of the learning part takes. The less steps are stored, the longer it will take before the right actions are learned. Trying to define the optimal amount of steps to be stored, should also be topic for further research.

Other parameters in the second experiment that should be tuned to fit real world environments, are the number of nodes that are involved in the SOM and the number of connections to sensor inputs from each node. The more nodes in the SOM and the more connections to sensor inputs, the longer it will take before they are all manipulated. But the less nodes there are, the less the resolution of the SOM is and the less it can react to specific situations. Yet another topic for more investigation.

8 Conclusion

Out of the results of the experiments described in this article, we can conclude that it seems impossible to get an operational control system for Lola from a straight-forward combination of two different but complementary existing control systems.

The reason of this failure are of two kinds. First, the used hardware does not allow to process many statements of the required complexity. Second, the implementation of the learning part was not tuned well enough for execution on a real mobile robot.

However, new insights are achieved that are valuable for further research on this topic. These insights indicate the direction of possible improvements on the approach of integration of reactive behaviour-oriented controls systems and learning mechanisms. One of the more promising directions can to be located in the distribution of processing power. Multiple processor units that each take care of a part of the job. The PDL framework is most fit to support distribution of

processing. Computational complex processes can be processed by a single powerful processor unit and other, less computational complex processes, can be grouped
and processed by another processor unit.

A final remark that we would like to make is that these experiments should not
be the end of the research, but should be the first phase of a more elaborated
investigation of this subject.

9 Acknowledgements

The research needed on this subject was sponsored by the ESPRIT basic research
project SUBSYM. We would like to thank Jose del R. Millan, Jukka Heikkonen,
Jari Mononen and Pasi Koikkalainen for the collaboration, help and remarks they
made on this experiment. We also would like to thank Luc Steels and Peter Stuer
for the comments they had on this text and helped to bring it into its current shape.

References

[1] Heikkonen, J., Koikkalainen, P., Oja, E., (1993), From Situations to Actions:
Motion Behavior Learning by Self-Organization, In proc. ICANN '93:
International Conference on Artificial Neural Networks, edited by Stan Gielen
and Bert Knappen, Springer-Verlag, pp. 262 - 267.

[2] Kohonen, T., (1989), Self-Organization and Associative Memory. Springer-
Verlag, Berlin, Heidelberg.

[3] Koikkalainen, P., Oja, E., (1990), Self-Organizing Hierarchical Feature Map,
In proc. IJCNN-90, International Joint Conference on Neural Networks, San
Diego, CA, 279-284.

[4] Koikkalainen, P., (1993), Fast Organization of the Self-Organization Map,
Symposium on Neural Networks in Finland, Abo Akademi, Axelia, Turku,
Finland.

[5] Millan, J. del R., (1992), Building Reactive Path-Finders through
Reinforcement Connectionist Learning" Tree Issues and An Architecture. In:
proc. of the 10th European Conference on Artificial Intelligence. Vienna,
Austria.

[6] Steels, L., (1992), The PDL reference manual. AI-memo 92-5 Free University
of Brussels.

[7] Steels, L., (1993), Building Agents with Autonomous Behavior Systems. In :
Steels, L., and R. Brooks (eds.) (1993), The 'artificial life' route to 'artificial
intelligence'. Building situated embodies agents. Lawrence Erlbaum Associates,
New Haven.

[8] Steels, L., (1994), The artificial life roots of artificial intelligence. Artificial
Life Journal, Vol 1,1. MIT Press, Cambridge.

[9] Sutton, T.S., (1988), Learning to predict by the methods of temporal
differences. Machine Learning, 216-224.

Distributed Reinforcement Learning

Gerhard Weiß

Institut für Informatik, Technische Universität München
D-80290 München, Germany

Abstract. In multi-agent systems two forms of learning can be distinguished: centralized learning, that is, learning done by a single agent independent of the other agents; and distributed learning, that is, learning that becomes possible only because several agents are present. Whereas centralized learning has been intensively studied in the field of artificial intelligence, distributed learning has been completely neglected until a few years ago.

This paper summarizes work done on distributed reinforcement learning. The problem addressed is how multiple agents can learn to coordinate their actions such that they collectively solve a given environmental task. Two learning algorithms called ACE and DFG are described that provide answers to the following two questions:

- How can multiple agents learn which actions have to be carried out concurrently?
- How can multiple agents learn which sets of concurrent actions have to be carried out sequentially?

Initial experimental results are provided which illustrate the learning abilities of these algorithms.

Keywords. Multi-agent systems, distributed reinforcement learning, activity coordination, ACE algorithm, DFG algorithm

1 Motivation

Multi-agent systems establish a central research area in distributed artificial intelligence (see, e.g., Bond & Gasser, 1988; Huhns, 1987; Gasser & Huhns, 1989; Brauer & Hernandez, 1991). The interest in these systems bases on the insight that many real-world problems are better modelled using a set of interacting agents instead of a single agent (Georgeff, 1987). In particular, multi-agent modelling allows to cope with natural constraints like the limited processing power of a single agent or the geographical distribution of data and to profit from inherent properties of distributed systems like robustness, parallelism and scalability. Various multi-agent systems have been described in the literature. According to the standard or "prototypical" point of view a multi-agent system consists of a number of agents being able to interact and differing from each other in their skills and their knowledge about the environment. Each agent is assumed to be composed of a sensor component, a motor component, a knowledge base and a learning component. An agent typically is restricted in its activity for three reasons:

– because of limitations imposed on the sensor component, it knows only a part of the environment (i.e., it is not "omniscient"),

– because of limitations imposed on the motor component, it is specialized in carrying out a specific action (i.e., it is not "omnipotent"), and

– its action can be incompatible with actions carried out by other agents (i.e., different actions may prevent each other from being executed).

This prototypical point of view also underlies the work described in this paper.

Two forms of learning can be distinguished in a multi-agent system (see also Shaw & Whinston, 1989). First, isolated or centralized learning, that is, learning that is done by a single agent and that does not require the presence of other agents (e.g., learning by creating new knowledge structures). Second, collective or distributed learning, that is, learning that is done by several agents and that becomes possible only because several agents are present (e.g., learning by exchanging knowledge). Whereas centralized learning has been intensively studied since the early days of artificial intelligence, distributed learning has been neglected until a few years ago. This is in contrast to the common agreement that there are two important reasons for studying this subject:

– to be able to endow artificial multi-agent systems, which typically are very complex and hard to specify, with the ability to improve their behavior on their own; and

– to get a better understanding of the learning processes in natural (human and animal) multi-agent systems.

Despite some initial research efforts (e.g., Brazdil & Muggleton, 1991; Shaw & Whinston, 1989; Sian, 1991; Sikora & Shaw, 1991; Tan, 1993), however, there is a great number of open problems and questions on distributed learning.

This paper summarizes work that focussed on the relation between learning and action coordination in multi-agent systems and that has been part of a broader research project aiming at a more unified perspective of learning in parallel and distributed systems (Weiß, 1995). The central problem addressed is how several agents can learn to coordinate their actions such that they collectively solve environmental tasks, even if the learning feedback is minimal and consists only of a scalar reinforcement value. Two algorithms called the ACE algorithm and the DFG algorithm are described which implement distributed reinforcement learning and endow multiple agents with the ability to generate appropriate sequences of sets of compatible actions.

2 The ACE Algorithm

2.1 Overview

The ACE algorithm (ACE stands for "ACtion Estimation") is designed to solve the problem of learning appropriate sequences of action sets in multi-agent systems (e.g., Weiß, 1993a)[1]. The working method of this algorithm can be overviewed as follows. Each agent estimates the usefulness of its action in different environmental states. Based on these estimates, in each environmental state the agents compete for the right to become active. Only the winning agents are allowed to perform their actions and, by the way, to transform the actual into the next environmental state. The agents learn by collectively adjusting and improving, over time, the estimates of their actions. This adjustment, which is also known as credit assignment or apportionment of credit, is done according to the action-oriented variant (Weiß, 1992) of a reinforcement learning model called bucket brigade (Holland, 1986) which originally comes from the field of classifier systems. All together, according to the ACE algorithm the overall behavior of the multi-agent system results from the repeated execution of the competition and the credit assignment activities. The next subsections give a detailed description of these two activities.

2.2 Competition

In each environmental state S_j a competition runs between the agents. Each agent A_i makes a bid B_i^j for the right to carry out its action and announces this bid to the other agents. This bid is calculated by

$$B_i^j = \begin{cases} (\alpha + \beta) \cdot E_i^j & , \quad \text{if } E_i^j > \Theta \\ 0 & , \quad \text{otherwise} \end{cases} , \tag{1}$$

where α is a small constant called risk factor, β is a small random number called noise term, Θ is a constant called estimate minimum, and E_i^j is A_i's

---

[1] Here a slightly different notation was used.

estimate of the usefulness of its action dependent on what it knows about S_j. (E_i^j is initialized with a predefined value E^{init}.) The α indicates the fraction of E_i^j the agent A_i is willing to risk for being allowed to become active. The β introduces noise into the competition process in order to avoid getting stuck into local learning minima. (In the literature on classifier systems various methods of introducing noise into the bidding process have been described; see, for instance, (Goldberg, 1989).) The Θ helps to prevent executing useless (low-estimated) actions. In the following, $\alpha \cdot E_i^j$ is called the deterministic part and $\beta \cdot E_i^j$ is called the stochastic part of B_i^j.

After the agents have announced their bids, they select the actions that are carried out concurrently. The agent having made the highest bid is allowed to execute its action, and each potentially active agent whose action is incompatible to the selected one withdraw its bid; this is repeated until no further action associated with a non-zero bid can be selected. Formally, action selection is described by

- $\mathcal{A}^{pot}[S_j] =_{def}$ set of agents that could become active in S_j,
  $\mathcal{A}^{act}[S_j] =_{def} \emptyset$

- until $\mathcal{A}^{pot}[S_j] = \emptyset$ do
  - select $A_i \in \mathcal{A}^{pot}[S_i]$ with $B_i^j > B_k^j$ for all $A_k \in \mathcal{A}^{pot}[S_j]$
  - $\mathcal{A}^{act}[S_j] = \mathcal{A}^{act}[S_j] \cup \{A_i\}$
  - $\mathcal{A}^{pot}[S_j] = \mathcal{A}^{pot}[S_j] \setminus (A_i \cup \{A_k \in \mathcal{A}^{pot}[S_j]:$
    A_k and A_i are incompatible$\})$

- Only the agents in $\mathcal{A}^{act}[S_j]$ become active

(Note that this kind of competition requires a rational or non-egoistic behavior of the agents in the sense that none of the agents insists the execution of a low-bid or an incompatible action.)

2.3 Credit Assignment

The agents assign credit to each other by adjusting the estimates of the usefulness of their actions. Informally, this is done as follows. The agents that won the actual competition reduce the estimates of their actions (the actual winners pay for their previlege to carry out their actions) by the amount of the deterministic part of their bids, and hand the sum of all reductions back to the agents that won the previous competition (the previous winners are rewarded for appropriately setting up the environment). The previous winners, in turn, add the received sum to the estimates of their own actions. Additionally, if there is an external reward from the environment, then it is distributed among the actual winners. This adjustment of the estimates is formally described as follows. Let S_j and S_l be the actual and the previous environmental state, respectively. The estimate E_i^j of each actual winner A_i is modified according to

$$E_i^j = E_i^j - \alpha \cdot E_i^j + R^{ext} / \left| \mathcal{A}^{act}[S_j] \right| \ , \tag{2}$$

where R^{ext} is the external reward (if there is any). The estimate E_k^l of each previous winner A_k is increased according to

$$E_k^l = E_k^l + B / \left| \mathcal{A}^{act}[S_l] \right| \ . \tag{3}$$

where $B = \sum_{A_i \in \mathcal{A}^{act}[S_j]} \alpha \cdot E_i^j$ is the sum of all reductions made by the actual winners.

The effects of this bucket-brigade-type credit assigment are as follows (see Holland, 1985). An agent's estimate of its action increases (decreases), if the agent pays less (more) than it receives. As a consequence, the estimates of actions that are involved in successful sequences of action sets (i.e., sequences that lead to external reward) increase over time and stabilize this sequence; and conversely, the estimates of actions that are involved in unsuccessful sequences decrease over time and destabilize this sequence.

3   The DFG Algorithm

3.1 Overview

Like the ACE algorithm, the DFG algorithm (DFG is an acronym for "Dissolution and Formation of Groups") implements distributed reinforcement learning of action sequences (e.g., Weiß, 1993b). In contrast to the ACE algorithm, however, the DFG algorithm explicitly distinguishes between single agents and groups of compatible agents as the acting units in a multi-agent system. Now agents as well as groups estimate the usefulness of their activities in different environmental states, and agents as well as groups compete for the right to become active. In particular, now learning encompasses two interrelated processes: first, bucket-brigade-type credit assignment; and second, group development, that is, the process of dissolving existing (useless) and forming new (useful) groups. The following subsections describe the DFG algorithm in detail.

3.2 Groups as Acting Units

As it is known from organization theory and management science, single agents typically serve as building blocks for more complex structured and

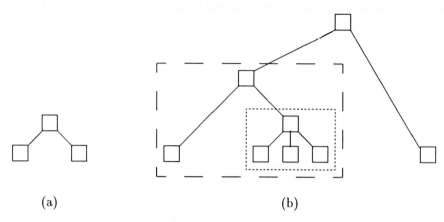

(a)   (b)

Fig. 3.1. Examples of group structures. *(a)* shows the most simple group which consists of a leader and two members each being a single agent. *(b)* shows a more complicated group which has two members, one being a group (dashed box) and the other being a single agent. The "dashed group" also has a group (dotted box) and a single agent as its members, where the "dotted group" has three members each being a single agent. (Legend: $\square$ single agent, / \ leader-member relations.)

autonomously acting units (see, e.g., Fox, 1981; Galbraith, 1973). The DFG algorithm adopts this point of view, and distinguishes between single agents and groups as the acting units in a multi-agent system. A group is considered to be composed of a group leader and several compatible group members, where a group leader is a single agent and a group member is either a single agent or another group. This recursive definition is rather general and covers both low and high structured groups; see figure 3.1 for an illustration. The task of a group leader is to represent the group's interests; this includes, for instance, to decide whether the group should persist as an autonomously acting unit, cooperate with another acting unit, or dissolve. The group members have to be compatible in the sense that the activity of no member leads to environmental changes that prevent the activity of another member.

The following simple notation is used in the following subsections. U_i refers to an acting unit, that is, either to a single agent or to a group. If U_i refers to a group, then $\overline{U_i}$ denotes the leader of this group; otherwise, if U_i refers to a single agent, then $\overline{U_i}$ simply denotes this agent, too. Finally, $[U_i, S_j]$ denotes the knowledge that the agents contained in U_i have about S_j; $[U_i, S_j]$ is called the knowledge context of U_i in S_j. (Note that $[U_i, S_j] \cap [U_k, S_j]$ may but need not be empty, and that $\bigcup_i [U_i, S_j]$ is not necessarily equal to S_j. Similarly, $[U_i, S_j] \cap [U_i, S_k]$ may but need not be empty; in particular, it may be the case that $[U_i, S_j] = [U_i, S_k]$, which means that a unit may be unable to distinguish between different environmental states.)

3.3 Competition and Credit Assignment

Competition and credit assignment is done analogously to the ACE algorithm. In each environmental state S_j the acting units compete for the right to become active. Each unit U_i calculates a bid B_i^j,

$$B_i^j = (\alpha + \beta) \cdot E_i^j \quad , \tag{4}$$

where α is the risk factor, β is the noise factor, and E_i^j is $\overline{U_i}$'s estimate of U_i's usefulness dependent on the knowledge context $[U_i, S_j]$. Only the unit making the highest bid is allowed to become active and, in this way, transforms the actual into a new environmental state. This selection of a single winning unit corresponds to the selection of the action set that is carried out in the actual state.

Credit assignment again is done in a bucket brigade style. Let U_i be the winning unit in the actual state S_j, and let U_k be the winning unit in the preceding state S_l. $\overline{U_i}$ reduces its estimate E_i^j by the amount of the deterministic part of its bid B_i^j, and hands this amount back to $\overline{U_k}$. $\overline{U_k}$, in turn, adds the received amount to its estimate E_k^l. Additionally, if the activity of U_i leads to an external reinforcement R^{ext}, then $\overline{U_i}$ adds this reinforcement to its estimate E_i^j. All together, credit assignment involves the following adjustments:

$$E_i^j = E_i^j - \alpha \cdot E_i^j + R^{ext} \tag{5}$$
$$E_k^l = E_k^l + \alpha \cdot E_i^j \quad . \tag{6}$$

Based on Grefenstette's (1988) convergence result it has been shown that under the DFG algorithm the estimates of successively active units tend to converge to an equilibrium level (Weiß, 1993b). Moreover, it can be shown that every solution path learnt under the DFG algorithm is cycle-free in the sense that no environmental state is involved in this path more than one time.

3.4 Group Development

The DFG algorithm distinguishes two contrary processes of group development, namely, group formation and group dissolution. Both processes largely depend on the past usefulness of the agents' and the groups' activities. In order to be able to decide about the formation of new groups and the dissolution of existing ones, each $\overline{U_i}$ calculates the mean values of its estimates over the previous episodes, where an episode is defined as the time interval between the receipts of two successive environmental reinforcements. More exactly, during each episode $\tau + 1$, $\overline{U}_i$ calculates the gliding mean value $M_i^j[\tau + 1]$ of its estimate E_i^j as

$$M_i^j[\tau + 1] = \frac{1}{\nu} \cdot \sum_{T=\tau-\nu+1}^{\tau} E_i^j[T] \quad , \tag{7}$$

where ν is a constant called window size and $E_i^j[T]$ is E_i^j at the end of episode T. Both group formation and group dissolution proceed in dependence on the gliding mean values of the estimates.

In the following, let S_j be the actual environmental state and let $\tau + 1$ be the actual episode. For each unit U_i that is potentially active in S_j, $\overline{U}_i$ decides that U_i is ready to cooperate and to form a new group with other units in the context $[U_i, S_j]$ if

$$M_i^j[\tau + 1] \leq \sigma \cdot E_i^j[\tau - \nu] \quad , \tag{8}$$

where σ is a constant called cooperation factor which influences the units' readiness to form new groups. In words, according to this decision criterion a unit intends to cooperate in a specific context, if the estimate of its usefulness tends to increase too slowly, to stagnate or to decrease. The units that are ready to cooperate in their activity contexts form new groups according to the simple principles "stronger units choose their cooperation partners first" and "stronger units are preferred as cooperation partners." Formally, group formation is done as follows. Let $\mathcal{U}$ be the set of all units that are ready to cooperate:

until $\mathcal{U} = \emptyset$ do

- Let $U_i \in \mathcal{U}$ be the unit with $E_i^j = \max_l \{E_l^j : U_l \in \mathcal{U}\}$. Then $\overline{U}_i$ announces a "cooperation offer" to the other units.

- For each unit $U_l \in \mathcal{U}$ that is compatible with U_i, $\overline{U}_l$ sends a "cooperation response" to $\overline{U}_i$.

- Let $\mathcal{U}^{resp} \subseteq \mathcal{U}$ be the set of all responding units. Then $\overline{U}_i$ chooses the unit $U_k \in \mathcal{U}^{resp}$ with $E_k^j = \max_l \{E_l^j : U_l \in \mathcal{U}^{resp}\}$ as the cooperation partner of U_i. $\overline{U}_i$ and $\overline{U}_k$ form a new group G that has U_i and U_k as its members and either $\overline{U}_i$ or $\overline{U}_k$ as its leader. (The leader initializes the estimate of the new group's usefulness with a predifined value E^{init}.)

- $\mathcal{U} = \mathcal{U} \setminus \{U_i, U_k\}$.

(Note that each group has exactly two members; this could be easily extended towards multi-member groups by allowing $\overline{U}_i$ to chose several cooperation partners.) There are two things that have to be stressed. First, if two units form a new group, then this occurs within the frame of the units' knowledge contexts. Second, the process of group formation does not require an exchange of environmental information – neither among group members, nor among group members and leaders, nor among leaders of different groups. With that, cooperation and group formation is a highly context-sensitive process, but does not require an exchange of environmental information among the units.

For each group U_i that is potentially active in S_j, $\overline{U_i}$ decides that U_i has to dissolve in its members if

$$M_i^j[\tau + 1] \leq \rho \cdot E^{init} \quad , \tag{9}$$

where ρ is a constant called dissolution factor which influences the robustness of the existing groups, and E^{init} is the initialization value of the estimates. According to this criterion a group dissolves, if the estimate of its usefulness, averaged over the previous episodes, falls below a certain minimum.

4 Experiments

4.1 Task Domain and Analysis

As a task domain the blocks world is chosen. This domain has been intensively studied in the fields of problem solving and planning, and is clear enough for doing the initial experimental studies in the unknown field of multi-agent learning. What has to be learnt by a given set of agents is to transform a start constellation of blocks into a goal constellation within a limited time interval. This paper summarizes the results on the task shown in figure 4.1. In this task, each agent is specialized in a specific action; for instance, agent A_1 is able to put block A on the bottom (symbolized by a $\perp$) and agent A_6 is able to put block D on block A. The precondition for applying an action $put(x, y)$ is that no other blocks are placed on x and y, i.e. x and y have to be empty. Each agent is assumed to have only minimal information about the environment: it only knows ("sees") whether the precondition of its action $put(x, y)$ is fulfilled. Because of this constraint, an agent is unable to distinguish between all different environmental states. In particular, an agent may be unable to distinguish between a state in which its action is useful and a state in which its action is not useful. (For instance, agent A_2 cannot distinguish between a "relevant state" in which block B is placed on the bottom and an "irrelevant state" in which block B is placed on block F. The fact that an action may be relevant in one state but irrelevant in another is sometimes called the Sussman's anomaly; see, e.g., (Ginsberg, 1986).) Two actions are considered to be incompatible if their concurrent execution is not possible. Examples of sets of incompatible

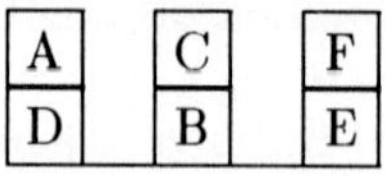

A	C	F
D	B	E

Start Constellation

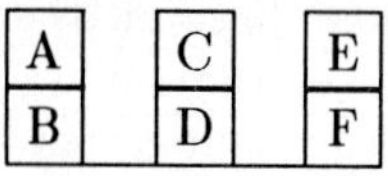

A	C	E
B	D	F

Goal Constellation

Agents: A_1: $put(A, \perp)$ A_2: $put(A, B)$

A_3: $put(B, F)$ A_4: $put(C, \perp)$

A_5: $put(C, D)$ A_6: $put(D, A)$

A_7: $put(E, F)$ A_8: $put(F, \perp)$

A_9: $put(F, E)$

Limited Time Interval: at most 4 cycles

Fig. 4.1. A basic blocks world task.

actions are $\{put(A, \perp), put(A, B)\}$ (i.e., a block cannot be placed on different positions at the same time), $\{put(B, F), put(E, F)\}$ (i.e., different blocks cannot be put on the same block), and $\{put(C, D), put(D, A)\}$ (i.e., a block cannot be put on a block which, at the same time, is put on another block). The transformation from the start into the goal constellation has to be done in at most four cycles.

There is one solution sequence (i.e., a sequence of action sets that transforms the start into the goal constellation) of length 2, 24 solution sequences of length 3, and 210 solution sequences of length 4. The solution sequence of length 2 is given by $\langle\{put(A, \perp), put(C, \perp), put(F, \perp)\}, \{put(A, B), put(C, D),$ $put(E, F)\}\rangle$. Because every solution sequence contains at least 5 actions, this task can not be solved by means of a sequential "one-action-per-cycle" approach. In the case of a random walk through the search space, the probability of finding the solution sequence of length 2 is less than 1 percent, the probability of finding a solution sequence of length 3 is less than 4 percent, and the probability of finding a solution sequence of length 4 is less than 5 percent. With that, the probability that a random sequence of maximal length four transforms the start into the goal constellation is less than 10 percent.

4.2 Results

The experimental setting was as follows. A trial is defined as any sequence of at most four cycles that transforms the start into the goal constellation

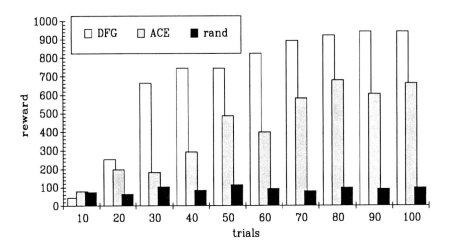

Fig. 4.2. Performance profiles.

(successful trial), as well as any other sequence of exactly four cycles that transforms a start into a non-goal constellation. Learning proceeds by the repeated execution of trials. At the end of each trial the start constellation is restored, and the agents again try to solve the task. Additionally, only at the end of each successful trial a non-zero external reward R^{ext} is provided. Parameter setting: $E^{init} = R^{ext} = 1000$, $\alpha = 0.15$, $\beta \in [-\alpha/5 \ldots + \alpha/5]$ (randomly chosen for every bid), $\nu = 4$, $\sigma = 1 + 3\alpha$, and $\rho = 1 - \alpha$.

Figure 4.2 shows the performance profiles of the ACE algorithm, the DFG algorithm, and a random-walk algorithm that randomly chooses an applicable set of compatible actions at the begining of each trial. (Further experimental results are described in (Weiß, 1993a, 1993b).) Each data value reflects the mean environmental reward per trial obtained during the previous 10 trials, averaged over 10 runs started with different random-number-generator seeds. There are several important observations. First, the learning performance of ACE and the DFG algorithm was significantly above the random performance level which is about 100. Both algorithms reached their highest average performance level after about 80 trials. This shows that under both algorithms the agents are able to learn useful sequences of action sets. Second, the DFG algorithm clearly performed better than the ACE algorithm; additionally, the DFG algorithm shows a more "'smooth"' performance than the ACE algorithm. The reason for that is that under the ACE algorithm the usefulness of an action is estimated only in dependence on the environmental state, whereas under the DFG algorithm it is estimated in dependence on both the environmental state and other actions carried out concurrently. As a consequence, the DFG algorithm does better cope with the fact that the usefulness of an action set does nothing say about the usefulness of a subset or a superset of this action set. Finally, as a consequence of the minimal-information constraint described above, the average performance of the ACE and the DFG algorithm remained below the maximum reward level (1000). This observation clearly shows the importance of information and information exchange mechanisms in the context of cooperating agents.

5 Critique and Future Work

This paper summarized work on learning and action coordination in multi-agent sytems. Two algorithms, ACE and DFG, were described which enable multiple agents to learn useful action sequences.

The experimental results are very encouraging, but they also showed that the ACE and the DFG algorithm leave room for improvement. A limitation of both algorithms is that their learning success relies on an explicit exploration of a sufficient number of state-action pairs. This leads to an impractical amount of time required for learning, if the state and action spaces are too large and complex. There are two standard methods which can be used in order to cope with this kind of limitation: to endow the agents with the ability to generalize over the search space; and to endow the agents with the ability to built up an internal world model which can be used for look-ahead and planning activities. The ACE and the DFG algorithm are very general learning schemes which both allow an incorporation of these methods. This incorporation has to be a topic of future research.

Other important research topics are, for instance, the investigation of alternative group concepts (e.g., groups with variable bindings between the members) and alternative strategies for the dissolution and the formation of groups. In attacking these topics, the artificial intelligence community should also take into consideration related work from other disciplines such as psychology (e.g., Guzzo, 1982) and economics (e.g., Argyris & Schön, 1978).

References

Argyris, C., Schön, D.A.: Organizational learning. Reading, MA: Addison-Wesley 1978

Bond, A.H., Gasser, L. (eds.): Readings in distributed artificial intelligence. San Mateo, CA: Morgan Kaufmann 1988

Brauer, W., Hernández, D. (eds.): Verteilte Künstliche Intelligenz und kooperatives Arbeiten. Berlin: Springer-Verag 1991

Brazdil, P., Muggleton, S.: Learning to relate terms in a multiple agent environment. In: Kodratoff, Y. (ed.): Machine learning – EWSL-91. Berlin: Springer-Verlag 1991, pp. 424–439

Fox, M.S.: An organizational view of distributed systems. IEEE Trans. on Systems, Man, and Cybernetics SMC-11(1), 70–80 (1981)

Galbraith, J.: Designing complex organizations. New York: Wiley 1973

Gasser, L., Huhns, M.N. (eds.): Distributed artificial intelligence (Vol. 2). London: Pitman 1989

Georgeff, M.P.: Many agents are better than one. SRI International, Menlo Park, CA, Technical Report 417, 1987

Ginsberg, M.L.: Possible worlds planning. In: Georgeff, M.P., Lansley, A.L. (eds.): Reasoning about actions and plans – Proceedings of the 1986 workshop. Morgan Kaufmann 1986, pp. 213–243

Goldberg, D.E.: Genetic algorithms in search, optimization, and machine learning. Reading, MA: Addison-Wesley 1989

Grefenstette, J.J.: Credit assignment in rule discovery systems based on genetic algorithms. Machine Learning 3, 225–245 (1988)

Guzzo, R.A.: Improving group decision making in organizations – Approaches from theory and research. Academic Press 1982

Holland, J.H.: Properties of the bucket brigade algorithm. In: Grefenstette, J.J. (ed.), Proceedings of the First International Conference on Genetic Algorithms and Their Applications. Hillsdale, NJ: Lawrence Erlbaum 1985, pp. 1–7

Holland, J.H.: Escaping brittleness: The possibilities of general-purpose learning algorithms to parallel rule-based systems. In: Michalski, R.S., Carbonell, J.G., Mitchell, T.M. (eds.): Machine learning: An artificial intelligence approach (Vol. 2). Morgan Kaufmann 1986, pp. 593–632

Huhns, M. (ed.): Distributed artificial intelligence. London: Pitman 1987

Shaw, M.J., Whinston, A.B.: Learning and adaptation in distributed artificial intelligence. In: (Gasser & Huhns, 1989, pp. 413–429)

Sian, S.S.: Adaptation based on cooperative learning in multi-agent systems. In: Demazeau, Y., Müller J.-P. (eds.): Decentralized AI (Vol. 2). Amsterdam: Elsevier 1991

Sikora, R., Shaw, M.J.: A distributed problem-solving approach to inductive learning. College of Commerce and Business Administration, University of Illinois at Urbana-Champaign. Faculty Working Paper 91-0109, 1991

Tan, M.: Multi-agent reinforcement learning: Independent versus cooperative agents. Proceedings of the Tenth International Conference on Machine Learning. 1993, pp. 330–337

Weiß, G.: Learning the goal relevance of actions in classifier systems. Proceedings of the Tenth European Conference on Artificial Intelligence. Chichester: Wiley 1992, pp. 430–434

Weiß, G.: Learning to coordinate actions in multi-agent systems. Proceedings of the 13th International Joint Conference on Artificial Intelligence. San Mateo, CA: Morgan Kaufmann 1993a, pp. 311–316

Weiß, G.: Action selection and learning in multi-agent environments. Proceedings of the Second International Conference on Simulation of Adaptive Behavior. Cambridge, MA: MIT Press 1993b, pp. 502–510

Weiß, G.: Distributed machine learning. Sankt Augustin, Germany: infix Verlag 1995

A New Three-degree-of-freedom Spatially Mobile Robot Topology

or

Kurt: A Mobile Robot for the Study of Prototypical Autonomy

Martin Nilsson

Swedish Institute of Computer Science, Box 1263, S-164 28 Kista Sweden
mn@sics.se

Abstract. Most autonomous mobile robots are constructed as vehicles with wheels, legs, or tracks, confining them to life in a two-dimensional world. This certainly limits the kinds of interesting behaviours that can be studied with these robots. Although there are also spatially mobile robots, constructed as helicopters, airplanes, or blimps, they usually have several serious disadvantages, such as very limited payload, high cost, difficulty of control, restriction to outdoor use, or high power consumption. In this paper, we propose a new topology for a spatially mobile robot which is simple, easy to control, inexpensive, and accepts relatively large payloads with relatively low power consumption.

Keywords. Robot, behaviour, space, mobile, autonomous, topology, kinematics.

Introduction

Although simulations can be very helpful for research in behavioural robotics, practical experiments on real robots are very important, as has been noted by many researchers, e.g. [5]. The capabilities of a robot dictate what behaviours can be studied, but unfortunately, very capable robots tend to become bulky, expensive, and complicated. Since robotic vehicles with wheels, legs, or tracks are relatively easy to build [3, 4], two-dimensional wall-following and obstacle-avoidance behaviours have dominated in behavioural robotics.

If this stereotypical robot could be generalised, say, to move autonomously in three dimensions, there is an abundance of new interesting behaviours that could be studied, not to speak of many potential practical applications. Spatially mobile robots have been thoroughly studied [2], but as they are constructed as helicopters, airplanes, or blimps, they usually have several serious disadvantages, including very limited payload, high cost, difficulty of control, restriction to outdoor use, or high power consumption. These problems become even more accentuated when considering multiple cooperating robots.

In this paper, we propose a new topology for a spatially mobile robot which is simple, easy to control, inexpensive, and accepts relatively large payloads with low power consumption. This topology allows a far greater range of interesting behaviours than a surface-bound vehicle robot.

Kurt resembles the NIST spider, a robot described by Albus, Bostelman, and Dagalakis [1]. This robot is a version of the Stewart platform, where the six links have been replaced by cables, controlling all six degrees of freedom, whereas Kurt only has three degrees of freedom. However, advantages with Kurt is that construction is considerably simpler, and that both the direct and inverse kinematics is very straightforward, as we shall show below. The remaining three degrees of freedom could in Kurt's case suitably be implemented on the platform with a local manipulator.

In section 2, we shall describe the basic structure of the robot. We suggest some examples of behaviours and applications that become possible with this robot in section 3. The kinematics for the robot is very simple, as we show in section 4. This section is particularly valuable for those who plan to design a similar robot. Section 5 concludes the paper.

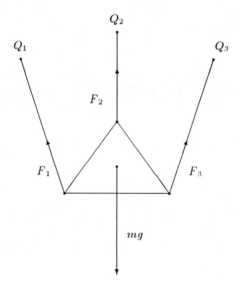

Figure 1: Robot structure

Robot Structure

As often happens in problem solving, ideas for solutions can often be found in Nature. In our case, the inspiration came from *spiders*, which have an elegant way of moving in space by hauling in and out wires. We named the robot

Kurt, from Japanese *kumo robotto*, meaning "spider robot." The heart of the robot is a platform, shaped as an equilateral triangle. The payload hangs down vertically from the centre of the triangle. There is one motor in each corner, and to each motor a wire is attached. The other ends of the three wires are attached to three different points on walls, ceiling, or poles (fig. 1).

By winding or unwinding wire, the robot can now move in space. Theoretically, every position under the attachment points can be reached by winding the appropriate amount of wire with the motors. Limited motor torque constrains the workspace somewhat, but not too seriously. In section 5, we show that if the robot is more than $s/7$ below the lowest attachment point, where s is the maximal distance between attachment points, the maximum load on any wire is always less than twice the robot's weight. This section also shows that both kinematics and inverse kinematics become very simple for this robot topology.

Behaviours and Applications

For Kurt, classical obstacle avoidance behaviour generalises to the spatial equivalent, and wall following corresponds to surface following. Other types of two-dimensional behaviours have corresponding three-dimensional behaviours, the main difference being that spatial mobility allows much larger workspace than surface mobility. For instance, hiding away can be done on a shelf somewhere, and finding a light source or beacon is done by moving in space. Object recognition becomes more interesting, as Kurt easily can also move above objects.

Most interesting are probably some behaviours made possible by Kurt's inherent features. For instance, resting on a shelf, Kurt can pull itself over the edge, and throw itself out very quickly to reach a desired position, when some work needs to be done. The wires can be used as touch sensors, detecting obstacles and other disturbances ("flies") in the workspace.

Two robots with this topology will be able to access each other not only from the sides, but also from above and below, and could thus be able to repair each other. Several robots could together lift and handle lengthy objects, which is hard with vehicle-type robots.

From a practical standpoint, Kurt could be used as an intelligent overhead crane. When idle, Kurt hides in a corner on a shelf. When Kurt's detectors are trigged, Kurt springs into action and moves over to the spot of action. Kurt could easily pick up things from floors, desks, or shelves that would be very hard for a floor-bound robot to reach. In this way, Kurt would be suitable as an office or home assistant robot.

Kurt could also be equipped with wheels or legs, so that it could move out of the wire workspace, under furniture, or deep inside shelves. An advanced version of Kurt could have climbing robots at the far ends of the wires instead of having them fixed. This would allow Kurt to travel to find some suitable climbable objects, and send up the climbing robots. After these robots have safely attached themselves at some height, Kurt could raise up and explore the space.

Kinematics

In this section we show that the proposed topology allows easy computation of the cartesian position and the wire forces, given the distance from each corner. We shall assume that wires have little mass compared to the robot, and that the diameter of the robot is small compared to the distance from the attachment points.

Since the robot's diameter is small compared to the distance from the corners Q_1, Q_2, Q_3, we can approximate it by a point mass m at a point P.

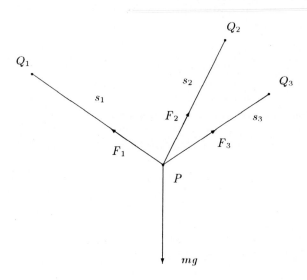

Figure 2: Wire forces

Assume that the corners $Q_i = (q_{x,i}, q_{y,i}, q_{z,i})$ are known, and that the distances from $P = (p_x, p_y, p_z)$ are measured to be s_i. Then for $i = 1, 2, 3$

$$\left| \overline{PQ_i} \right|^2 = s_i^2 \tag{1}$$

i.e.

$$\begin{cases} (p_x - q_{x,1})^2 + (p_y - q_{y,1})^2 + (p_z - q_{z,1})^2 = s_1^2 \\ (p_x - q_{x,2})^2 + (p_y - q_{y,2})^2 + (p_z - q_{z,2})^2 = s_2^2 \\ (p_x - q_{x,3})^2 + (p_y - q_{y,3})^2 + (p_z - q_{z,3})^2 = s_3^2. \end{cases} \tag{2}$$

Subtracting the first equation from the others results in

$$\begin{cases} (p_x - q_{x,1})^2 + (p_y - q_{y,1})^2 + (p_z - q_{z,1})^2 \quad = \\ \quad s_1^2 \\ p_x(2q_{x,1} - 2q_{x,2}) + p_y(2q_{y,1} - 2q_{y,2}) + p_z(2q_{z,1} - 2q_{z,2}) = \\ \quad s_2^2 - s_1^2 + q_{x,1}^2 - q_{x,2}^2 + q_{y,1}^2 - q_{y,2}^2 + q_{z,1}^2 - q_{z,2}^2 \\ p_x(2q_{x,1} - 2q_{x,3}) + p_y(2q_{y,1} - 2q_{y,3}) + p_z(2q_{z,1} - 2q_{z,3}) = \\ \quad s_3^2 - s_1^2 + q_{x,1}^2 - q_{x,3}^2 + q_{y,1}^2 - q_{y,3}^2 + q_{z,1}^2 - q_{z,3}^2. \end{cases} \tag{3}$$

From the latter two equations we can extract p_y and p_z as linear expressions in terms of p_x. These expressions substituted into the first equation gives a quadratic equation for p_x, from which easily two solutions can be obtained, one of which corresponds to P, and the other one to the mirror image of P in the triangle $\triangle Q_1 Q_2 Q_3$.

Theoretically, the workspace of the robot is the whole volume under the $\triangle Q_1 Q_2 Q_3$. This is constrained by the maximal allowable wire forces.

Let us compute the force in wire $\overline{PQ_1}$. We have

$$\boldsymbol{F}_1 + \boldsymbol{F}_2 + \boldsymbol{F}_3 = mg\boldsymbol{e}_z. \tag{4}$$

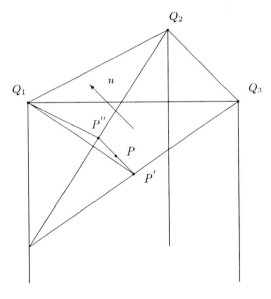

Figure 3: Maximum forces occur on the sides

Let $n = (n_x, n_y, n_z)$ be a vector normal to the plane through P, Q_2, and Q_3. For instance, we could choose $n = \overline{PQ_3} \times \overline{PQ_2}$. If we multiply by n on both sides,

$$n F_1 = mg e_z n = mg n_z. \tag{5}$$

so

$$|F_1| = \frac{mg n_z}{|n| \cos(F_1, n)} = \frac{mg \hat{n}_z}{\sin \varphi}, \tag{6}$$

where φ is the angle between $\overline{PQ_1}$ and the PQ_2Q_3 plane and $\hat{n}_z = n_z/|n|$.

Let v_i be the vector from the centre of the platform to corner i. The triangle $\triangle P_1 P_2 P_3$ is equilateral, so

$$\begin{cases} v_1 + v_2 + v_3 = 0 \\ |v_1| = |v_2| = |v_3|. \end{cases} \tag{7}$$

If the diameter of the platform is small compared to the distances from Q_1, Q_2, and Q_3, we can use the values of the forces calculated in the previous section. Torque equilibrium around the centre of $\triangle P_1 P_2 P_3$ requires

$$F_1 \times v_1 + F_2 \times v_2 + F_3 \times v_3 = 0. \tag{8}$$

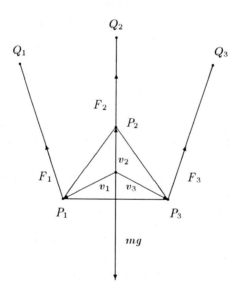

Figure 4: Calculating platform attitude

Using (7) we get

$$\begin{cases} \boldsymbol{v}_3 = -\boldsymbol{v}_1 - \boldsymbol{v}_2 \\ |\boldsymbol{v}_1| = |\boldsymbol{v}_2| = |\boldsymbol{v}_1 + \boldsymbol{v}_2| \\ (\boldsymbol{F}_1 - \boldsymbol{F}_3) \times \boldsymbol{v}_1 + (\boldsymbol{F}_2 - \boldsymbol{F}_3) \times \boldsymbol{v}_2 = 0. \end{cases} \quad (9)$$

From the last equation of (9) we can express $\boldsymbol{v}_2$ as a linear function of $\boldsymbol{v}_1$, $\boldsymbol{v}_2 = \mathbf{A}\boldsymbol{v}_1$. The remaining equations are

$$|\boldsymbol{v}_1| = |\mathbf{A}\boldsymbol{v}_1| = |(\mathbf{A} + \mathbf{I})\boldsymbol{v}_1|. \quad (10)$$

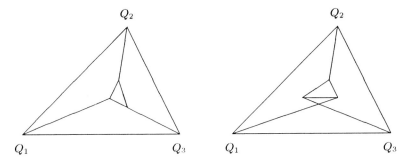

Figure 5: Correct and tangled configurations (top view)

In order to find the direction of $\boldsymbol{v}_1$, we can assume that one of the three cases $v_{1,x} = 1$, $v_{1,x} = 0$, or $v_{1,x} = -1$ hold, and solve the two remaining coupled quadratic equations for $v_{1,y}$ and $v_{1,z}$. There will be several solutions; some of these will correspond to configurations will tangled wires. The correct solution is the one such that projected in the horizontal plane, each P_i is nearer Q_i than any P_j is, $i \neq j$.

It is helpful to have some idea of how the wire forces change as the robot moves. Intuitively, the forces should increase as the robot rises. An obvious question is whether there is an upper bound for the wire forces, provided the robot is at least a certain distance d below the lowest corner. In this section we shall show that in fact for the maximum force F_{max} in any wire,

$$d \geq \frac{s}{4\sqrt{k^2 - 1}} \Rightarrow F_{max} \leq kmg, \quad (11)$$

where $k > 1$, and s is the diameter of the projection of $\triangle Q_1 Q_2 Q_3$ on the horizontal plane.

In order to derive this result, we shall first show the following maximum principle. *The maximum wire forces for a given height always occur on the vertical sides of the workspace.*

Again, consider the PQ_2Q_3 plane. The horizontal plane through P intersects this plane along a line $P'P''$, where P' and P'' lie on the sides of the workspace. We notice that $\hat{n}_z$ is constant along this line, but φ changes, so equation 6 implies that $|\boldsymbol{F}_1|$ will be maximal when $\sin\varphi$ is minimal. Clearly, φ, and thus $\sin\varphi$, will be minimal at the point most distant from Q_1. This must happen in either P' or P''. In other words, $|\boldsymbol{F}_1|$ is maximal only on the sides.

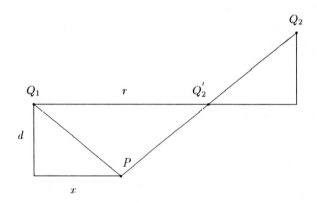

Figure 6: The robot on the Q_1Q_2 side of the workspace

In order to find an upper bound on the wire forces, the maximum principle allows us now to assume that P lies in the vertical plane through Q_1 and Q_2. Suppose that Q_2 is positioned higher than Q_1. The wire forces are parallel with the respective wires, so $\boldsymbol{F}_1 \parallel \overline{PQ_1}$ and $\boldsymbol{F}_2 \parallel \overline{PQ_2} \parallel \overline{PQ_2'}$. Denoting $|\overline{PQ_2'}|$ by r,

$$\begin{cases} \boldsymbol{F}_1 = \overline{PQ_1} \cdot c_1 = (-x, d) \cdot c_1 \\ \boldsymbol{F}_2 = \overline{PQ_2'} \cdot c_2 = (r - x, d) \cdot c_2 \end{cases} \tag{12}$$

Equilibrium requires

$$\boldsymbol{F}_1 + \boldsymbol{F}_2 = (0, -mg) \tag{13}$$

from which we get the equations

$$\begin{cases} -xc_1 + (r - x)c_2 = 0 \\ dc_1 + dc_2 = -mg \end{cases} \tag{14}$$

with the solution

$$\begin{cases} c_1 = -mg/d \cdot (r - x)/r \\ c_2 = -mg/d \cdot x/r. \end{cases} \tag{15}$$

We have

$$|\boldsymbol{F}_1| = |(-x, d)| \cdot |c_1| = \frac{mg}{d} \frac{r - x}{r} \sqrt{x^2 + d^2}. \tag{16}$$

If the premise of (11) holds,

$$d \geq \frac{s}{4\sqrt{k^2 - 1}} \geq \frac{r}{4\sqrt{k^2 - 1}} \geq \frac{r^2}{4r\sqrt{k^2 - (1 - x/r)^2}} = \tag{17}$$

$$\frac{r^2/4}{\sqrt{r^2 k^2 - (r - x)^2}} \geq \frac{x(r - x)}{\sqrt{r^2 k^2 - (r - x)^2}}. \tag{18}$$

Squaring both left and right hand sides,

$$d^2 \geq \frac{x^2(r - x)^2}{r^2 k^2 - (r - x)^2}. \tag{19}$$

Extracting k^2,

$$k^2 \geq \frac{1}{d^2} \frac{(r - x)^2}{r^2} (x^2 + d^2). \tag{20}$$

Taking the square root and multiplying by mg, finally

$$kmg \geq \frac{mg}{d} \frac{r - x}{r} \sqrt{x^2 + d^2} = |\boldsymbol{F}_1|. \tag{21}$$

The same relation is derived similarly for $\boldsymbol{F}_2$, which concludes the proof.

As a curiosity, if Φ denotes the "golden ratio," $(\sqrt{5} - 1)/2$, we can show that $d \geq \Phi^{5/2} s < 0.3003s$ implies that $F_{max} \leq mg$, by setting $k = 1$ and maximising the right hand side of equation (19). This bound is better than (11) for k less than approximately 1.3.

Discussion and Conclusions

We have described a robot topology for movement in three dimensions. The design uses only three motors in a mechanically simple symmetric configuration. The system can easily be built so that power is not consumed when the robot maintains its position. Both direct and inverse kinematics is very simple. A drawback is that the robot must be attached to three points by wires. However, considering that a floor-bound robot is very restricted vertically, as well as usually walled-in by thresholds, doors, etc, this drawback doesn't seem to be all that serious in comparison. This robot could without problems cooperate with conventional vehicle-type robots in order to combine the advantages of both.

Acknowledgments

We are indebted to Luc Steels, the other organisers, and the participants of the Nato Advanced Sciences Institute on the Biology and Technology of Intelligent Autonomous Agents, for a workshop we found tremendously stimulating and inspiring. We are grateful to Gunnar Sjödin for making us pay attention to the bound for the special case $k = 1$.

References

1. J. Albus, R. Bostelman, N. Dagalakis, The NIST spider, A robot crane, J. Res. Inst. Stand. Technol. 97, 373–385, 1992
2. M. Arbib, G. Bekey, and T. Lewis, Brain simulation and autonomous robots. In B. Svensson, Proc. Int. workshop on mechatronical computer systems for perception and action, Centre for Computer Architecture, Halmstad University, Sweden, 1–10, June 1993
3. A.M. Flynn, R.A. Brooks, W.M. Wells III, D.S. Barrett, Intelligence for miniature robots, Journal of Sensors and Actuators, Vol. 20, 187–196, Also appears as Squirt, the prototypical mobile robot for autonomous graduate students, MIT AI Memo 1120, July 1989
4. J.L. Jones and A.M. Flynn, Mobile robots — inspiration to implementation, AK Peters, Wellesley, MA, 1993, ISBN 1-56881-011-3
5. T. Smithers, Taking eliminative materialism seriously, a methodology for autonomous systems research, Handout of the Nato advanced sciences institute workshop on the biology and technology of intelligent autonomous agents, Levico, Italy, February 1993

How Swarms Build Cognitive Maps

Dante R. Chialvo[1] and Mark M. Millonas[2]

[1] Fluctuations & Biophysics Group, The Santa Fe Institute, 1399 Hyde Park Road, Santa Fe, NM 87501 and Ecology & Evolutionary Biology and ARL, Neural Systems, Memory & Aging, University of Arizona.
[2] Theoretical Division and CNLS, MS B258 Los Alamos National Laboratory, Los Alamos, NM 87545 and Fluctuations & Biophysics Group, The Santa Fe Institute.

Abstract. Swarms of social insects construct trails and networks of regular traffic via a process of pheromone laying and following. These patterns constitute what is known in brain science as a cognitive map. The main difference lies in the fact that the insects write their spatial memories in the environment, while the mammalian cognitive map lies inside the brain. This analogy can be more than a poetic image, and can be further justified by a direct comparison with the neural processes associated with the construction of cognitive maps in the hippocampus. We investigate via analysis and numerical simulation the formation of trails and networks in a collection of insect-like agents. The agents interact in simple ways which are determined by experiments with real ants.

1 Introduction

The self-organization of neurons into a brain-like structure, and the self-organization of ants into a swarm are similar in many respects. The former, for obvious reasons, has received more attention recently. However, the basic ideas of connectionism and mass action can be traced to the earlier work of Wilson on social insects.[3] Wilson defines communication as "action on the part of an organism that alters the probability pattern of behavior in another organism in a fashion adaptive to either one of both of the participants."[1, 2] Collective communication, such as that found in social insects, occurs when this communication leads to an emergent behavioral structure (mass action) which is adaptive. This organization, which has been called "mass communication",[3] occurs in spite of the fact that the individual organisms neither posses, nor are able to communicate, the complex global types of information these structures are able to transmit. Thus, in a certain sense, while the ants are able to construct a cognitive map for their behavior, a good part of this map lies outside of the individual ants.

Unlike simple direct action, which can often be studied by observation of the actual organisms, and broken down into its component parts, collective behavior involves statistical effects of a large number of individual agents. The behavioral coding is expressed in collective rather than in individual terms, and it is sometimes difficult in the most complex cases to determine which behavioral signals in combination are responsible for a particular collective behavior, and which are sufficient. How then are we to dissect the swarm and determine how individual behavioral components interact, and how they function in the whole? What we would like to do here it to show that it is possible to do theoretical and numerical analysis of this situation. In addition, we point out the similarities between the underlying dynamics, and the emergent properties of the swarm to the organization of cognitive maps in the hippocampus.

In *The Insect Societies* Wilson forecasted the eventual appearance of what he called "a stochastic theory of mass behavior" and asserted that "the reconstruction of mass behaviors from the behaviors of single colony members is the central problem of insect sociobiology." He forecasted that our understanding of individual insect behavior together with the sophistication with which we would be able to analyze their collective interaction would advance to the point where we would one day posses a detailed, even quantitative, understanding of how individual "probability matrices" would lead to mass action on the level of the colony. By replacing *colony members* with *neurons*, *mass behaviors* or *colony* by *brain behavior*, and *insect sociobiology* with *brain science* the above paragraph could describe the paradigm shifts in the last twenty years of progress in the brain sciences.[4]

In an attempt to realizes some of Wilson's vision we focus here on what we will call "proto-swarms". A proto-swarm is a minimal combination of behavioral signals, observed and isolated *in vivo*, which are sufficient to form some rudimentary cooperative structure observed in nature. We hope that in this way it might be possible to isolate and study one or more of the underlying mechanisms that allow real organisms to act collectively. We believe that a fuller understanding of the the amount of physiological evolution required to produce a behavioral change can be achieved in this way since it often possible to scan the physiological space of possibilities in rather comprehensive way.

2   The Stochastic Transition Probabilities

Here we will be concerned with the basic behavioral components of the pheromone trail laying and following behavior of individual organisms, and the resulting self-organization of regular patterns of flow, such as trails and networks of regular traffic. Clearly such behavior has proved adaptive, and is widespread amongst the social insects today.

The systematic analysis of the chemical signals of social insects in terms of it component chemicals, their effects on the individuals, and their glandular sources began with the program of Wilson. It has been shown that ants make use of (at least) two different types of sense-data processing, *osmotropotaxis*, a kind of instantaneous pheromonal gradient following, and *klinotaxis*, a sequential method.[6] Here we will be interested only in the former of these methods.

The basic osmotropotaxic sense datum is a measurement of the difference in pheromone concentration between the antennae. This sense datum is translated onto a response of the organism. Phenomenological forms for this response in various controlled settings have been determined recently.[7, 8, 9] It has been shown that, at least on the phenomenological level, there is a significant amount of noise in the response function. Subsequently, the description we use must be stochastic. This noisiness, as we shall see, is fundamental to the behavior of the swarm as a whole. Whether or not it is fundamental on the level of the sensory processing system is not known, though this is a likely possibility. Already there is some evidence that noise on the intrinsic level plays an important role in information processing in sensory neurons. The relationship is not entirely frivolous, and pheromonal communication has been called the "giant synapse". In addition, there are a number of fairly obvious physiological and perhaps even evolutionary parallels between pheromones and neurotransmitters. Real ants are able to emit a number of types of chemical signals, and might be understood as "a walking secretory gland". In the same sense it is now understood that neurons are able to release a number of types of neuro-transmitters.

The state of an individual ant can be described by a "phase variable" containing its position $\mathbf{r}$, and orientation θ. Noisy pheromone gradient following is formally equivalent to noisy potential gradient following in statistical physics, an analogy which has already been justified in some detail elsewhere.[13, 14, 15] Since the response at a give time is assumed to be independent of the previous history of the individual, it is sufficient to specify a transition probability from one place and orientation $(\mathbf{r}, \theta)$ to the next place and orientation $(\mathbf{r}', \theta')$ an instant later. At this level of description, the motion of an individual organism is rigorously equivalent to a continuous Markov process whose probabilities are determined at every moment in time by the instantaneous distribution of pheromone $\sigma(\mathbf{x}, t)$.

In [13, 14], the behavior of all possible proto-swarms of this type of swarm was classified and generalized as far as possible via theoretical considerations. In that previous work transition rules were derived and generalized from the noisy response function of Refs. [10, 11, 12], which in turn were found to reproduce a number of experimental results with real ants.[7, 8, 9] The response function can effectively be translated into a two-parameter transition rule between the cells by use of a pheromone weighting function

$$W(\sigma) = \left(1 + \frac{\sigma}{1 + \delta\sigma}\right)^{\beta}. \tag{1}$$

This function measures the relative probabilities of moving to a cite $\mathbf{r}$ with pheromone density $\sigma(\mathbf{r})$ as discussed in [13, 14]. In order to illustrate this, we consider the following situation pictured in Fig. 1. For simplicity the ant is constrained to walk along a path untill it come to a fork in the road. At this point the ant chooses according to the weights given in Eq. 1. We fix the pheromone on segment two, $\sigma_2 = 1$, and measure the probability for the ant to choose segment one, W_1. We can illustrate the degree of response of the ant to differences in scent by varying σ_1 as illustrated in Fig. 2.

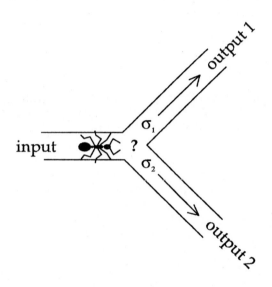

Fig. 1.

The parameter β is associated with the osmotropotaxic sensitivity. Also it can be seen as a physiological inverse-noise parameter or gain. This parameter controls the degree of randomness with which the ant follows the gradient of the pheromone. For low values of β the pheromone concentration does not greatly affect its choice, while high values cause it to follow pheromone gradient with more certainty. Here we call $1/\delta$ the sensory capacity. This parameter describes the fact that the ant's ability to sense pheromone decreases somewhat at high concentrations. This gives rise to a peaked function for the average time an ant will stay on a trail as the concentration of pheromone is varied–a fact which has been observed repeatedly experimentally[16, 17, 18]. This is perhaps significant in light of one of the results presented here: trails and networks form more easily in the presence of this saturation effect. This saturation also can be found in first order sensory neurons in which there is a limited "dynamic range" in which sensory inputs are encoded.

In addition to $W(\sigma)$ there is a weighting factor $w(\Delta\theta)$, where $\Delta\theta$ the change in direction at each step. More work need to be done to determine the precise form of this factor, which takes into account the sensory anisotropy the organisms due to the orientation and motion of the insects. Everything else being equal, very sharp turns are much less likely than turns through smaller angles, thus $w(\Delta\theta)$ is a decreasing (and symmetric) function of the magnitude of $\Delta\theta$. Stated in another way, they have a probabilistic bias in the forward direction. The actual transition probability rate can be calculated after setting the step size, and normalizing.

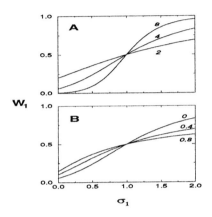

Fig. 2. Response function for (A) $\delta = .2$ and values of β indicated in the figure and (B) $\beta = 4$ and values of δ indicated in the figure.

3 Learning

For the picture of the swarm to be complete it must be supplemented by a description of how pheromone evolves. We assume that each organism emits pheromone at a given rate η, and that this pheromone remains fixed at the point of emission (no spatial diffusion). As the pheromone evaporates at rate κ from the surface it creates a pheromone vapor, in proportion to its quantity, above the surface which the insects sense. If not constantly replenished the pheromone field and the vapor slowly disappear due to evaporation.

Note that there is no direct communication between the organisms but a type of indirect communication through the pheromonal field. The ants are not allowed to have any memory and the individual's spatial knowledge is restricted to local information about the pheromone density. The pheromonal field contains information about past movements of the organisms, but not arbitrarily far in the past, since the field "forgets" its distant history due to evaporation in a time $\tau \approx 1/\kappa$. Therefore the distribution of pheromone represents the memory of the recent history of the swarm, and in a sense it contain information which the individual ants are unable to hold or transmit. In a gross analogy, memories are believed to be written as a stronger coupling among individual, or groups of neurons. These couplings are strengthened by neural co-activity[19, 20] much in the same way that the pheromonal field is preserved or strengthened by coherent frequent ant traffic.

4   Mathematical Analysis

Since a detailed mathematical analysis of this type of system has already been made[13, 14], we will confine ourselves here to a brief statement of the pertinent results. In addition to the two parameters discussed above another important parameter is the number of ants present, or alternatively, the mean density of ants ρ_0. The scent decay rate κ and the emission rate η and ρ_0 can effectively be combined into a single stress parameter, the average pheromonal field $\sigma_0 = \rho_0 \eta/\kappa$. The most important theoretical result is the location of the second order phase transition, that is the location of the boundary separating totally random behavior from ordered behavior of various types in the physiological phase space. As shown in [13, 14] ordered behavior sets in when the inequality $\sigma_0 f'(\sigma_0)/f(\sigma_0) - 1/\beta > 0$ holds, where $f(\sigma) = (1 + \sigma/(1 + \delta\sigma))$. This criterion is true for any behavioral function $f(\sigma)$, and allows us to calculate the physiological phase boundaries in $(\sigma_0, \beta, \delta)$ space. For the particular behavioral function used here the transition lies along the curve

$$\beta_c(\delta, \sigma_0) = 1 + 1/\sigma_0 + 2\delta + \delta\sigma_0 + \delta^2\sigma_0. \tag{2}$$

The symmetry which allows for a mean-field type solution for the location of the transition line case is maintained up to the point of the transition and is effectively spontaneously broken at that point. This means the points of transition from disorder to order can be determined theoretically, even in the more general case, but not the resulting patterns, which must be determined via simulations.

In addition to being the major landmark in the physiological parameter space, we believe both on general grounds, and because of the results presented in the next section that the location of this line has significant behavioral implications. Clearly the behavior of groups of real ants is ordered, and this order plays an important role in the functioning of the swarm Just as importantly, the behavior should not be too rigid and ordered since fluctuation and instabilities might increase the flexibility of response of the mass action. Thus, if there are significant fluctuations in the patterns of mass action this might aid the swarms in responding to a changing environment. We might conclude that the ordered region is a "good" place to be, but near to the transition line in such a way as to optimize the conflicting tendencies of controlled ordered behavior versus flexible random behavior. We will also see in the next section that large fluctuation actually serve to stabilize some of the important pattern of collective behavior of the system.

5 In Numero Swarm Behavior

For the purposes of simulation it is necessary to introduces some discretization of space and time, and to translate the noisy behavioral function observed experimentally into transition rules on this discrete space[13]. The ants will be allowed to move from cite to cite on a square lattice. These discrete rules are merely tools for the approximation of a continuous model, and other discretizations are possible. We allow each ant to take one step on the lattice of points (cells) at each time step. As a result of discretizing the space an individual ant at each time step finds itself in one of these cells, and its sensory input is influenced by the concentration of pheromone in its own cell and each of the eight neighboring cells. In addition, each ant leaves a constant amount of pheromone at the node in which it is located at every time step. This pheromone decays at each time step at a rate κ. Toroidial boundary conditions are imposed on the lattice to remove, as far as is possible, any boundary effects. The normalized transition probabilities on the lattice to go from cell k to cell i are then given by

$$P_{ik} = \frac{W(\sigma_i)w(\Delta_i)}{\sum_{j/k} W(\sigma_j)w(\Delta_j)}, \qquad (3)$$

where the notation j/k indicates the sum over all the cells j which are in the local neighborhood of k. Δ_i measures the magnitude of the difference in orientation (direction) for the previous direction the last time the ant moved. Since we are using a neighborhood composed of the cell and its eight neighbors on a square lattice, Δ_i can take only the discrete values $0-4$, and it is sufficient to assign a number w_i for each of these changes of direction. Here we used weights of (same direction) $W_0 = 1$, and $w_1 = 1/2$, $w_2 = 1/4$, $w_3 = 1/12$ and $w_4 = 1/20$ (u-turn). Once the parameters β, δ and w_i are set, a large number of ants (here we used 307) can be placed on the lattice at random positions. The random movement of each ant is determined by the probabilities P_{ik}. We usually take the initial condition of the pheromone to be zero at every point on the lattice. Every time step each ant leaves a quantity η (here $\eta = 0.07$) of pheromone in each cell, and the total amount of pheromone σ_i in each cell is decrease at a rate κ (here $\kappa = 0.015$) at the end of each time step.

In Fig. 3 we show the evolution of the distribution of the ants on a 32X32 lattice as time progresses. As early as a hundred time step a spatial structure begins to emerge consisting of a network of trails. Later on some of the trails are consolidated, and some are lost. This dynamics can also be visualized in Fig. 4 where in the bottom panel the occupancy of a vertical cross-section

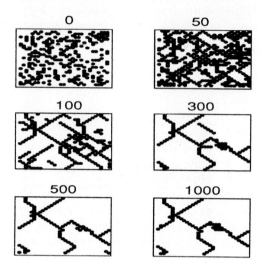

Fig. 3. Snapshots of the evolution of the distribution of ants for the times indicated $(\beta = 3.5 \quad \delta = .2)$.

of the middle of the lattice is plotted. The degree of order can be quantified in many ways. In the top panel of Fig. 4 the dashed line is the coefficient of variation (standard deviation over the mean) of the direction of crossing a given cell averaged over the whole lattice. The continuous line is the ratio of number of cells which have greater concentration than the equilibrium mean σ_0 over the number of cells with pheromone less than σ_0.

Fig. 5 shows the behavior of the system in $\beta - \delta$ parameter space. There are three types of behavior: disorder, patches and trails. Within the latter two regions there are a range of patch sizes, and a range of typical line shapes depending on the precise position within each of the respective regions. Well defined trails form in the region above the phase transition line (ordered phase), but near the transition from disorder. Further away from the order-disorder line the clumping tendency overcome the directional bias, and no lines form. Near the border between lines and patches we observed some hysterysis effects, and preformed lines persist into the region of patches. In the trail forming region of the parameter space shown in the phase diagram in Fig. 5, the ants form well-defined trails.

This is the simplest (local, memoryless, homogeneous and isotropic) model which leads to trail forming that we know of, and the formation of trails and networks of ant traffic is not imposed by any special boundary conditions, lattice topology, or additional behavioral rules. The required behavioral elements are stochastic, nonlinear response of an ant to the scent, and a directional bias. Furthermore the parameters of the system need to be tuned

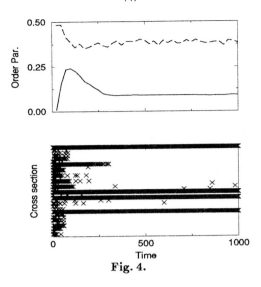

Fig. 4.

somewhat to the appropriate region, and not any nonlinear rule of the above type will do. If the nonlinear response or the directional bias are removed no lines form, and lines that are already formed do not persist. Notice that the role of the decreasing sensory capacity in this region of the physiological parameter space it to broaden the range of gain at which ants will be able to form trails. We suspect that the ability of neurons in the hippocampus to form plastic networks will be affected in a similar way.

6 Some Conclusions

The main conclusion to be drawn here is that osmotropotaxic scent following of the very simple kind described above is *sufficient to produce evolution of complex pattern of organized flow of social insect traffic all by itself*, and not just sufficient to allow for trail *following* behavior. We believe this observation may be of some help both in classifying different types of recruitment behaviors, in understanding how this type of behaviors evolved, and its relation to other factors such as tandem running which exist in various mixtures of importance with the types of behavior discussed above.

Since the self-organizing properties of the swarm are instability driven, the structures that form have some very interesting properties with respect to large perturbations. We performed the following experiment: the system was tuned to a region of the phase plane where lines form, and a network of traffic was allowed to form. Then β was decreased so that system is tuned below the transition line ($\beta < \beta_c$). One observes that the ants fall away from their orderly patterns and immediately start executing random walks on the lattice. As a result the pheromone distribution starts to fluctuate more and more randomly. If β is then tuned back to its original value at some time later the line will eventually reform with little or no change. This occurs even if the randomization is allowed to proceed to the point that the pheromonal field is almost totally randomized. For times up to about the decay time $\tau = 1/\kappa$ even small, virtually undetectable memory effects of the field can be amplified causing the patterns to reform without significant changes. We can not resist comparison of this amplification phenomenon with recent results of Wilson and McNaughton[20] in which rat memories seem to be consolidated during sleep states in which the noise is substantially larger than in the walking state.

There are two, seemingly conflicting functional abilities this amplification phenomenon allows the swarm to have. Firstly, organized patterns of behavior are really quite stable with respect to large perturbations which might have an obvious usefulness to operation in a changing and unpredictable environment. Secondly, because the patterns are due to the formation of an initially weak cooperative structure of the right kind, the swarm can act as an information amplifier, and even a weak external perturbation (such as the presence of a food source) might lead to a significant response. Thus the swarm posses both a long memory, and the ability to learn.

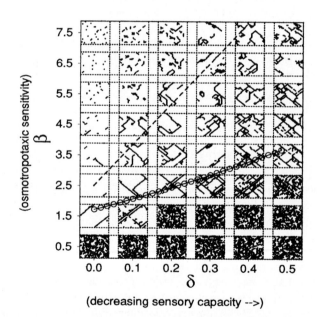

Fig. 5. Physiological Phase Plot: A snapshot of the last two iteration after 1000 time steps starting from a random initial distribution of ants for each δ and β value indicated on the axis. The circles joined by a line is the mean field prediction of a second order phase transition given by Eq. 2. The dashed line is the approximate location of the transition between trails and patches.

As shown in Fig. 5, lines form a little above the second order phase transition line. In this region there is a cooperative effect between the fluctuations and the ordering effects. Below the transition line, or just above it, the fluctuations are too great, and no cooperative structures form. However, too far above the transition line the fluctuations are suppressed, and the order dominates overwhelmingly. This means that the ants are basically induced to "turn around" with sufficient probability that only patches form. The role of such "U-turns", has been investigated experimentally and theoretically in [21], where it was also suggested that these u-turns might play an important role in the self-organization of ant traffic.

The region in which the lines form shown in Fig. 5 represents a cooperative effect between the fluctuations and the ordering behavior. The formation of lines can also be understood in terms of the theory in [13, 14]. In the region where lines form the clumping effect is both unstable in the longitudinal direction and stable in the transverse direction of the ant motion, and it is this fact which give rise to the formation of stable line of traffic. The result is a fairly robust region of line formation.

Since information (in the form of the orderly movement of ants from one place to another) can be said to flow only in the region of stable line formation, which lies above the disordered region and below the very ordered region, these results might be thought similar to the hypotheses of "complexity at the edge of chaos" which asserts that complex behavior emerges in the vicinity of a marginally stable state.[22, 23] However we point out that the system, as far as we know, does not "self-organize" to this region unless it may be said that on the evolutionary time scale the organisms found such behavior adaptive. In addition there is not an "edge" but rather a large robust region where the most complex structures form, and this region is *above*, not *at* marginally stable state. Lastly we do not believe that the reason for this type of behavior has anything to do with the very speculative ones which are sometimes suggested by some researchers.[24, 25]

DRC is supported by NIMH. The Santa Fe Institute receives funding from the John D. and Catherine T. MacArthur Foundation, the NSF (PHY9021437) and the U.S. Department of Energy (DE-F605-88ER25054). The authors would like to acknowledge helpful conversations with W. Schaffer and the initial contributions of E. Rauch to the work presented here.

References

1. E. O. Wilson, The Insect Societies, Cambridge: Belknam Press, 1971.
2. E. O. Wilson, Sociobiology, Cambridge: Belknam Press, 1975.
3. E. O. Wilson, Animal Behavior **10**, 134-164 (1962).
4. W. J. Freeman, Mass Action in the Nervous System, New York: Academic Press, 1975.
5. B. Hölldobler, and E. O. Wilson, The Ants, Cambridge: Belknap (1990).
6. W. Hangartner, Z. vergl. Physiol. **57**, 103 (1967).

7. R. Beckers, J.-L. Deneubourg, S. Goss. and J. M. Pasteels, Insectes Soc. **373**, 258 (1990).

8. J.-L. Deneubourg, S. Aron, S. Goss and J. M. Pasteels, J. Insect Bchav. 32, 159 (1990).

9. S. Goss, R. Beckers, J.-L. Deneubourg, S. Aron, and J. M. Pasteels, In: Behavioral Mechanisms of Food Selection, Nato ASI Series G20 (Hughes, ed.) Berlin Heidelberg: Springer-Verlag (1990).

10. V. Calenbuhr and J.-L. Deneubourg, In: Biological Motion (W. Alt, and G. Hoffmann, eds.) 453, Berlin: Springer-Verlag (1990).

11. V. Calenbuhr and J.-L. Deneubourg, J. Theor. Biol. **158**, 359 (1991).

12. V. Calenbuhr, L. Chrétien, J.-L. Deneubourg and C. Detrain, J. Theor. Biol. **158**, 395 (1991).

13. M. M. Millonas, J. Theor. Biol. **159**, 529 (1992).

14. M. M. Millonas, In: Artificial Life III(C. G. Langton, ed.). Santa Fe Institute Studies in the Sciences of Complexity, Reading, Massachussetts: Addison-Wesley, (1994).

15. M. M. Millonas, In: Pattern Formation in Physical and Biological Systems(P. Claudis, ed.). Santa Fe Institute Studies in the Sciences of Complexity, Reading, Massachussetts: Addison-Wesley, (1994).

16. R. P. Evershed, E. D. Morgan and M. C. Cammaerts, Insect Biochem. **12**, 383 (1981).

17. C. Detrain, J. M. Pasteels, and J.-L. Deneubourg, Actes coll. Insectes Sociaux **4**, 87 (1988).

18. S. Gérardy and J. C. Verhaeghe, Actes coll. Insectes Sociaux **4**, 235 (1988).

19. M. A. Wilson and B. L McNaughton, Science **261**, 1055 (1993).

20. M. A. Wilson and B. L McNaughton, Science **265**, 679 (1994).

21. R. Beckers, J.-L. Deneubourg and S. Goss, J. Theor. Biol. **159** , 397 (1992).

22. P. Bak, C. Tang and K. Weisenfield, Phys. Rev. A **38**, 364 (1988).

23. P. Bak, K. Chen and M. Creutz, Nature **342**, 780 (1989).

24. N. H. Packard, In: Complexity in Biological Modeling (S. Kelso and M. Shlesinger, eds) (1988).

25. C. G. Langton, Computation at the edge of chaos: Phase transitions and emergent computation, Ph. D. Thesis, University of Michigan, 1991.

Multiple Neural Experts for Improved Decision Making

Ethem Alpaydın
Dept of Computer Engineering, Boğaziçi University, TR–80815 Istanbul, Turkey
alpaydin@boun.edu.tr

Abstract. There are many choices that should be made before a neural network can be used in an application. These include the network architecture, learning method, training data, feature representation, initial parameter values, training set order, cost function. A number of alternatives are generated and tested and the one that performs best on a separate test set is adopted. Instead of discarding the rest, we propose here to combine them by taking a vote. First the approach is advocated and the existing literature is surveyed. Then we put the approach into a Bayesian framework where the weights in votes are taken as Bayesian priors computed based on model complexities. Initial results on classification of handwritten numerals are promising. The voting approach can be generalized to regression where the function approximated is continuous as opposed to discrete 0/1 of classification.

1 Learning and Generalization

Learning can be defined as a function approximation problem: We are asked to map a multi-dimensional input vector x to one output $f(x)$. We do not know $f(x)$ but we are given a set of examples $\{x, y\}$ on which to train our approximation function. Generally, there is error in reception in that $y = f(x) + \eta$ where η is generally assumed to be zero mean, input independent "noise." Classification is a special case where $f(x)$ is a 0/1 function that denotes membership of class C, i.e., $f(x)$ is 1 if $x \in C$, and 0 otherwise. This framework can be generalized to the multiple output case by defining separate $f^k(x)$ for different outputs. The function we use for approximation is of the form $\mathcal{F}(x)$ where $\mathcal{F}$ is the model whose parameters are computed so as to minimize the error, e.g., in the mean square sense, on a given training set, T.

$$E = \frac{1}{|T|} \sum_{\{x_i, y_i\} \in T} [y_i - \mathcal{F}(x_i)]^2 \tag{1}$$

In the case of neural networks, $\mathcal{F}$ is defined by the network architecture, i.e., the number of hidden layers, the number of hidden units, connectivity between layers, etc. Its parameters correspond to the modifiable weights of the connections of this network. Each choice made by the network designer, e.g., the training sample used, the learning method used, parameters of the learning method, or the hyperparameters (architecture of $\mathcal{F}$) is one additional source of bias that causes convergence to a different set of parameter values and thus to a different performance on data unseen during training.

The approach generally taken is trial and error. We generate a number of models $\mathcal{F}_i(x)$ as alternatives and of these, the one that best generalizes on a seperate data set, C, is chosen and is subsequently used:

$$k = \text{ArgMax}_j \left[\frac{1}{|C|} \sum_{\{x_i, y_i\} \in C} [y_i - \mathcal{F}_j(x_i)]^2 \right] \tag{2}$$

There are two problems here:

1. The cross-validation set C contains noise just like the training set T and the best performing on C, $\mathcal{F}_k$, need not be the best one in real-life.
2. The best performing model's performance is not necessarily a superset of the performances of the discarded models. There may be patterns for which the best one fails but one of the others succeed.

It is for this reason that we do not want to discard other models but combine them suitably so as to improve performance. Examples are "synergy" of clustering multiple networks [8], "portfolios" [10], "multiple" networks [1], "consensus" theory [3], "stacked" generalization [16], "ensembles" [12].

1.1 Generating a Separate Test Set

In this article we assume in several cases that we have a data set C, distinct from T for cross-validating models trained on T. This is generally done by taking a large training sample and using one half as T and the other as C. In certain applications, the data set may be so small that we do not want to lose half but use the whole for training; after all this is non-parametric estimation and we need all the data we can lay our hands on. In such a case, we can use one of a number of techniques to get a separate data set from the training set. One technique is "leave-one-out." Here given a training set of m patterns, we train the model on $m-1$ patterns and compute the error on the mth. By cycling this one pattern, it is as if we have a separate set as big as the training set at the expense of m separate trainings. "Jackknife" is leave-n-out. "Bootstrap" is generating new data sets by sampling randomly from the training set *with* replacement.

2 Simple Voting

When a decision maker cannot be sufficiently trusted, getting fault tolerance by introducing redundant, multiple copies and taking a vote is a trick that has been used very frequently[1]. In the absence of a priori information as to the reliabilities of the voting models, the best strategy is to give equal say to all the models, which we call *simple voting*. Then when we have N models $\mathcal{F}_i$, the output voted, R, is:

$$R(x) = \frac{1}{N} \sum_{i=1}^{N} \mathcal{F}_i(x) \qquad (3)$$

The idea is that any of these decision makers has more probability of getting the right answer and thus overall, the more systems one has, the higher is the probability that a voting scheme will get the right answer. This assumes that the networks have enough variance that they will generally fail under different circumstances. This is demonstrated in the context of neural networks by Hansen and Salomon [6] who have shown how local minima differ in the case of a one hidden layer feedforward network. They also showed that, if each network can get the right answer more than half of the time and if network responses are independent, then the probability of error by a majority rule decreases as the number of networks are increased.

Voting can also be seen as a way of getting rid of randomness introduced in learning. Averaging decreases variance, i.e., noise. It is known that the best predictor of a random variable in terms of minimizing its mean square error is the mean whose unbiased estimator is the sample average. In signal processing a well-known method to eliminate noise is using a low-pass filter which is basically local averaging.

Each learning rule has a number of parameters which affects generalization and they are chosen arbitrarily. The additional bias introduced by choosing one specific value or range of values can effectively be eliminated by training multiple models with alternative values and taking a vote. In the case of Pearlmutter and Rosenfeld [11], this parameter is the set of initial random weights in backpropagation. In the case of Hampshire and Waibel ([5], reported in [11]), it is the objective function. In Alpaydın's case [1], this parameter is the order of patterns in the training set. One can also divide the training set into N and train N separate models and take a vote.

3 Weighted Voting

When we take a simple majority voting, each network has equal "weight" on the result of voting. The disadvantage here is that when there are a number of similar agents whose predictions generally coincide, there is no way of allowing a more expert to overrule them. We would normally like better networks to have more weight. The goal of consensus theory [3] is to get a consensus among experts where experts are weighted according to their goodness:

---

[1] This is one of the reasons why democracy is better than monarchy where there is only one decision maker.

$$R(x) = \sum_{i=1}^{N} \mathcal{F}_i(x)\mathcal{W}_i \qquad (4)$$

where the weights sum up to one:$\sum_i \mathcal{W}_i = 1$. There are a number of ways by which these weights can be computed:

1. In the case of *reflective agents*, each network, together with its output, indicates its "certainty" on its output. In classification, this can simply be the difference of values of the two highest active units. In regression, this requires a separate *monitoring* network [15] and a separate data set on which the monitor is trained.
2. These weights can be thought of as a second learning model combining the output of lower-level models and can thus be trained like any other on a separate data set. This method, as proposed by Wolpert [16], is called *stacked generalization*.
3. This certainty may be dependent on the network architecture. We know for example that complex models with many free parameters overfit on a small sample. Thus we may like to have the tendency to give them less weight unless they are more successful than simpler models, i.e., their complexity is justified [2]. Here we interpret weights as Bayesian priors with which we modulate likelihoods.

Regardless of the way we combine the models, there is no reason to have multiple models if they are quite similar. One can only get fault tolerance if the models fail under different circumstances[2]. In neural networks, choosing parameters randomly like initial weights and hyperparameters like network architecture independently, one guarantees this independence to a certain extent. Actually to get minimum error, we do not want independence; we want them to be inter-dependent. When we have one model that has a certain success, we want to add a second model that succeeds best for inputs on which the first one fails; we do not care about the new one's overall performance. Assuming simple voting [10, 12]:

$$\mathrm{Var}(R) = \mathrm{Var}\left(\frac{1}{N}\sum_{i=1}^{N}\mathcal{F}_i\right)$$
$$= \frac{1}{N^2}\left[\sum_i \mathrm{Var}(\mathcal{F}_i) + 2\sum_i\sum_{j\neq i}\mathrm{Cov}(\mathcal{F}_i,\mathcal{F}_j)\right] \qquad (5)$$

[2] This is the reason why democracy is better than oligarchy where there is a group of decision makers with very similar trainings and experiences. Performance can be improved as in UK by incorporating a House of Commons from a different background.

If the models are independent, the second term cancels and variance (and Mean Square Error) decreases, with increasing N. However variance can be decreased further by choosing mutually "orthogonal" models. For example if we have two models, instead of having them independent, we would want them to be negatively correlated.

4   Reflective Agents

In this approach, together with each model's output, the voting system also is given a value about how much that model can be trusted or the model's certainty of its output. This value can be fixed or may change from one input to another.

In the case of classification, the "certainty" of a classifier can be computed as the difference of the values of the two highest active units [1]. When one has cross-validation data, a separate data set, one can assess the performance of the trained model on this data set and use the success achieved as the measure of "belief" in this model [17].

It is also possible that this monitoring network is also given a copy of the input. In such a case an input-dependent assessment of the monitored model is possible. Then we need a competitive strategy by which different models become expert on different parts of the input space to best utilize the computational resources and the voting mechanism works as a "gating" model [7].

5   Stacked Generalization

We can treat the voting mechanism as a separate generalizing layer which should be treated using a different learning set. This is named "stacking" generalizers one on top of another [16].

6   Weights as Bayesian Priors

In the case of classification, we have separate R^k for each class output and the response of the system is the code of the class, R^c, which has the highest R^k:

$$R^c(x) = \max_k \left[R^k(x)\right] = \max_k \left[\sum_{i=1}^{N} F_i^k(x) W_i\right] \tag{6}$$

In a Bayesian setting, this can be written as:

$$Pr(C^c|x) = \max_k \left[Pr(C^k|x)\right] = \max_k \left[\sum_{i=1}^{N} \underbrace{Pr(C^k|x, \mathcal{F}_i)}_{F_i^k(x)} \underbrace{Pr(\mathcal{F}_i)}_{W_i}\right] \tag{7}$$

The W_i values may be interpreted as a Bayesian prior indicating the "plausibility" of the corresponding model. This prior may also be interpreted as a regularizer [13] or a complexity term preferring less complex models [14]. We can define an error function that takes into account both the success and the model complexities and minimizing this function will give us the desired W values.

$$E(W) = \sum_{\{x_i, y_i\}} \left[y_i - \sum_i^N \mathcal{F}_i(x_i) W_i \right]^2 + \beta \sum_i^N W_i |\mathcal{F}_i| \qquad (8)$$

with the additional constraint that $\sum_i W_i = 1$. The first term is the usual sum of square of errors on the training set. The second term penalizes big networks. $|\mathcal{F}_i|$ denotes the complexity of model i, e.g., the length of its description or the cardinality of its parameter set. The relative importances of the two terms is given by the value of β.

One other possibility is to add a separate bias unit indexed 0 with null complexity, i.e., $|\mathcal{F}_0| = 0$, and train also its weight, W_0. Then the response is computed as follows:

$$R(x) = \sum_{i=1}^N \mathcal{F}_i(x) W_i + W_0 \qquad (9)$$

In such a case, W_i can be negative too, thus in the cost functional equation (8), W_i in the second term should be squared. Weights no longer sum up to one. Due to the one more free parameter added, this mechanism is more powerful than the previous one. This corresponds to using a hyperplane for combining the models.

6.1   Simulation Results

This scheme is applied to the problem of recognition of handwritten digits. The training set and test set each contains 600 digits [4]. We have used multiple networks trained with the back-propagation learning rule to train our models. For the same problem, we have used four networks all with one hidden layer with 5, 15, 35, and 75 hidden units. All the networks are trained on the same training set. Network complexities, $|\mathcal{F}_i|$, are taken to be equal to the number of hidden units as they are the indicators of complexity and everything else is equal. The regularization parameter, β, is set to be $0.0005/m$ where m is the network with the largest number of hidden units, here 75. The results given in table 1 are the averages of 20 runs after 50 epochs on the training set.

Table 1. Results of simple, weighted, and weighted with bias voting types.

Voters	Hidden Units	Success	Error	Reject
	5	60.66, 7.07	3.67, 5.13	35.67, 11.44
	15	90.42, 1.57	2.24, 0.67	7.34, 1.54
	35	90.85, 1.34	1.09, 0.44	8.06, 1.32
	75	90.52, 1.34	0.58, 0.34	8.91, 1.36
Voting type				
	Simple	97.17, 0.63	2.22, 0.55	0.61, 0.24
	Weighted	97.23, 0.61	2.17, 0.54	0.61, 0.24
	Weighted / bias	97.26, 0.59	2.74, 0.59	0.00, 0.00

6.2 Discussion

Success mean increases and variance decreases by voting, which is an indication of better generalization. Note that the approach taken here is different from "adaptive mixtures of local experts" [7] in that here, participating networks essentially learn the same task but converge to different solutions due to differences in their structure or parameter settings. We can say that the weights denote relative performances of the networks over the whole input space as opposed to some part of it. MacKay [9] cites Patrick and Wallace's work where, "studying the geometry of ancient stone circles, [they] discuss a practical method of assigning relative prior probabilities to alternative models by evaluating the lengths of the computer programs that decode data previously encoded under each model. This procedure introduces a second sort of Occam's razor into the inference, namely a *prior* bias against complex models" (emphasis in the original). Of course, nothing refrains one from also using the first order Occam's razor, i.e., add a regularization term also when training the individual networks. One may also envisage a combination of systems based on different training mechanisms, neural and nonneural, e.g., fuzzy or programmed, systems in such an architecture. The relative complexities of quite different systems, $|\mathcal{F}_i|$, may however be very difficult to assess. One can always use simple voting where weights are equal if such an assessment is not possible; this corresponds to perfect a priori ignorance.

7 Acknowledgements

The digit database is provided by Isabelle Guyon. Francesco Masulli of University of Genoa helped in collecting related literature. This work is supported by Grants EEEAG–41 from TÜBİTAK, the Turkish Scientific and Technical Research Council and Grant 93HA0132 from Boğaziçi University Research Funds.

References

1. Alpaydın, E. (1991) "GAL: Networks that grow when they learn and shrink when they forget," ICSI, TR-91-032, Berkeley, CA.
2. Alpaydın, E. (1993) "Multiple networks for function learning," *IEEE International Conference on Neural Networks*, March, San Francisco, 9–14.
3. Benediktsson, J.A., Swain, P.H. (1992) "Consensus Theoretic Classification Methods," *IEEE Transactions on Systems, Man, and Cybernetics*, **22**, 688–704.
4. Guyon, I., Poujoud, I., Personnaz, L., Dreyfus, G., Denker, J., le Cun, Y. (1989) "Comparing different neural architectures for classifying handwritten digits," *International Joint Conference on Neural Networks*, Washington DC.
5. Hampshire, J., Waibel, A. (1989) "A novel objective function for improved phoneme recognition using time delay neural networks," CMU, TR CS-89-118.
6. Hansen, L.K., Salamon, P. (1990) "Neural Network Ensembles," *IEEE Pattern Analysis and Machine Intelligence*, **12**, 993–1001.
7. Jacobs, R.A., Jordan, M. I., Nowlan, S.J., Hinton, G.E., (1991) "Adaptive Mixtures of Local Experts," *Neural Computation*, **3**, 79–87.
8. Lincoln, W.P., Skrzypek, J. (1990) "Synergy of Clustering Multiple Back Propagation Networks," in *Advances in Neural Information Processing Systems 2*, D. Touretzky (Ed.), Morgan Kaufmann, 650–657.
9. MacKay, D.J.C. (1992) "Bayesian Interpolation," *Neural Computation*, **4**, 415–447.
10. Mani, G. (1991) "Lowering Variance of Decisions by using Artificial Neural Network Portfolios," *Neural Computation*, **3**, 484–486.
11. Pearlmutter, B.A., Rosenfeld, R. (1991) "Chaitin-Kolmogorov Complexity and Generalization in Neural Networks," in *Advances in Neural Information Processing Systems 3*, R. Lippmann, J. Moody, D. Touretzky (Eds.), Morgan Kaufmann, 925–931.
12. Perrone, M.P. (1993) "Improving Regression Estimation: Averaging Methods for Variance Reduction with Extensions to General Convex Measure Optimization," PhD Thesis, Department of Physics, Brown University.
13. Poggio, T., Torre, V., Koch, C. (1985) "Computational vision and regularization theory," *Nature*, **317**, 314–319.
14. Rissanen, J. (1987) "Stochastic Complexity," *Journal of Royal Statistics Society B*, **49**, 223–239, 252–265.
15. Beyer, U., Smieja, F. (1993) "Learning from Examples, Agent Teams, and the Concept of Reflection," GMD, TR-93-766, St Augustin, Germany.
16. Wolpert, D.H. (1992) "Stacked Generalization," *Neural Networks*, **5**, 241–259.
17. Xu, L., Kryzak, A., Suen, C.Y. (1992) "Methods of Combining Multiple Classifiers and Their Applications to Handwriting Recognition," *IEEE Transactions on Systems, Man, and Cybernetics*, **22**, 418–435.

Understanding Complex Systems:
What Can the Speaking Lion Tell us?

Erich Prem

The Austrian Research Institute for Artificial Intelligence
Schottengasse 3, A-1010 Vienna, Austria
erich@ai.univie.ac.at

Abstract. The rebirth of complex systems in several distinct domains of research has posed new epistemic questions. Self-organizing systems as well as autonomous robots have a tendency to not only behave in an unpredictable way, they are also extremely difficult to analyse. In this paper we discuss three problems with neural networks that are important for complex systems in general. They are related to the proper design of a self-organizing system, to the role of the system engineer, and to the proper explanation of system behavior. We present a generally applicable solutions, which is based on a "symbol grounding" neural network architecture. We also take a look at an implemented network which hints at the fact that grounding in this case must involve teleological terms. We then discuss the relation of this approach to the measurement problem in physics and point out similarities to existing positions in philosophy. However, it should be noted that our "solution" of the explanation problem may be judged as being a very sceptic one.

Keywords. Self-organization, complex system, symbol grounding, epistemology, explanation

1 Introduction

1.1 Understanding and self-organization

Scientists not only search for solutions to problems, they also seek to construct a solid basis upon which their results can be justified and explained. That such an absolutely secure ground of science is impossible to be found within empirical fields is one of the truisms of our time. The recent rebirth of complex systems in many different domains confronts scientists with new epistemic problems. Complex systems have a tendency to not only behave in an unpredictable way, they are also extremely difficult to analyse.

Both, rebirth and the epistemic problems, are manifest in computer science, for instance, through the discussions that center around artificial neural networks or, as it is called, connectionism. Such "newly" developed techniques are usually accompanied by theoretical arguments around their usefulness and drawbacks. In the case of self-organizing neural networks there has been much discussion about the virtues and possible advantages of emergent properties. In this paper we want to put our finger on three problems with neural networks that are important for self-organization and for complex systems in general. These problems are related to the proper design of a complex system, to the role of the system engineer, and to the proper explanation of system behavior. The problems are different aspects of one central theme: *Only that can be designed which has been understood and only that can be understood which has been designed.*

In the case of an artefact this means that it is usually understood what it does and how it does it because the elements of which the artefact has been constructed are well understood and have been arranged in such a way that they work together in a well-defined way. Things soon become more difficult in the case of natural systems. The above characterization, of course, is not to say that we could never arrive at an understanding of some natural object out there in the world. It means, instead, that during the process of scientific inquiry such natural objects undergo an in-depth analysis. Experiments are conducted and a conceptual framework is constructed so as to understand what is going on in and about the object of study. This framework, in an admittedly imprecise sense of the word, is *constructed* and therefore understood. "Understanding" in this case refers to the designed explanatory framework.

This paper addresses questions concerning understanding the responses of a complex system. We shall try to present a generally applicable solution, one that will also be implementable for neural networks. It is based on a "symbol grounding" architecture. The increasing importance of autonomous systems which are put in a physical environment (cf. [4, 2]) leads us to assume that these issues will be even more important in the future. However, our approach will also be a sceptic solution of the problem. In the light of the extremely wide applicability of the principle of self-organization, we admit that this paper barely can scratch the surface of epistemic questions dealing with emergence, design, measurement, indeterminacy of translation, semantics, and other problems related to complex systems.

1.2 Terminology

In this paper we distinguish between physically and non-physically (information-ally) self-organizing systems. They differ with respect to their epistemic qualities. A *physically self-organizing system* is open to flow of energy or material. Its behavior is fully predictable at a low influx of energy by means of laws which describe the behavior of the system's atomisms ("equilibrium state") [14]. At a higher influx of energy, through the interaction of elements, the behavior of the system is not fully accounted for by initial and boundary conditions [15]. To successfully predict it, one needs to develop a new way of describing the system. In these descriptions sets of atomic states map on a state in the new descriptive frame (the macro description). This "change of view" means the development of a new observable of the system, which is important for successfully predicting it. We say that a new phenomenon has emerged.

In an *informationally self-organizing system* no physical emergence takes place, since the system is not necessarily open to flow of energy. Taking artificial neural networks as an example (which we will do throughout the paper), such systems share some properties with their physical pendants. This correspondence consists in the fact that (i) different levels of descriptions are possible to explain the system and (ii) the new observables appear through the interaction of many elements which can be assumed to be atomic. As in the former case the new observables are mappings from the "atomic" state space to a macro description of the system. The main difference to physical self-organization consists in the fact that, when simulated on a conventional computer architecture, the number of possible observables of the system is fixed. (P.Cariani has called this "computational emergence" [5].) Therefore, the change to the new description frame is not in the same sence necessary as in the physical system (see below).

A well-known socio-biological example of self-organization happens in large populations of nest building insects, e.g. termites [14, 15]. A single termite can be described as a simple autonomous agent who reacts upon a specific scent so that material is deposited immediately. If many insects interact through similar deposit of building and odor emitting material, this simple mechanism results in the construction of an emergent, seemingly planful behavior—the construction of a termite dome. Investigating the system by only considering individual termites cannot reveal this emergent phemonenon. The adequate description level of this behavior is the emergent phenomenon and the interaction of the individuals. It is important to realize that emergence only appears because of many interacting termites. Macro-desriptions like "The insects are constructing an arch." are important in describing and predicting the system of termites, whereas the micro-descriptions of single termites do not possess this predictive value (for a detailed discussion see [15, p.10]).

A good example for computational emergence in neural networks has been studied by Elman [9]. He has trained a recurrent neural network with sequences of patterns which are supposed to represent simple sentences. The task of the network was to predict the next word in the sentence. A cluster analysis of hidden unit activation shows that the network has formed representations which groups verbs, nouns, and certain object categories in clusters. In the language which we have just been using this means that for the network the concept of a verb or person has emerged in the hidden unit representation. This is something that could not be desribed at the level of the weights or single units alone. Explaining the network at this higher levellof linguistic structures makes it possible to predict what will happen to novel input patterns. Of course, such a prediction could also be derived from a mathematical analysis of network parameters, but the higher level description is understood much more easily.

2 System design problems

2.1 Conventional systems, rule-based systems

In what follows I will call a conventional system a system which can easily be explained by only looking at the elements of which it is constructed and their simple interaction. In this sense, a classical rule-based expert system in AI is conventional, whereas a self-organizing neural network is not.

A conventional system can be designed to behave in a well-defined way, because the internal dynamics of the system are highly constrained. No self-steering interaction between the programmed entities takes place, therefore there is no emergent behavior of the system (for a more mathematical formulation see e.g. [16, pp.31–39]). A mechanical example for such a system is the motor of a car, where the parts of the engine constrain the turbulances of exploding gas to guide the forces in order to operate the wheels. Another example is a rule-based system in AI, where the system designer tries to capture the mapping a problem space to a space of solutions by means of a partition of the former. This can be done, because the computer program ensures that the rules are mapped on the physical state space of the computer system in a way that the system is kept at a low energy level. Therefore, no physically relevant interactions between the rules (resp. the physical states of the computer) take place and no emergent properties of the system arise. Thus, the rules specify the physical state transitions and the transitions in the problem-solving state space.

The conventional approach to designing a solution with such systems consists in classifying "situations", i.e. input data, and connecting them to "actions". In immediately grounded connectionist systems or autonomous systems the classification refers to the data which arrives from the sensoric input. Such an input can be regarded as a measurement of the physical environment, like e.g. a video-camera recording. The system designer can classify the data according to what is happening at her own sensory "device" by means of the concepts which she has aquired. Being in possession of a relevant conceptual framework enables the designer to develop the rules and to *understand* what the rules do. When later explaining an action of the system, the programmer can give explanations in the sense that a specific set of data items "means" this or that and consequently leads to the action in question (cf. [24]). Explaining and understanding the system therefore crucially depend on the fact that the system actually *uses* the same conceptual framework as we do. Another reason, why rules are so useful in describing what a system does, is based on the correspondence of state-transition sequences and the derivation process in rule-bases. States, as "clearly identifiable situations" in which a system is (cf. [5]), can be compared to (sets of) concepts which have a clear and distinct meaning to humans and which clearly identify features of a system. Rules which connect such concepts can be compared to state-transitions.

For several reasons (pointed out e.g. by [17, 8]) the usage of a designer's own concepts in the constructon of these rules imposes severe limitations, even in the case of a machine-learning system. The influence of the designer restricts the system not only in the way that the mapping from situations to actions is being designed. The more profound problem is the pregiven conceptual basis of the designer. This problem has originally been one of the main arguments for artificially neural systems and is now extended through the importance of autonomously acting systems in physical environments. We shall call it *the designer problem*.

The problem with designing a conventional system in AI goes a little deeper. As S. Harnad has continually pointed out [10, 11], the symbols which are used by the programmer of such systems do, of course, not have the same meaning to the machine than they have to her. (This is one of the reasons why it is so difficult to write a good expert system: one easily forgets that all those connotations, which words have for the programmer, are not in the program.) Moreover, the concepts which a programmer uses to solve the task, simply may be an inefficient means of solving the problem; other concepts (the system's "own" ones) may be better suited.

2.2 Artificial neural networks

In an artificial neural network, the overall behavior of the system is a consequence of the interaction of many units. Therefore, the appropriate level of describing and explaining the system is often the level at which the behavior emerges and not at the level, where it is described, what a unitt does. I.e. we usually do not use "weight-and-unit"-talk when explaining the features or the "knowledge" of a net. (But we *do* use the symbolic rules of an expert system in the explanation of how it works.) With respect to section 1.2 it must be noted that the emergent character of neural networks is a function of the system taken into account. If one views the computer together with the neural network simulator as the system in question, than no *new* states emerge. It is only with respect to the designed entities that we can talk of "emergent properties".

A great part of the neural network literature only deals with the appropriate design of these entities. The question which confronts the designer is: "How can a network be designed, so that it shows the desired emergent behavior". This question is dual to: "What sort of behavior will emerge from that system?" and can therefore be labelled *the inverse emergence problem*. This problem can be translated to our termite example as: Which behavior should a single termite have, so that instead of the dome the insects build a bridge?

In neural networks there certainly exists an important influence in designing the architecture (number of units, learning parameters and rules, etc.). But it can be said that this is not comparable to the aforementioned strong influence of the designer. The hope at least is that the *relevant* system behviour only emerges from the interaction of many units. Therefore, the decisions made through the network do not directly depend on the conceptual framework of the net designer, especially if the input to the system is "immediately grounded" in the physical environment [3] (and not of the symbolic sort as e.g. in NETtalk [27]). The designer's strategy consists in enabling the system to self-organize to a satisfactory degree.

One of the well-known drawbacks of this technique is the *explanation problem.* (See e.g. [18].) The representations (or better: the models realized in hidden units) are *conceptually opaque.* This is why it is now very difficult to explain the system behavior within our own conceptual framework, i.e. with words containing personal experiences, sensual impressions etc. and not purely mathematical terms. We shall refer to this way of explaining a network as "understanding". One of the procedures that has been suggested and often used for this is to investigate the hidden units' representation through the construction of some kind of correlational semantics. This can be done by activating input units for which a predication e.g. "there-is-a-dog" can be given. A statistical relation between this input and an activation of a hidden unit can lead to assertions of the kind: "Hidden unit 7 is activated for dogs." However, this approach often does not work, because the hidden units simply do not represent something that would be easily expressible in natural language (within our conceptual framework). The more complicated the network architecture, the more implausible this approach becomes. It is especially difficult in cases where hidden units are connected to other hidden units or where recurrent connections exist in the system. For systems with a manifold of inputs (e.g. many and/or complex sensors with many states that receive data directly from the physical environment) it may become technically impossible to trace statistical relations between the parts of a self-organizing system and its emergent properties.

A solution to this problem is not just an epistemic need of the philosophically eager scientist, but a technical imperative. Without even a rudimentary approach to predicting the system, to evaluate it there can be no confidence in its results. It is also for reasons of maintanance that the explanation problem must be addressed. Note that even if we would be satisfied with a merely mathematical solution to the problem, a *physically* self-organizing system develops *new* observables. Therefore, understanding of such systems without the endeavour of applying difficult techniques of analysis is not feasible. The "explanation"-problem thus is a principal limitation of any self-organizing system. It coexists with a dualtity between design-based limitations and understandability.

2.3 Behavior-based robots

There is a perspective from which it makes sense to include behavior-based robots or autonomous systems in this discussion. Proponents of this kind of embodied AI have continually pointed out that a concrete behavior of a behavior-based robot is created through the interaction of many independent behavioral modules and through the robot's interaction with the physical world. For several reasons, such robots are a good example for the systems which we have in mind in this paper. The great number of interacting modules and their sometimes complex way of interaction makes it difficult, if not impossible, to describe an observed robot behavior in terms of the modules which generate it. Secondly, the high interaction dynamics of the robot with its environment and the large amount of sensor data make a direct trace of internal system variables impossible. In effect, the observer of such a robot is forced, as in the previous examples of neural networs or termites, to change the level of description from entities which generate the behavior to rules which contain "higher", in this case behavioral, concepts. Very often goal-oriented intentional talk is used (although there have been arguments that this can and should be avoided [28].)

The *problem of inverse emergence* in this case refers to the question which behavioral modules should interact at which dynamics and in which way with each other so as to generate the desired behavior. The *explanation problem* refers to the need for generating descriptions of the robot behavior in other terms than those which directly describe internal system variables. Traditionally, the *designer problem* is not as important in robotics, as in the other examples. However, there is an increasing number of architectures in which behavior-based and self-organizing (neural network) approaches are combined.

3 Making the system speak

3.1 Law-like rules and concepts

If we want to explain a self-organizing system, we are searching for rule-like descriptions of the system behavior. Since we hope that the system is not using rules which we designed (because of the designer problem), we must find another way to guarantee that the rule by which we describe the system actually is one the system follows.

Consequently, our first step will consist in a technique of attributing rule-like descriptions to the self-organizing system. The basis of this description must involve an understandable characterization of the situation in which the system finds itself. To understand what these situations for the system are, we need a possibility to make the system use one of our concepts without actually using it for decisions. This paradox requirement can be overcome by forcing the system to translate between the "concepts" it "uses" and the ones which we (the system observers) easily understand. What we need is a mapping of the system's states that result from self-organization to the concepts which we understand. An example architecture for achieving this is shown in fig. 1.

This architecture is supposed to achieve the following. It is able to "mimic" the designer's response to a given situation. When the system sees a flower, it responds with "flower!". In order to make the system answer with symbols which are understandable, there will have to be a training process through which the system builds its translation function. (See figure 2.) We do *not* assume to know this translation function. We only make the system respond in a plausible way to the situations in which it finds itself. Moreover, by assuming that the system is self-organizing and dealing with immediately grounded sensory data, we assume

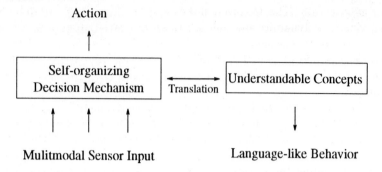

Fig. 1. Generating descriptive explanations: A translation mechansim is used which ensures that the system's own concepts are translated into understandable ones, which generate some kind of language like behavior (cf. [10, 24]).

that no analysis of the translation function is possible. Strictly speaking, we do not even have to know the state space of the decision making action system as long as the descriptions are consistent with what we expect them to be.

Fig. 2 shows how such a system could be trained to produce the correct descriptons, which would make sense to a human observer. It also shows what a correctly functioning system should produce.

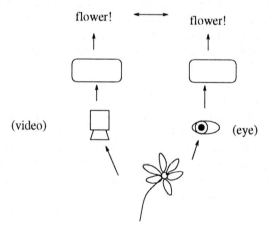

Fig. 2. Supervising the system: By teaching the system to produce the right symbols in the right situation the system generates the desired translation function.

3.2 Automated generation of descriptions

Such a mapping does not necessarily have to be constructed by the system. In our termite example scientists have found an explanation of the system through careful investigation. The termites, however, form a relatively simple system because of the low dynamics and the easy to observe interacting units. Note that this approach to understanding the system is not limited to neural networks, but can be regarded as generally applicable to complex systems. All we need is a consistent language-like behavior that is achieved through a translation mechanism.

The point which we would like to make here is that such a system can be explained by using *its own* descriptions of the situations. We can construct predictive rules based on the system's own situation descriptions. When it senses a "dog", it reacts accordingly. These attribution of situations shall form the basis of our understanding the self-organizing mechanism. But can we be sure that this is a useful description of what the system does?

3.3 Objections

The difference between (i) investigating the termites and (ii) the mimicry of a conventional conceptual frame seems to be a very profound one. In the case of the termites (i) the concepts which explain the behavior *really* describe what happens. In (ii) the concepts are just classifications of the system's states that do not *really* describe what is going on in the system.

What is obviously meant with such an objection is that in (i) we have grasped "nature" with our concepts and describe what is causally happening and could not otherwise be explained satisfactory. In (ii) the system's translation is just a mapping on a set of words that could well be *wrong*, i.e. not what the system *really* does. The system could, e.g. if we taught it wrong usages of words, lie to us and truly do something else. Can this be a serious doubt?

Of course, we must assume that the system has learned to use the words it learned correctly. Its description of situations should be comparable with the way we classify them and with the actions which the system undertakes. This question, we propose, must be formulated in the following way: Which fact can guarantee that the system *means* (through parroting) what I mean with this same word? The answer is that if the system is truly complex (i.e. not simply analyzable), then the only fact that does guarantee this can be found in the system's behavior. If it behaves as if it would use my terms correctly, we not only can but must say that it does.

This problem has been extensively discussed within a philosophical debate around the later works of Ludwig Wittgenstein [29]. In Kripke's interpretation [13], the main achievement of Wittgenstein was to have shown that there is no other guarantee for what somebody *means* with a word but some outer criterion to be found in his or her behavior. According to this position it is useless to search for facts within the speaker which could guarantee what is meant with a word. And it is also impossible to ensure that the word will always be used correctly in the future.

With respect to our problem here this position holds that as long as the system correctly uses my *words* we have to say that the system means the same as we do. In this sense do we *understand* the situations in which the system is. Additionally, as long as the system behaves according to rules which are based on these situations and their descriptons, we have properly designed the state space which is necessary to describe the system. We can then say to understand the system's actions in terms of our rules. For the sake of clarity, we shall reformulate this position in section 5. Let us first present a neural network architecture which is capable of implementing the necessary translation function for the case of an informationally self-organizing system.

4   A symbol grounding example

4.1   Symbol grounding

The architecture, which we have described so far, is essentially a "symbol grounding" architecture. Symbol grounding tries to answer the question as to how it is possible for a computer program to use symbols which are not arbitrarily interpretable [10]. The idea is to directly connect symbols to sensory experience and thereby enrich the meaning of these signs. Neural networks seem particularly suited for fulfilling these requirements for the following reasons: They are able to form categories or clusters in an unsupervised fashion. They can also be trained to label inputs in a supervised way. And they seem suited for directly dealing with immediately grounded sensory input, i.e. input which comes directly from a measurement device.

Typically, the models possess two types of input: one for (simulated) sensory data, another one for symbolic descriptions of the data. The networks are usually chosen so that the sensory input is categorized through un- or selfsupervised categorization algorithms like Kohonen networks [12]. The resulting category representations are then associated with some symbolic description of the static input.

4.2   Example

The two main possible architectures which could be used for achieving the desired functionality are shown in fig. 3.

In the left variant of the architecture the self-organizing decision mechanism which generates the desired behavior is independent from a component which categorizes and symbolically describes the input. In version (b) the decision mechanism which generates the output is also used to describe what is going on in the system, it is thus more similar to the architecture in fig. 1.

Details of this architecture are described in [25]. Here, it suffices to know that in the left version, the generation of the network's output is completely independent from the categorization which happens in order to have symbols grounded. Categorization is achieved through a Kohonen-network [12] and the learning of symbols is done via a simple supervised Hebbian learning scheme.

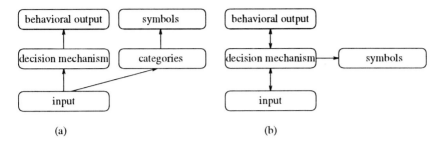

Fig. 3. Neural network architectures for symbol grounding: (a) shows a grounding module which generates symbols independently from the network's decision mechansim. (b) tries to ground the symbols in the system's own representational frame.

In the case of architecture (b) the network is trained to produce the desired (behavioral) output by means of supervised backpropagation learning. In this case, the categorization component does not directly find clusters in the input space as in version (a), but in the state-space of the hidden layer of the backpropagation network. These states are again categorized through a Kohonen network and then labeled by the same technique as in network (a).

The practical upshot of all this is that with this architecture symbols can, in principle, be grounded in sensory data. These networks succeed in correctly "uttering" a set of symbols in specific situations. However, they are very different with respect to what can be grounded, see below.

It must be added here that the generation of the correct symbols is, of course, not an easy task; quite the contrary: it is so difficult that whole areas of computer science are involved in finding ways to generate these mappings (e.g. vision, image understanding).

5   Understanding and measurement

One possible interpretation of our approach is to compare it with a measuring device and a physical process (figure 4). What we have introduced above is a system that responds with "flower!" to describe its own state. Now imagine a physical system and a meter, e.g. a thermometer that delivers numbers. We usually assume to understand the physical system if we can use the measurement in a rule which predicts another aspect of the physical system (of course, another measurement). If correct prediction fails, we can either change the rule, change the meter or change the explanation of the meter. Note that the system being measured can have (and presumably always has) additional states, which are not captured by the meter. But unless this does not result in wrong predictions, we are not only satisfied with our description, we say that we have *understood* the system. We can even say that physical events correspond to changes of state which are only specified by the evaluation of observables on states [26]. This means—as long as our predictions are correct and we have no need to introduce or to assume the existence of additional states—that all there is about physical systems (i.e. all there *is*) is that for which an observable exists.

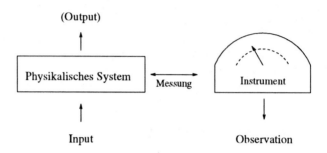

Fig. 4. A physical measurement.

Compared to our neural network example it now becomes clear in which sense we understand what the network does: As long as its own descriptions of the situations in which the system finds itself are consistent with what we expect these descriptions to be like, there is no reason whatsoever to say that we do not *really* understand what is going on in the system. In the example architecture all there is to explain the behavior producing neural network (the left part of the systems in figure 3) is what the network's translation function (the right part in figure 3) tells us about it. This forms the *relevant set of states* of the connectionist system.

Assume now that we have a physically self-organizing system which can develop new interesting states. Imagine further that, since we cannot predict what the relevant states will be, we invent a procedure to build new measuring devices which capture the newly developed observable (cf. [20, pp.105–108], [26, p.90]). This new meter maps the states of the system to numbers or symbols. In a way, this would be a better way of explaining the system, since we now have a direct meter for interesting phenomena. However, it is not clear how we can understand the new measurement. Note, that we do not know what the meter is actually measuring, we can only see its output, which consists in "meaningless" numbers or symbols. Trying to find out what an unknown meter measures, results in extensive experiments where one compares the results of the measurement with well-known situations measured by other devices or sensory organs.

It has been pointed out by Pattee [21, 22] that all this are typical features of measuring devices, since we do not know the internal dynamic constraints of the meter. Nevertheless, we can understand new measurements, because someone can explain them to us. I can, for example, be told that a bat can use its subsonic sensors to detect objects in the night. We understand the bat's measuring device through a comparison with our eyes and think that subsonic "hearing" is in a way like "seeing" (mainly of paralleling their *functions*.

This sort of translation is exactly what happens in the proposed architecture. The system is constructing a meter and automatically "explaining" it by using words the way we would do. In the case of the simulated neural network the meter construction is not a principal problem, because the state space of the computer is fixed. Therefore, we can theoretically construct a meter of the system's state space onto any desired scale. This cannot be done easily in the case of a physically self-organizing system [5]. Although one such self-building measuring device has been described and built by Gordon Pask [19], the applicability of the proposed method in physically self-organizing systems depends on physical methods for the construction of the translation function from "internal states" to symbols.

6 Practical problems of this approach

Experiments with both architectures that have been described above (fig. 3) show that the actual set of descriptions which can be grounded, i.e. which can be consistently generated by the architecture, is quite different from each other. (Details of the experiments are described in [25].)

In the experiments, feature vectors with random noise are presented to the networks. In the case of the architecture, where the input is directly categorized, it easy to "ground" symbolic descriptions of the input data. I.e. to generate labels for the inputs.

In the case of the architecture which categorizes the hidden states of the backpropagation network, grounding of the very same set of symbols fails (if the categorization component is forced to find a small set of categories, otherwise some symbols may be successfully grounded.) In this second case, only descriptions of the *output* can be grounded, i.e. only labels for the output of the backpropagation network can be correctly produced.

What these experiments show is that in the case of using the concepts which the system has built it can be very difficult to generate the right symbols. The reason lies in the fact that the *representation* which the system has constructed is more a model of the output than of the input. Consequently, in our neural network approach it is only possible to make the network describe its *output*, not its *input*.

We therefore face something similar to the original *designer problem:* Because of the fact that the network has generated a representation which is useful for itself, it has now become difficult to use these encodings of the input as the generator for symbolic descriptions which would make sense to the human observer. This, on the other hand, can also be of some practical value. If the desired translation function cannot be generated, this can mean that the information in the domain of the translation mapping is not sufficient to generate such a map. This, in turn, would be an indication that this term is not a good choice for a concept which explains the system behavior and the system's representational framework. Therefore, this technique also show us a way to the explanation of neural network models via some kind of "infering statistics".

Of course, simple extenstions of the architecture are possible, so as to categorize input as well as hidden layer states. In such an approach both kinds of labels, input and output descriptions, can be grounded. However, all approaches imply that the relevant information in order to generate the right set of symbols is in the state set which is taken as the basis of grounding.

We have seen that in the approach, where the system's own state-space is used, the goals of the network, i.e. what it is supposed to achieve, become more important to the formation of hidden unit representation. Consequently, more descriptions which contain terms like "in order to" or "so as to" can be securely grounded. This brings us to another, completely different meaning of "understanding", which is explained in the next section.

7 Understanding and goals

There is a second level at which the question of understanding what a system does can be answered. In what we said above, we tried to reduce the problem to the question of properly *discovering* the state space of the system, i.e. the situations which are important to predict its behavior. This was supposed to help us to discover the predicting *rules*. However, understanding can also mean to recognize the purposiveness of a specific behavior. In this case the *rules* which the system follows are themselves proven to be useful, meaningful, or goal-achieving. It has been previously pointed out by the author [24] that neural networks in general, but especially together with their appearance in autonomous systems or "artificial life" models tend to be explained in a teleological i.e. goal-oriented way.

In the case of our self-organizing system, however, we cannot be sure about the goals which the system will try to achieve. Because it is a technical system, we assume that it has been designed to fulfill one of our goals. This is why it may be a useful technical system. The experiments with the neural network architecture showed that it becomes difficult to explain the architecture with our own terms, since we could only ground concepts which describe the system's goals. A strategy for grounding must therefore be based on knowledge of the system's set of goals. This would then allow us, in a limited sense, to have the system speak about the situations, in which it finds itself. This would help in explaining and maintaining it. The following example is taken from [7].

Consider, for example, the case of an autonomous agent which has learned to correctly use a set of symbols for objects and locations (for example, to solve the famous "fetch and find" task). We oberve the robot in action. For some unknown reason it cannot find its way out of the kitchen, instead it always bumps into the table. Asking the agent about its location, i.e. having the agent use one of its grounded location symbols, it could e.g. answer with the "bathroom"-symbol. In this case, we could easily discover that the system wrongly "believes" that it currently is in the bathroom, maybe because of similar sensor readings. (It bumps into the table instead of leaving the bathroom through the door.)

But would the proposed mimicry, as a tool for explaining complex system behavior, be enough if the goals of the system would be different from our own goals? First, as we have seen before, it would be technically very difficult to ground correct descriptions of situations. But secondly, would these terms mean what we intended them to mean? Wittgenstein would deny this, since we can only understand what a system says and does, because of our socially shared way of living, our common "Lebensform." *If a lion could speak, we could not understand him.* [29, p.568]

8 Acknowledgements

The author wishes to thank Robert Trappl for having made this research possible as well as Georg Dorffner for continuous discussions. The Austrian Research Institute for Artificial Intelligence is supported by the Austrian Federal Ministry of Science and Research.

References

1. Bertalanffy L.von: The Theory of Open Systems in Physics and Biology. Science, **111** (1950), pp. 23 ff.
2. Braitenberg V.: Vehicles. Experiments in Synthetic Psychology, MIT Press, Cambridge, MA (1984)
3. Brooks R.A.: Elephants Don't Play Chess, in Maes P. (ed.), Designing Autonomous Agents, MIT Press, Cambridge, MA, Bradford Books (1960) pp. 3–16
4. Brooks R.A.: Intelligence without Reason, in Proceedings of the 12th International Conference on Artificial Intelligence, Morgan Kaufmann, San Mateo, CA (1991) pp. 569–595
5. Cariani P.: Implications from Structural Evolution: Semantic Adaptation, in Caudill M. (ed.), Proceedings of the International Joint Conference on Neural Networks (Winter Meeting), Washington D.C., Lawrence Erlbaum, Hillsdale, NJ (1990) pp. 47–50
6. Diederich J.: Explanation and Neural Computation, GMD, Nr. 458 (1990)
7. Dorffner G., Prem E.: Connectionism, symbol grounding, and autonomous agents, Proceedings of the 15th Annual Meeting of the Cognitive Science Society, Boulder, CO (1993) pp. 144-148
8. Dreyfus H.L.: What Computers can't do. A critique of artificial reason, Harper & Row, New York (1972)
9. Elman J.L.: Finding structure in time, Cognitive Science **14** (1990) pp. 179–211
10. Harnad S.: The symbol grounding problem, Physica D, 42 (1990), pp. 335–346
11. Harnad S.: Lost in the hermeneutic hall of mirrors, JETAI Journal of Experimental and Theoretical Artificial Intelligence **2** (1990) pp. 321–327
12. Kohonen T.: Self-organized formation of topologically correct feature maps, in Anderson J.A. (ed.), Neurocomputing, A Bradford Book, MIT Press, Cambridge, MA (1988) pp. 511–522
13. Kripke S.: Wittgenstein on rules and private language, Harvard University Press, Cambridge, MA (1982)

14. Kugler P.N., Turvey M.T.: Information natural law and the self-assembly of rhythmic movement, Lawrence Erlbaum, Hillsdale, NJ (1987)

15. Kugler P.N., Turvey M.T.: Self-Organization, flow fields, and information, in Kugler P.N., Turvey M.T. (eds.), Selforganization in biological work spaces, North-Holland, Amsterdam (1989) pp. 1–33

16. Lange O.: Wholes and parts. A general theory of system behaviour, Pergamon, PWN-Polish Scientific Publishers, Oxford (Warszawa 1962), 1965

17. Lischka C.: Ueber die Blindheit des Wissensingenieurs, die Geworfenheit kognitiver Systeme und anderes ..., KI **4** (1987) pp. 15–19

18. Partridge D.: What's wrong with neural architectures, Computer Science Dept., Univ. of Exeter, Research Report 142 (1987)

19. Pask G.: Physical analogues to the growth of a concept, Mechanization of thought processes: Proc. of a symposium, National Physical Laboratories, November 1958, HMSO, London

20. Pask G.: An approach to cybernetics, Hutchinson, London (1961)

21. Pattee H.H.: Universal principles of measurement and language functions in evolving systems, in Casti J.L., Karlqvist A. (eds.), Complexity, language and life: Mathematical approaches, Springer-Verlag, Berlin (1986) pp. 268–291

22. Pattee H.H.: The measurement problem in physics, computation, and brain theories, in Carvallo M.E. (ed.), Nature, cognition and system II - current systems-scientific research on natural and cognitive systems, Vol.2: On complementarity and beyond, Kluwer, Dordrecht (1992) pp. 197–192

23. Perlis D.: How can a program mean?, in Proceedings of the 10th International Joint Conference on Artificial Intelligence (IJCAI-87), Morgan Kaufmann, Los Altos, CA (1987) pp. 163–166

24. Prem E.: Aspects of rules and connectionism, in Trappl R. (ed.), Cybernetics and Systems '92, World Scientific Publishing, Singapore (1992) pp. 1343–1350

25. Prem E.: Symbol Grounding: Die Bedeutung der Verankerung von Symbolen in reichhaltiger sensorischer Erfahrung mittels neuronalen Netzen, PhD Thesis, University of Technology, Wien, Austria, 1994

26. Rosen R.: Fundamentals of measurement and representation of natural systems, North-Holland, New York, 1978

27. Sejnowski T.J., Rosenberg C.R.: Parallel networks that learn to pronounce english text, Complex Systems, **1** (1987) pp. 145–168

28. Smithers T.: Taking eliminative materialism seriously: A methodology for autonomous systems research, in: Varela F.J. et al., Towards a practice of autonomous systems (1991) pp. 31–40

29. Wittgenstein L.: Philosophische Untersuchungen, Suhrkamp, Frankfurt/Main, Werkausgabe Bd.1 (1986), 1953

Multi-Modal Active Sensing for a Simple Mobile Agent

Kristian T. Simsarian

Swedish Institute of Computer Science, Box 1263, S-164 28 Kista Sweden
kristian@sics.se

Abstract

This work presents a methodology for the combination and use of many simple features of differing modalities that are extracted from the environment. There are two main ideas presented here, first that it is important to consider all modalities for extracting sensory features and second, that useful information can be extracted without resorting to computationally expensive methods. Many useful features can be extracted easily and cheaply and recognition robustness could be achieved through the combination of many such simple features. Examples of such feature parameters are based on color, texture, motion, range, sound, and magnetism. A design is given for a small, simple, and fast embodied agent employing the ideas contained herein.

Introduction

Many researchers in the field of robot perception have provided various vision techniques for recognizing and building a structure model of a previously experienced object. Much of this work centers on building better techniques for feature extraction. Examples of such features are texture, color, edge, shading, and motion flow. Researchers have also combined many of these feature types into a single representation similar to what David Marr described as the $2\frac{1}{2}D$ sketch[12]. This multi-feature work generally falls into the category of "sensor fusion." The work in this paper provides an alternative to that traditional view and proposes the combination of many simple versions of different feature extraction techniques assembled into a feature vector, which can then be used to index into a set of feature-based models or directly into a set of actions for a task. The target application is a small autonomous robot with limited on-board computing. Thus, the idea is not only to combine many different types of features into one indexing vector, but to also make each component very simple to compute. One point which this research wishes to assert is that sensing in the various modalities is important and that not all the necessary information can be easily extracted from any one sensor. Another point is that in order to develop and test these theories the agent must be embodied in the real world.

Related Research

The sensing in this work follows closely along the line of work known as *qualitative vision*. Qualitative vision methods do not attempt to extract elaborate computational models from the environment, but rather extract qualitative features which require less computation, have better response time, focus on salient and relevant aspects of the environment and use environmental constraints more effectively [11]. A good example of this qualitative approach is the detection of object motion parameters. An agent does not have to build an accurate model of the motion structure to answer the question "Is this thing moving and is it going to hit me?" Instead, a much simpler calculation can be performed to find the focus-of-expansion (FOE) and then based on the FOE location (i.e. whether it is inside or outside the image), make predictions about object motion. A catalogue of such methods are well summarized in a paper by Nelson[17].

Ian Horswill tries to explicitly identify such environmental constraints for use in navigation[10]. For example, in trying to identify other independent agents, a rule such as "moving objects indicate agency" can be used. This rule, however, is not universal and can depend strongly on the environment that the agent is assumed to occupy. For this reason Horswill calls such rules "habitat constraints." This methodology is embodied in Poly, a mobile robot designed to give tours of the MIT AI Lab that can travel at 1 meter/second. This high speed is not achieved by using more computing power but instead is achieved by extracting features from the environment that are better suited for the task. A mainstay of Poly's corridor navigation is a simple algorithm that quickly extracts the corridor's vanishing point (the basic vanishing point extraction algorithm originally appears in [4]). This demonstrates that general purpose tasks could be achieved by combinations of simple task-specific skills.

Engleson describes a robot self-localization technique that uses simple vision cues to form an "image-signature" of a location[9]. Most of these features are quite inexpensive to compute and in combination are described to provide reasonably robust recognition in many cases. Variations of some of those features are described again in this paper and are suggested as simple vision parameters.

Nelson describes a method of using simple features, such as dominant edge orientation, extracted from a low resolution spatial image to locate a specific location for the task of visual homing[16]. Work on homing has also been done by Nehmzow who used a solar based compass[15]. There is evidence that shows that different animals employ many different methods of direction finding and even that a single animal can employ redundant methods (*e.g.* pigeons). Examples of such tracking and locating methods used in homing are light-gradient following, solar and lunar tracking, visual recognition, magnetic field tracking, electrical potential detection, and odor and pheromone detection[25]. The work in this paper suggests that many of these features in combination should be explored for the task of robot homing.

Swain demonstrated that simple color histograms can provide surprisingly accurate object recognition at low computational cost[23]. However under varying lighting conditions where the spectral and illumination qualities are subject to change, color does not maintain its invariance and is therefore not always a robust feature. The idea of histogramming features is extended in this present work to encompass other vision and non-vision based features. The computational cost of adding parameters is not great (especially at reduced resolution) while the increase in robustness is expected to be high.

Rimey identifies what he calls "sufficing-vision" which is similar to the notion of extracting just enough information for a task[18]. Rimey also mentions that the features that fulfill this notion of sufficing are often meaningless without context. In other words, the features can be powerful for a given task and context but do not constitute a general world model. This methodology falls into the work called "active/animate vision"[1],[3].

By analogy with active vision, the work described in this paper is termed "active sensing".[1] In addition to the visual sensors being active, the agent itself is active and has the capability to investigate a stimulus in order to improve and extract sensor information. Active sensing requires that an agent be embodied and have the ability to explore and exploit features in its environment.

Extracted Features

This section provides a description of the qualitative features that this work proposes to use. Each feature parameter is meant to be as simple as possible while supplying independent information about the environment.

While it is important to take advantage of any information that a sensor can provide, it is also important that sensors be used for their strengths and not over extended when viewed within the context of a given robot. A use of a certain sensor should not be discarded because it cannot provide all information. Instead simple sensors should be viewed in combination with other sensors and with agent morphology. An autonomous system has a specific kind of interaction with the world and this interaction needs to be taken into consideration when designing a robot's sensory system. A fit system in the biological world has tuned its sensors to suit its survival needs. Likewise we should, as robot designers, take a robot-centered view on the sensation constraints and allowances when designing and tuning sensors. This idea of robot-centered design is similar to Smithers' arguments for the importance of exploring the "interaction space" of an autonomous system [20].

In this description of extracted feature parameters, it is assumed that the agent has already fixed its attention to some location in the scene and that

---

[1] The term "active sensing," as used here, should not be confused with the term "active sensor" (devices that transmit a signal in order to make a reading).

the sensor used is employing a variable resolution imaging scheme such as [19][2]. The eye movements that shift the focus of attention, called *saccades*, can be guided by low-resolution visual cues, i.e. motion or color as described in [24]. Once the agent's attention has been drawn to a part of the scene, the next step is to transform the sensory information into a robot response. Based on what has been foveated, the agent must identify what it is looking at (if object identification is the task) or what action it should take. Henceforth the sensory information acquired from the environment (and the environment itself) will be called *the stimulus* and the action the robot takes *the response*. Attending to a more limited part of the scene, and the cue that guided attention to that point, is treated here as a form of segmentation. The idea is to fit a number of easily computed features into a vector that describes the overall characteristics of this attended portion of the scene.

Since this technique would be for a small autonomous robot, the features should be easily computed. Each of the following vision based features will be applied to each low-resolution picture element contained in the fixated region of interest of a course image. Thusly the result encodes coarse spatial information in the feature vector..

The following is a catalogue of features.

Edges Calculated edge-based measures are Y-gradient magnitude, X-gradient magnitude, overall edge density (number of zero-crossings), and magnitude and number of corresponding X and Y edges (these correspondences often indicate object boundaries);

Motion From two time varying images a simple measure of global motion direction and magnitude can be obtained;

Color With a color-sensitive sensor, the global amount of color in each independent color parameter can be calculated;

Texture With the appropriate hardware (i.e. a digital signal processing chip), simple texture information can be calculated, such as dominant gradient direction, intensity, and frequency. Such texture information can be hashed into low, medium and high frequencies. If this hardware is not available simpler measures can be used, for example a count of the number of neighboring pixels that differ;

Specularity When specularities are present they often appear as saturated sensor values. A count of saturated pixels serves as a very crude specularity measurement;

Sound If the agent is equipped with a pair of sound sensitive devices it is possible to calculate simple sound parameters such as intensity, frequency and source direction;

Compass Directional input from a compass can provide information on viewing direction and location;

Range A spatially distributed range sensor array, such as that provided from sonar or infrared can give scale information.

Speed and inclination Features about the agent's movement (agent kinesthetics) are important to give context to other extracted features. Such kinesthetic parameters are extracted from devices such as wheel shaft encoders and inclinometers.

After the feature vector is calculated it should be matched to a response in such a way that the weighting of individual features can be altered. It should also be possible for other factors such as recent experience and task goals to influence the weighting. For example, if the agent is searching for metallic objects the specularity measurement should be given relatively large weighting. The ability to emphasize and deemphasize certain features is meant to prime the sensors, thus giving the agent the ability to have context for a task. It is well known that context can strongly and decisively alter categorization in humans. See Bruner and Postman [7] for a seminal demonstration of this. Weights matrices are meant to give this ability.

In the following, v is an n-dimensional feature vector, where n is the total number of parameters extracted from the stimulus.[2] W is a nxn matrix of weights. f is an indexing function that selects out an action based on the transformed input. The action is another vector for commanding the response (i.e. the actuators).

$$f\,(v\;W)\;=>\;response$$

While the matrix of weights transforms the input linearly, f is allowed to be a non-linear computation such as a neural net. f is then, for a given size of stimulus/response pairs, a constant time indexing function. There is also a mechanism to select a context dependent weight matrix. The function f, the weights-matrix mechanism, and the matrix itself could all be learned functions.

The proposed advantage to using this method would be the acquisition of many different types of information. This model also contains a notion of robustness in the sense that if one feature is a weak indicator in a particular situation, the information from the other sensors can compensate. Given the context of using a small fully autonomous robot, it is important that each of these measurements be cheap to compute. Most of the above features outlined above are just that.

In order to achieve more robust light invariance the color space employed is HSV (hue, saturation, value). Using just the chroma components, HS, provides a reasonable robustness under varying illumination that has constant spectral but variable intensity components [22].

In this framework, in addition to the parameters coded explicitly, there is also information encoded implicitly. For example, range information gives stimulus scale and the color information contains the spectral characteristics of lighting. If the stimulus or the agent is moving this movement will be

---

[2] For visual features there is one parameter of each type extracted from the different spatial locations in a 3x3 subsampled array representing the image.

embedded in the visual and audio cues, as well in the kinesthetic sensors of the agent, i.e. shaft encoders and inclinometers. In addition, information about ego-motion can be used to better measure other movement in the environment.

For more robust segmentation in an object-identification task, in addition to using foveation, a more sophisticated method to separate out the stimulus from the background can be employed. The correspondences of X and Y direction spatial derivatives can be calculated and then the convex hull of these correspondences could be called the outline of the stimulus. Since these locations in the image can often define object boundaries, features extracted within these boundaries will have a greater correspondence with the stimulus.

Fetching Task

Tom Mitchel [14] proposed a new challenge to behavior-based robotics[6]. This challenge was posed in terms of a new task: "Fetch X from Y and bring it to Z." Inherent in this problem is stimulus recognition. The stimulus X is an object and both Y and Z could be objects, agents or locations. To achieve this task, the fetching agent must be able to match a stimulus to a response. Thus, this kind of stimulus recognition is important. The work in this paper bears this new task in mind and proposes that through the combination of the simple features presented here, such stimulus recognition can be achieved in practical computational time. The subject of planning and control for such a fetching agent is beyond the scope of this paper.

An example of such a fetching task is the robot acquisition of particular toy wind-up creatures which have been released into an environment with distractors[3]. In addition to varying colors, such toys often have varying behaviors, i.e. hopping, jumping, walking, crawling, and thus emit different sounds. Therefore sound, as well as color, size and motion are clues to identification. If it is motion that attracted the focus of attention, and the stimulus is directly in front of the agent, then this information could be present in data from the range/proximity sensors. Thus a stimulus that is distant gives a different feature response than a target stimulus which is close. Features such as motion would thus not be viewed alone. While simple motion parameters suffer the traditional ambiguity of: *big motion far away* versus *small motion close by,* the likelihood of such ambiguities is reduced when combined with other sensor information.

Embodiment

Embodiment and operation in the real world are critical to discover the real behavior of the agent and the sensors. In this work the discovery and use

---

[3] Distractors could be other moving objects or other such toys that the robot is not supposed to acquire.

of the real-world behavior of a sensor, i.e. the sensor "bugs," is important. For example, it does not matter if a magnetic compass is subject to local fluctuations as long as that sensor behavior is relatively reliable in a given location. Likewise the ultrasonic detection of a corner, such as that by a Polaroid range sensor [5], does not have to be identical to a corner detected by a human. Such a measurement should only be repeatable to be useful. For further discussion on similar sensor use, see work by Mataric [13] and Connell [8].

Sensors should be considered in combination with other sensors. For example infrared and sonar range sensors compliment each other well since sonar can give decent coarse range measurements over long distances, while infrared can be engineered to give fine range information over short distances. Much of the subtleties of sensor and agent interaction with the environment can not be determined *a priori*, but must be discovered by running tests in the real-world with an embodied agent. Recently several researchers have made strong arguments for embodiment, these arguments will not be repeated here and the reader is instead referred to [6][20][21].

A small fast mobile testbed is being developed for this work. It initially consists of a toy R/C car base with a microcontroller to coordinate the sensing, acting and additional computation, i.e. DSP and communication. Range, magnetic, light, sound, proximity and tilt sensors will be employed and combined in the scheme described above. The goal is to enable this agent to run untethered with the ability to explore its environment. When possible, the sensory processing, i.e. the computation of vision based spatial and temporal derivates, will be done in analogue or digital hardware.

Simulations and Surrealism

When the artist René Magritte wrote "Ceci N'est pas une pipe" under his painting of a pipe, he was drawing attention to the fact that it is not a real pipe being viewed, but merely a representation. Though we as embodied human viewers realize this and would not try, for instance, to grasp Magritte's pipe, we have not allowed our agents this basic distinction when we run simulations that present the sensors with crafted images. We know a pipe not by experiencing it through images, but by experiencing real pipes in previous episodes. Training a system to answer "pipe" to a presentation of an image of a pipe is not a real test to a system that needs to deal with such objects in the real world. It is also denying the agent the opportunity to experience that object. In the real world, humans have a rich interaction with objects. In addition to tactile manipulation, humans have the ability to view objects from varying angles, and depth, (not to mention sound, taste and smell). Not only does this active approach give the agent more information, it allows the agent the ability to choose its own input. Which, after all, is one basic goal of artificial intelligence as a whole.

References

1. Yiannis Aloimonos, Isaac Weiss, and Amit Bandyopadhay, Active vision, In Proceedings of the Image Understanding Workshop, pp. 552–573, Los Angeles, CA, February, 1987

2. X. Arreguit, Analog VLSI artificial neural networks, In Proceedings of the first IFIP Workshop, Grenoble, March 1992

3. Dana H. Ballard, Animate vision, Artificial Intelligence, 48: 57–86, 1991

4. P. Bellutta, G. Collini, A. Verri, and V. Torre, Navigation by tracking vanishing points, In AAAI Spring symposium on robot navigation, AAAI, 1989

5. C. Biber, S. Ellin, E. Shenk, and J. Stempeck, The polaroid ultrasonic ranging system, In Audio engineering society 67th convention, 1980

6. Rodney A. Brooks, Knowledge without reason, MIT AI Technical Report, number 1293, April 1993

7. J.S. Bruner and L. Postman, On the perception of incongruity: A paradigm, Journal of personality, 18: 206–223, 1949

8. Jonathan H. Connell, Minimalist mobile robotics: a colony architecture for an artificial creature, AP, 1990

9. Sean P. Engleson and Drew V. McDermott, Image signatures for place recognition and map construction, In Proceedings SPIE symposium on intelligent robotic systems, sensor fusion IV, November 1991

10. Ian Horswill, A simple, cheap, and robust visual navigation system, In From animals to animats II: Proceedings of second international conference of adaptive behavior, 1993

11. W. Lim, A. Blake, P. Kahn, and D. Weinshall (eds.) Workshop on qualitative vision, AAAI, 1990

12. David Marr, Vision, W.H. Freeman, 1982

13. Maja J. Mataric, Integration of representation into goal-driven behavior-based robotics, IEEE Transactions on robotics and automation, 8(3): 304–312, June 1992

14. Tom Mitchel, Lecture: A new challenge to behavior-based robotics, In NATO Advanced Study Institute: The biology and technology of autonomous agents, 1993

15. Ulrich Nehmzow and Brendan McGonigle, Robot navigation by light, In Proceedings of european conference on artificial life, Brussels, 1993

16. Randal C. Nelson, Visual homing using an associative memory, In Proceedings of the DARPA image understanding workshop, pp. 245–262, 1989

17. Randal C. Nelson, Qualitative detection of motion by a moving observer, International journal of computer vision, 1991

18. R. D. Rimey, Studying control of selective perception with t-world and tea, In Proceedings: IEEE workshop on qualitative vision, IEEE press 1993

19. Alan S. Rojer and Eric L. Schwartz, Design considerations for a space-variant visual sensor with complex-logarithmic geometry, In Proceedings of the international conference on pattern recognition, 1990

20. Tim Smithers, Taking eliminative materialism seriously: A methodology for autonomous systems research, In NATO advanced study institute: The biology and technology of autonomous agents, 1993

21. Luc Steels, Building agents out of autonomous behavior systems, In L. Steels and R. Brooks (eds.), The artificial life route to artificial intelligence, Lawrence Erlbaum, 1993
22. Markus Stricker, Color and geometry as cues for indexing, Technical Report CS92-22, Department of Computer Science, University of Chicago, November 1992
23. Michael J. Swain and Dana H. Ballard, Color indexing, International journal of computer vision, 7: 11–32, 1991
24. Michael J. Swain, Dana H. Ballard, and Roger E. Kahn, Low resolution cues for guiding saccadic eye movements, In Proceedings of IEEE conference on computer vision and pattern recognition, pp. 737–740, 1992
25. Talbot H. Waterman, Animal navigation, Scientific American Library, 1989

Evaluating an Active Camera
Controlled by a Subsumption Architecture

Claudio S. Pinhanez *

University of São Paulo, Institute of Mathematics and Statistics
Present address: MIT, Media Laboratory, E15-368C
Cambridge, MA 02139, USA

Abstract. The quantification of the performance in highly reactive systems is the main topic of this paper. A simulator of an active camera, controlled by a subsumption architecture, was designed, built, and tested with different real-world images. Some special characteristics of the system, called *active eye*, permit the study and presentation of the system's behavior. Maps displaying the paths followed by the active eye can be visualized, providing a better understanding of the effects of the incremental building method typical of subsumption architectures. A method is then developed to obtain a numerical measure out of path histograms, based on the ratio between true positives and false positives. This measure is consistent with experienced variations of performance during the building process, and to the system's robustness to noise. The results of the active eye's higher level ability of detecting straight edges can also be displayed as images, which bear an interesting resemblance to hand-drawings.

1 Introduction

Most of the literature on behavior-based systems and subsumption architectures seems to fail to provide information about the actual patterns of
behavior exhibited by the artificial creatures, or numerical measures on improvement or robustness. The reader often faces sentences like *"... The robot
has spent a few hours wandering around a laboratory and a machine room.
..."* (Brooks, [1]), or *"... the robot sits still until a person, say, walks by,
and then moves forward a little. ..."* (Brooks, [2]), which mention only the
authors' visual impressions.

We are not questioning the results of the previous and on-going research,
but pointing out that sentences like those are interpreted differently by different readers. For instance, Brooks & Flynn, in [3], describe a robot's behavior

---

* This work was supported by the scholarship for research students (*kenkyu-sei*)
of the Ministry of Education of Japan (*Mombushō*), complemented by FAPESP
(grant number 90/2940–7d), and, later, by CNPq (grant number 203117–89.1)

(Squirt Jr.'s) as *"... his normal mode of operation is to act as a 'bug', hiding in dark corners and venturing out in the direction of noises, only after the noises are long gone, looking for a new place to hide near where the previous set of noises came from. ..."*. Of course it is difficult to explain in a paper how a creature reacts to its environment, specially when dealing with the uncertainties of the real world. But surely someone who has never watched Squirt Jr. in action can hardly imagine its actual movements. How does the behavior look like? Is Squirt Jr. like an insect? Or more like an intentional toy car?

Some very simple presentation of the accomplished behaviors can be found in the very first article on subsumption architectures (Brooks, [1]), in pictures portraying the paths followed by a simulated robot walking through corridors. Numerical evaluation of robot behavior has been even more rare in the subsumption architecture's literature, though we can cite the article of Mahadevan & Connell, [4], describing some measures of learning in a robot's performance.

The purpose of this article is to examine the behavior of a subsumption architecture-based system, by developing a numerical measure of performance. The measure is shown to be consistent to the visual impressions, and helps on the evaluation of the incremental method used when building the system, and also makes addressable, to some extent, the problem of robustness evaluation.

2 Basics of Subsumption Architectures

The *subsumption architecture* is a model for integrating sensors, actuators and control in a single framework. Here, only some essential ideas are examined: a good account can be found in the works of Brooks and his colleagues at MIT ([1], [2], [5]).

The basic unit of the subsumption architecture is a device consisting of a finite state machine with *timers*, called an *AFSM* — Augmented Finite State Machine. An *AFSM* is activated whenever it receives a message in one of its *registers* or a timer is triggered. The registers of different *AFSM*'s can be interconnected by *wires*; special *connectors* can be used to inhibit or suppress the information flowing through the wires.

We can view the subsumption architecture as a way to build highly reactive systems: first, the fundamental *behaviors* are implemented, tested in the real world and debugged. After that, a layered repertory of behaviors is built incrementally upon the previous network, but not necessarily in a hierarchical way. Ideally, no representation of the world is employed, neither a central control device, and the *AFSM*'s run in parallel (see Brooks, [6]).

Our research used a discrete simulator of a subsumption architecture(a more detailed description can be found in [7]). We call *1 unit of time*, or *1 ut*, the time needed by the simulator to run one cycle of all active *AFSM*'s, and to deliver the produced messages.

3 The Active Eye

We performed our experiments in a simulator of active cameras, using a high resolution image and a program, named *active eye*, which moves a group of *sensors* around the image, trying to locate interesting features without sensoring the whole image. Only the essential characteristics of the system are described bellow; a detailed description of the structure of the active eye can be found in [8] and in [9].

We define *the concentration of the active eye's presence on the areas of the image richest in edges* as the objective of our system. This must be accomplished using only local information, by concentrating the sensors in a *fovea*-like structure, and moving this apparatus throughout the image. The active eye must be able to find edges wherever they are, even if emersed in uniform areas, and also to treat images with different subjects, *without calibration*.

The second objective is *the following and detection of long straight edges*. The idea here is putting the system to follow edges, as a way both to naturally increase the activity in edge regions, and also to ease the finding of the straight lines. It must be noted that the detection of lines is a higher level task than the concentration of attention, and usually requires more complex algorithms. This second objective was set as a way to evaluate the use of behavior-based models for different levels of task complexity.

The term *sensor* is used here in a particular sense, meaning simply a method of examining the matrix of intensities of the image. The sensors are contained in an area of approximately 100 pixels of diameter, and are heavily clustered around the center. Although the whole system is quite simple compared to an actual foveated active camera, the structure provides enough complexity and difficulty in terms of camera control.

According to the principles of behavior-based systems, it is necessary to perform tests in the real world, what means, in our case, real images. Figure 1 shows the four images we use throughout this paper to illustrate our results. The first two — called *block1* and *block2* — depict outdoor scenes, and the other two — *lena* and *kim* — are indoor images, with people and objects. They are all 512×480 pixels large, with 256 levels of grey.

3.1 The Building of the Active Eye

The *active eye* can be understood as a simulator of an active camera which searches, detects, and follows edges. We tried to have our system built according to the principles of subsumption architectures, and, particularly, each behavior was exhaustively tested and debugged before the next was added, by observing its result on test images like the ones shown in Fig. 1.

Some of the *AFSM*'s are connected to *sensors*, and it is necessary to take in account, and to explore, if possible, the movement of the camera when designing the sensors. It is also desirable, in order to achieve real-time

Fig. 1. The four test images: block1 (upper-left), block2 (upper-right), lena (lower-left) and kim (lower-right); all images are 512 × 480 pixels large, 8 bits per pixel.

performance, that the sensors examine only a few pixels per unit of time, and based on simple calculations, guess whether something of interest was found. Accuracy proved to be not so important, since the actual control of the eye is the result of the complex interaction of the whole network of *AFSM*'s.

Figure 2 shows a simplified diagram of the system used in the active eye, whose structure and construction we briefly describe in this section. A better explanation of the symbology can be found in the works of Brooks (for example, in [2]). In this diagram the big rectangles indicate the *AFSM*'s, the small vertical rectangles mark the registers, and the arrows represent the wires which interconnect them; a number x multiplying an *AFSM* shows that x copies of it are used. Black dots represent a connector type **Bisector**, an i within a circle means a connector type **Inhibitor**, and an s indicates a connector type **Suppressor**. A triangle in the upper-left (respectively lower-right) corner is used to mark *AFSM*'s directly connected to sensors (respectively actuators).

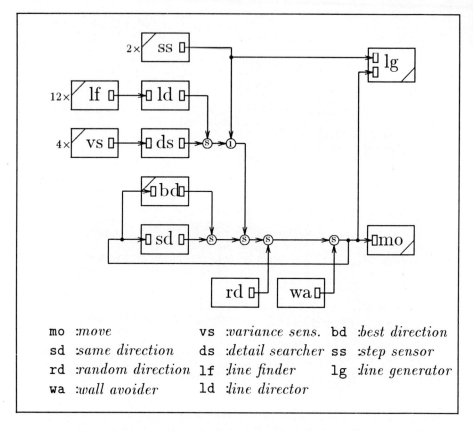

Fig. 2. Simplified diagram of the subsumption architecture used for controlling the active eye; for clarity of presentation, identical *AFSM*'s are grouped, and parallel wires are represented by only one arrow.

The following behaviors are currently implemented in the active eye. For each behavior added, we recorded the plot of the path followed by the center of the eye. The kind and intensity of the improvement caused by the layering of each new behavior is shown in Fig. 3, which was obtained by having the active eye scrutinizing the image block1 (upper-left image of Fig. 1), in a short run of 50,000 uts.

Random Walk Generator : The most basic behavior of the active eye is implemented using the *AFSM*'s **mo** (*move*), **sd** (*same direction*), **rd** (*random direction*) and **wa** (*wall avoider*). A simple feedback loop produces a regular straight movement of the center of the eye, changed randomly every 1,000 units of time. It is completely independent from the image in use, since no sensing is done, as shown in Fig. 3 (upper-left).

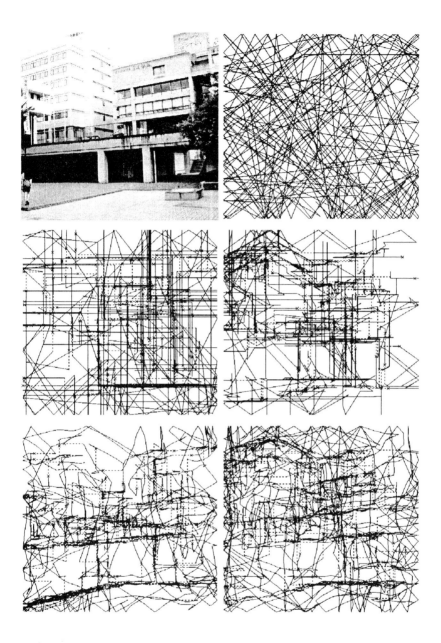

Fig. 3. The paths followed by the center of the active eye under the control of different sub-systems, for 50,000 units of time: the random walk generator (upper-left); the movement generated by the attractor to high variance areas (upper-right); the saccadic movements caused by the attractor to high variance areas, where a cross marks the arrival point (middle-left); the inclusion of the edge detector (middle-right); the addition of the corrector of direction (lower-left); and the whole system (lower-right).

Attractor to High Variance Areas : The first behavior which senses the image has the objective of pushing the focus of attention towards areas with high local variance (where edges can be presumely easierly found). Composed by four **vs**-type *AFSM*'s (*variance sensor*), each of them calculating the variance of four different areas located to the left, to the right, above and below the center; and by **ds** (*detail searcher*) which determines if a good direction was found. If this is the case, **ds** sends to **mo** a special message which generates a 50-pixel jump of the center of the eye towards the high variance area, something that can be compared to the *saccadic movements* of the human eyes. The effect of adding this behavior to the random walk generator is shown in the upper-right picture of Fig. 3. The saccadic movements are represented by dashed lines ending in small crosses which mark the arrival position, and, for clarity, are displayed isolated in the middle-left picture of Fig. 3. Note that the path followed by the center of the eye starts concentrating in the edge-rich areas of the image.

Edge Detector : In order to detect edges and change the direction appropriately (for following them), a behavior consisting of 12 *AFSM*'s of type **lf** (*line finder*) and a selector **ld** (*line director*) was added, suppressing, when active, the attraction to high variance areas. Each **lf** scans a different direction, looking for uniform areas with huge gradients. The 12 **lf**'s are arranged radially around the center of the eye, covering every direction which is a multiple of $30°$. The value of the gradients are sent to **ld** which selects the direction with maximum gradient. The resulting effect can be appreciated in the middle-right picture of Fig. 3, where we can see that the system is already able to find line edges, without losing its previous abilities of wandering or discovering high variance areas.

Corrector of Direction : This behavior aims to smooth the paths followed by the center, and is implemented using a very simple *AFSM*, **bd** (*best direction*). The effects of tiny corrections in the actual direction are calculated every unit of time; if there exists a correction which results in a higher gradient in the direction perpendicular to the path being followed, **bd** pumps it into **mo**. The behavior of this module is critically dependent on the movement of the camera. The addition of this module produces a dramatic change in the behavior of the active eye, as can be seen in Fig. 3 (lower-left picture).

Sensor for Movements over Edges : Two **ss**-type *AFSM*'s (*step sensor*) check if an edge is being followed, and, if so, inhibit both the edge detector and the attractor to high variance areas, though keeping **bd** alive. Some sensor characteristics of **ss** also add some *inertia* to the eye, enabling it to follow broken lines, and even to detect some types of *subjective contours*. The lower-right picture of Fig. 3 exhibits clearly the concentration of activity in the edge areas. As the system is now more able to follow edges, the area covered by the active eye is enlarged, maintaining, though, most of the abilities showed so far.

Segment Generator : The detection of a long edge segments is done by `lg` (*line generator*), based on the edges found by `ss`. The internal structure of the segment generator is explained bellow in this article.

The path maps shown in Fig. 3 clearly provide a better understanding of the incremental method of building behaviors. However, they are obtained using only 50,000 units of time — what means that the center of the active eye visits at most 50,000 different pixels, about 1/5 of the total number of pixels. And above 50,000 units of time, those maps turn out to be an almost meaningless collection of random lines.

4 Developing a Measure of Evaluation

Although the results of Fig. 3 show that our system exhibits an interesting behavior, we must address now the problem of measuring the activity of eye. The bad point about path maps is that they only show whether the active eye has passed over a pixel, but not how many times. If the eye visits a pixel 10 times, this means that it has been concentrating its activity on that pixel. By counting the number of times the center of the active eye has passed over each different pixel, we obtain what we call the *path histogram* of the active eye on the image.

Figure 4 shows the path histograms of the images of Fig. 1, considering the whole active eye system working during 245,760 units of time (the number of pixels in each image). As can be seen, the blackest areas coincide with the most distinct edges. We develop our evaluation method based on those path histograms comparing the histogram with "desired" results.

Our numerical evaluation method is based on these path histograms, by comparing the histogram with "desired" results.

4.1 The $T+/F+$ Ratio

What can be a numerical definition of "concentrating the activity in the edge areas"? A path histogram really provides information about concentration of activity, but tells nothing about the actual edges of the image. Therefore, it is necessary to compare the path histogram to an image which reflects the edge areas. Our approach applies a common edge detector on the images (in this case, the maxima of the first derivative of a symmetric exponential filter, according to [10]), producing a image with black lines of 1-pixel width, and enlarges those edge lines by applying a morphological dilation on the edge image.

Now that we have defined the target pixels of our system, we need to determine when the eye has accomplished its task. It seems reasonable to say

Fig. 4. The path histograms of the four test images, for 245,760 units of time.

that whenever the system has passed more than twice over the same pixel,
then this pixel is concentrating the activity of the eye (for a 245,760 units of
time run): indeed, in order to visit a pixel three or more times, the eye must
give up passing over at least two other pixels. Besides that, the center of the
eye can mistakenly visit a pixel when moving from one point to another, or
visit twice when two directions of movement cross each other; we think that
three times is a situation much more difficult to happen just by chance.

Thus, the path histogram is thresholded at 2, resulting in a binary image
with black pixels indicating three or more visits. We then perform a logical
AND between the thresholded histogram and the dilated edge image, obtaining
the desired points which are really detected by the active eye. The number of
black pixels resulting from this operation reflects how successful the system
is in concentrating its activities in edge areas.

Table 1. Values of $T+$, $T-$, $F+$, $F-$, and $T+/F+$ for the four test images.

	block1	block2	lena	kim
$T+$	20498	19748	17552	17866
$T-$	116718	116206	153682	148130
$F+$	3659	4449	6947	6143
$F-$	104884	105357	67579	73621
$T+/F+$	5.60	4.44	2.53	2.91

However, a good system must also be able to avoid uniform areas. This reasoning leads to the definition of the following quantities:

$T+$: *true positives* — the number of black pixels (enlarged edges) on the desired image which concentrate the activity of the eye (more than two visits in the path histogram);

$T-$: *true negatives* — the number of white pixels (non-edges) on the desired image which do not concentrate the activity;

$F+$: *false positives* — the number of white pixels on the desired image which erroneously concentrated the activity;

$F-$: *false negatives* — the number of black pixels on the desired image which must, but do not concentrate the activity.

Figure 5 shows the meaning of $T+$, $T-$, $F+$, and $F-$ in the case of the block1 image, for 245,760 uts. Here, the considered pixels are displayed as black dots. Table 1 displays the actual number of pixels counted on the four test images.

We propose to use the ratio between $T+$ and $F+$ as the numerical measure of the behavior of the active eye, since it reflects both the successes and the failures. The ratio increases as the number of successfully detected points increases, and the division by $F+$ avoids randomness as a good policy. Considering, for instance, the case of block1, where the $T+/F+$ ratio is 5.60, we have the active eye visiting 5.6 desired pixels every wrong pixel. The last line of Table 1 contains the ratio between $T+$ and $F+$, for the four test images.

It is reasonable to believe that the $T+/F+$ ratio produces a number which can be associated to the intrinsic difficulty of the image, in terms of edge finding. However, the real utility of the $T+/F+$ ratio is when comparing the performance of different systems working in the same conditions, *i.e.*, on the same image.

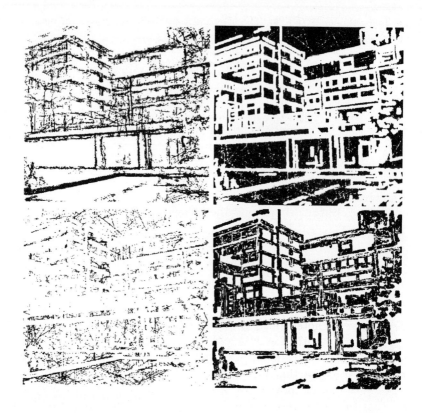

Fig. 5. The $T+$ points (upper-left), the $T-$ points (upper-right), the $F+$ points (lower-left) and the $F-$ points (lower-right), for the image block1, with 245,760 units of time and a path histogram threshold at 2; the points are displayed in black.

5 Evaluating the Building of the Active Eye

Having defined a way of measuring the behavior of the active eye, let's take a look of its values during the implementation process. Table 2 contains $T+$, $F+$ and $T+/F+$ for the five different phases of behavior implementation, for 245,760 units of time. The word "random" stands for the system composed only by the random walk generator, "ds–vs" represents the addition of the attractor to high variance areas ($AFSM$'s **ds** and **vs**), and so on.

The results show that the system greatly enhances its performance with the addition of the line detector behavior, and largely due to reductions in $F+$: the active eye improves not by finding more edges, but by avoiding non-edge areas. Comparing these results to the path maps of Fig. 3 we can conclude that, in our system, the fundamental point is the ability of running away from uninteresting areas.

Table 2. Values of $T+$, $F+$, and $T+/F+$ in the five different phases of the building of the eye, using the four test images.

		random	ds–vs	ld–lf	bd	ss
	$T+$	10687	18425	19960	18723	20498
block1	$F+$	10049	9541	4069	3779	3659
	$T+/F+$	1.06	1.93	4.91	4.95	5.60
	$T+$	11081	19066	16328	20719	19748
block2	$F+$	9655	9279	4184	4249	4449
	$T+/F+$	1.15	2.06	3.90	4.88	4.44
	$T+$	135	15647	16178	17161	17552
lena	$F+$	193	13545	6116	5628	6947
	$T+/F+$	0.70	1.16	2.65	3.05	2.53
	$T+$	8126	15659	16952	18639	17866
kim	$F+$	12610	13176	7364	5491	6143
	$T+/F+$	0.64	1.19	2.30	3.39	2.91

It can also be noted that, in three of the images, the value of $T+/F+$ without ss is greater than with it, and, in lena's case, even the system lacking bd and ss is better than the complete active eye. The $T+$ and $F+$ values tell almost nothing about this apparent contradiction with the path maps, but when we examine the movements of the active eye on the computer screen, we see that this is the result of an undesirable behavior: the eye without ss sometimes follows a short edge, then re-detects the already pursued edge, turns its direction 180° degrees, follows the edge again, and loops in this cycle for a while. The adding of ss's inhibits this waste of effort, but make thougher lf's task of finding edges.

In a later work, [9], we developed a slightly different numerical measure of performance, which ended up being more successful in penalizing the undesirable loop behavior.

6   Evaluating the Robustness to Noise

In order to evaluate the robustness to noise of the active eye, we performed a very simple experiment. The four test images were corrupted with white noise, at different rates — 5%, 10%, 20%, 50% and 100%. The images are then used as input to the eye system, and the $T+/F+$ ratios are computed for runs 245,760 units of time long. Table 3 displays the results.

We can see that, even for a noise rate of only 5%, the performance of the system is severely impaired. As before, the main cause is a large degradation on the $F+$ values. Comparing with the $T+/F+$ ratios for the building of the eye (Table 2), the addition of 5% of noise is enough to bring the value of $T+/F+$

Table 3. Values of $T+$, $F+$, and $T+/F+$ for the four test images corrupted with noise, for 245,760 units of time.

		0%	5%	10%	20%	50%	100%
block1	$T+$	20498	16030	15041	11904	10717	10320
	$F+$	3659	7211	7536	9910	10190	10700
	$T+/F+$	5.60	2.22	2.00	1.20	1.05	0.96
block2	$T+$	19748	15634	14510	13111	11688	10744
	$F+$	4449	7351	8832	9021	9674	10276
	$T+/F+$	4.44	2.13	1.64	1.45	1.21	1.05
lena	$T+$	17552	14375	10997	9572	7866	7851
	$F+$	6947	9420	11417	12246	13211	13169
	$T+/F+$	2.53	1.53	0.96	0.78	0.60	0.60
kim	$T+$	17866	13962	11957	9804	8022	7873
	$F+$	6143	10245	10796	12330	12980	13147
	$T+/F+$	2.91	1.36	1.11	0.80	0.62	0.60

down to ds–vs levels (i.e., the active eye composed only by the random walk generator and the attractor to high variance areas). Incrementing the amount of noise worses even more the results, and from 20% to 50% the system acts almost like it is sensing a completely random image (100%).

Those results could be considered as arguments against the general thesis that the systems built upon subsumption architectures are robust to noise. But we must remember that the claims about subsumption architectures are concerned about *noise found in the real world*. White noise, in the way it was added to the test images, seems to be far from "real noise": there is no correlation to the intensity of the pixels of the neighborhood, no pattern, etc. Besides that, the design of the sensors did not take in account very noisy images, and specially, white noise.

We think the $T+/F+$ ratios on Table 3 must be understood not as a proof neither as a counter-example of the robustness of subsumption architectures, but as a warning that subsumption architecture-based systems must be built using the real and ultimate environment. If noise happens in the "real world", we must start testing and debugging with noisy images from the very beginning of the implementation process.

7 A Higher Level Skill

The active eye has also the objective of detecting long, straight segments on the image. As said before, this ability is of higher level than that of just concentrating the eye activity. We can compare with the difference, in terms of complexity, between the problem of identifying high gradient areas and the problem of matching straight lines to a gradient map.

The observation of the eye activity on the screen of the computer confirms that the capacity of following edges was obtained as a consequence of the implementation and debugging of **ss**. However, it is not easy to show this result to someone who is not seeing the screen, and even for us it is difficult to precisely determine the behavior, after some time has passed and the window is full of pathmarks. But if the active eye is really following edges, it is not difficult to detect straight lines, and to generate the coordinates of their extremities.

The generation of long edge segments is thus implemented by **lg** (*line generator*), based on the edges detected by **ss** (see Fig. 2). The **lg** *AFSM* monitors the output of the **ss**-sensors: if they become active, **lg** records the position of the center of the eye in an internal register. When they cease to send messages, the present position is checked against the one stored in the register: only the coordinates of segments with length of more than 20 pixels are generated, provided that the direction of the eye has not changed in the meantime (what is verified by monitoring the input of **mo**). Since the generated segments are a kind of output of the whole system, we can view **lg** as if it was connected to some "actuator", what is indicated in the diagram of Fig. 2 by the triangle in the lower-right of **lg**.

The line generator was tested on the images shown in Fig. 1, and the generated segments, for 245,760 units of time, are plotted in the Fig. 6. The results were found to be quite interesting. Although some spurious edges are generated, the segments are similar to those obtained by common line-detecting algorithms. Of course, due to the structure of the active eye, the same segment is detected many times, what can be viewed as a sign of robustness. An extraordinary and quite unexpected result is that the plotting of the generated segments produces pictures similar to hand-drawings: this is one more evidence that our active eye has a non-trivial behavior.

In spite of being a visual — and thus, empirical — evaluation of performance, the pictures clearly confirm that the active eye is able to follow and detect edges.

8 Other Features of the Active Eye

We want to point out here some characteristics of the active eye that made us believe that its behavior is comparable to those obtained by other subsumption architecture-based systems, and, therefore, that our evaluation results somehow reflect those of previous works.

First, good results were obtained by the employment, without calibration, of the active eye on images portraying different subjects (buildings, people), what signals that we can expect a reasonable robustness concerning multiple types of images. It was found interesting that tests performed on synthetic

Fig. 6. The segments generated by 1g, for 245,760 units of time.

images showed bad results, with repetition of movements, cycling, etc. As put by Brooks, in [11], *"The world grounds regress."*, and, like Simon's ant, ([12]), the richness of the behavior of our system is in great part due to the varietiness of the real images.

Another aspect is the dynamic characteristic of the active eye: the sensing and moving are done in parallel, and the results of one affect the other. In some sense, our simulator is situated and embodied in its "real world", bearing a kind of on-line reactiveness with the sensed images. The behavior of the active eye is only due to local sensing, and the same is valid for the detection of lines. Moreover, just from the beginning, the detected segments — distributed all around the field of sight — can be delivered to actuators or higher order systems, and, as times goes by, the results are improved, by the generation of new lines. Even the fact that the same area is prospected many times, what is bad in terms of computational efficiency, can be used in an on-line system as a kind of positive redundancy.

9 Conclusions

Throughout this paper we presented some results on the evaluation of our active eye's performance. The path maps of Fig. 3 seem to express the idea of incremental building of behaviors, though still crude. The visualization is improved using path histograms, and their combination with maps of "desired images" (enlarged edge images) makes possible the definition of a numerical measure of performance.

Table 2 shows that the $T+$, $F+$, and $T+/F+$ values can mirror, quite successfully, our visual impression about the behavior of the active eye during the implementation phases. And, therefore, the figures on sensitivity to noise exhibited in Table 3 can be accounted to really reflect the deep depreciation in the activity of the eye under noisy conditions; this fact does not necessarily contradict subsumption architecture's claims on robustness, according to the reasons stated above.

The plots of the detected segments, displayed in Fig. 6, are amazing. They stress our view about the importance of the behavior evaluation, and also seem to show that, sometimes, an interesting behavior has already been acquired by a system, but an improper visualization can impair its perception.

We think that our work on the visualization and measuring of the active eye's behavior can be similarly applied to other subsumption architecture-based systems, since the reactive behavior of our system seems to resemble those of previous works in the field. *"Intelligence is in the eye of the observer."* (Brooks, [11]), and, thus, we hope the above presented results help to clarify the degree of intelligence (understood as complexity of behavior) behavior-based creatures can exhibit.

Acknowledgements

This work was started at the Dept. of Computer Controlled Machinery, Faculty of Engineering, Osaka University, Japan (*Denshi Seigyō Kikai Kōgakka, Kōgakubu, Ōsaka Daigaku*). The author wishes to thank Prof. Yoshiaki Shirai and the staff of Prof. Shirai's Laboratory for the kind cooperation.

References

1. Brooks, R.: A robust layered control system for a mobile robot. IEEE Journal of Robotics and Automation. RA-2(1) (1986) 14–23
2. Brooks, R.: A robot that walks: emergent behavior from a carefully evolved network. Neural Computation. 1 (2) (1989) 253–262
3. Brooks, R., Flynn, A.: Fast, cheap and out of control: a robot invasion of the solar system. Journal of the British Interplanetary Society. 42 (1989) 478–485
4. Mahadevan, S., Connell, J.: Automatic programming of behavior-based robots using reinforcement learning. Artificial Intelligence. 55 (1992) 311–365

5. Flynn, A., Brooks, R.: Building robots: expectations and experiences Proc. of IEEE/RSJ International Workshop on Intelligent Robots and Systems'89. Tsukuba, Japan (1989) 236–243
6. Brooks, R.: Elephants don't play chess. Robotics and Autonomous Systems. 6 (1990) 3–15
7. Pinhanez, C.: Um simulador de subsumption architectures (A simulator of subsumption architectures). University of São Paulo, Computer Science Dept. Technical Report, RT–MAC–9204. (1992)
8. Pinhanez, C.: Controlling a highly reactive camera using a subsumption architecture. Proc. of Applications of AI 93: Machine Vision and Robotics. Orlando, USA (1993) 100–111
9. Pinhanez, C.: Behavior-based active vision. IJPRAI. 8 (6) (to appear)
10. Shen, J., Castan, S.: An optimal linear operator for edge detection. CVPR'86. Miami, USA (1986)
11. Brooks, R.: Intelligence without reason. A.I. Memo No. 1293 (1991)
12. Simon, H.: The sciences of the artificial. MIT Press, Cambridge, USA. (1969)

AMOS: Basic Autonomy via Integrating Symbolic and Subsymbolic Mechanisms

Christian Schlegel and Manfred Knick

FAW Ulm, Postfach 2060, D-89010 Ulm, Germany
{schlegel,knick}@faw.uni-ulm.de

Abstract. In the AMOS project, the FAW uses a mobile robot to study questions related to the deep integration of symbolic and nonsymbolic information processing. AMOS aims at methods for autonomously acquiring new concepts from the environment [3]. This paper presents an architecture which integrates both symbolic planning as well as nonsymbolic reactive mechanisms. In particular, the concept of a plan break down and a region of interest plays a fundamental role. Implemented on a mobile platform, real world experiments show that by a suitable interaction of different modules basic autonomy can be obtained by using relatively simple and well-known algorithms.

1 Introduction

When an autonomous agent operates in real world it has to cope with a variety of problems which all have to do with the dynamic nature and complexity of the world. This requires real-time capabilities and makes it impossible to always have a complete and correct model of the environment. For example, it is not recommendable to have an offline planning module which is separated from plan execution because current sensor information must be considered to detect whether the current plan can be executed any longer or whether there are such changes in the world requiring a fast decision not to go ahead as planned before. Some of the occuring problems can be treated more efficiently with symbolic reasoning processes than with nonsymbolic reactive mechanisms and vice versa. For example, classical planning should not be used for computing the movement of a robot in terms of steering angles but at the level of rooms which have to be traversed for reaching a goal location. Nonsymbolic reactive mechanisms are very powerful to cope robustly with a changing world as long as for instance unknown obstacles have to be avoided.

Instead of using a strict hierarchical approach where nonsymbolic algorithms are only for executing the symbolically planned actions a different view is proposed. In this view, the symbolic level is not a command generator but can be seen as a symbolic information processor which is used whenever symbolic information processing has to be done. Within this view, the focus of attention is laid onto the problem solving capability of the different approaches. Therefore, nonsymbolic approaches are used for those tasks which require robustness and low cycle times and symbolic approaches are used whenever planning or reasoning processes are adequate. This allows the autonomous agent to benefit from both kinds of information processing.

As a consequence of the former view, for example, only those parts of the agent's environment are symbolically represented which are of relevance for the problems to be solved symbolically. As detailed in the following sections, an obstacle is of no relevance for an autonomous system on the symbolic level as long as nonsymbolic algorithms are able to bypass it. This will change drastically if an obstacle blocks up a door passage because then this has to be taken into consideration within the knowledge about passable links between rooms. Therefore, an obstacle carries a different semantics depending on the tasks the autonomous system has to fulfil and depending on the obstacle's location.

It is one of the aims of the AMOS project to have a system being able to learn from such situations. On the basis of the interaction between symbolic and nonsymbolic modules, it is possible to get closer hints to where (related to the autonomous agent semantics) interesting situations are located. These regions of interest are one of the key ingredients which allow classification [2] of such situations. The classification scheme combined with the detection of the region of interest will serve as a basis for inventing new concepts into the symbolic world model to improve performance.

In the following sections, details of the implementation of this architecture on a mobile platform to setup basic autonomy as well as real world experiments are described. Further details can be found in [10].

2 Architecture

The experimental platform for AMOS is a three-wheeled industrial vehicle which is driven and steered by a single front wheel. The system is equipped with sensors, such as odometers and a gyrus which e.g. allow to detect the vehicle's position. A 2-D laser scanner supplies relative distance information in a horizontal plane about 60 cm above the ground. The platform is controlled by a transputer system together with a pc-notebook, which are all onboard.

The whole system consists of different modules being connected through a message based communication system [9]. Each module provides server functionality for some information and is at the same time client to other servers. The different modules are permanently active, running on the same priority level and are able to communicate asynchronously with each other without any central supervisory instance. Each module is able to access all services available by servers. The dynamics of the interaction between the modules is not precoded but depends on the current sensor information and in some sense reflects therefore the dynamic of the interaction of the robot with its environment. Each module is involved in handling the current situation as long as it is able to provide useful contributions, that is communication between modules depends on the problem currently to be solved by the system. In some situations, a solution can be found by a symbolic planner by generating a totally new plan. In other situations, it is sufficient that a nonsymbolic module generates a new path to avoid an unknown obstacle.

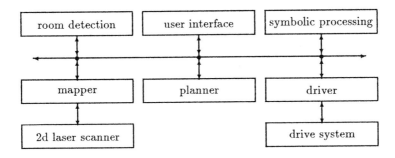

Fig. 1. Modules of the autonomous vehicle

The modules which at least are reasonably needed on an autonomous mobile system are shown in figure 1.

Within the subsymbolic modules, the mapper is responsible for map building. All the distance information from the 2D laser range finder is mapped into a grid, where each cell can be free or occupied by an obstacle [6]. The main advantage of this kind of representation is the simple update process. Furthermore, different grids can be fused in a single grid by simply combining corresponding cells. Within the mapper, a longterm map represents structures of the environment which are stable during a certain period of time whereas the current map holds the latest scanner data. The current map can be used in different modes. In one mode it is always cleared before a new scan is mapped into it. In the accumulative mode the map contents are kept so that new scanner data only update the map. In another mode parts of the longterm map are copied into the current map, which afterwards is updated by current scanner data. Finally, knowledge from the symbolic model about areas the robot is not allowed to enter can be used for labeling cells as being occupied by obstacles.

The current map, obtainable from the mapper, serves as the basis for the path planning module. In a first step, each free cell belonging to the goal area and the cell representing the current robot position are marked. Then a breadth first search, which directly operates on the grid, propagates the front of a wave from the goal area [5]. Each free cell discovered by the wave is attributed one of four directions (east, west, south and north) to point to the neighbour cell from which it was discovered. In each step the wave propagates by one cell-layer (figure 2) until the current robot position is reached or until there are no more free cells reachable. In the first case, the computed path can be found by simply following the direction entries in the cells. In the second case, no path exists. The wave is propagated from the goal area because then all entries in the grid guide to the goal area. The path found is only optimal in the sense of L_1 metrics. For further shortening, only the most distant cell of the current path, which under the current obstacle configuration is just reachable by a straight line, is considered as an intermediate goal point. In most cases, this heuristic, in combination with the permanent active mapping and planning process, yields to optimal paths in the sense of L_2 metrics as shown in figure 3. The shaded cells are occupied through obstacles as for example walls. The arrows show the path as it is generated by the breadth first search, and the dashed lines mark the path after applying the shortening heuristic. When the robot reaches the cell marked with a filled circle, the next part of the originally planned path can be shortened. This results in a new goal point for the driver, where it is important to recognize that there always is only one goal point available to the driver which can be updated at any time. Especially, there is no restriction to exchange the goal point before it is reached. Usually, only the last goal point, which marks the final goal, is really reached by the robot.

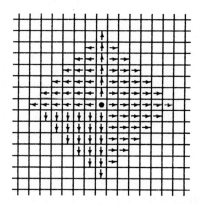

Fig. 2. Propagating the wave

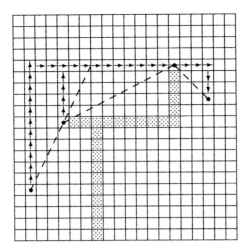

Fig. 3. Continuous path optimization

The driver is responsible for generating the final path from the current position to the intermediate goal point determined by the path planner. For this task a fuzzy controller is used [8] [4], which continually computes a suitable steering angle. The applied strategy is comparable to a rubber band pulling the robot to the destination point. A major advantage is that the inverse kinematic problems do not have to be solved explicitly. Continuous motion is performed by iterative downloading intermediate goal points from the path planner to the fuzzy controller.

The room detection is able to recognize whether the robot's position is within a certain area. These areas are described by a closed polygon and can be named. Therefore, room detection reduces to the point location problem [7]. The management of the polygons is done within the symbolic model of the environment. Because the polygonal descriptions of certain areas usually correspond to rooms, this module is able to generate a message whenever a room is entered or left. In addition, the room detection sends messages about rooms which can be seen from the current room. This is easily detected by exploiting information from the laser scanner.

The structure of the symbolic level is shown in figure 4. The knowledge level is based on a frame system which is implemented within prolog. The knowledge level contains a very simple environmental model, a representation of the robot status and knowledge about actions the robot can perform. Because at the moment the only performable action is driving around, the operators mainly represent knowledge about which rooms on principle are connected with each other. The operators have a STRIPS-like notation [1].

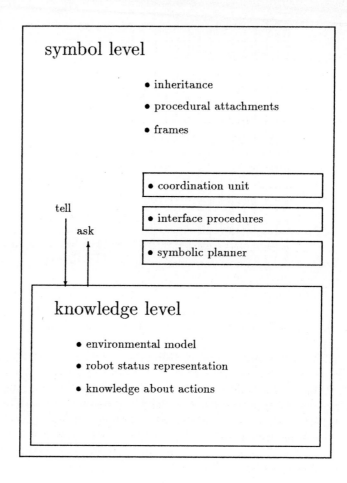

Fig. 4. The structure of the symbolic level

The environmental model consists of polygonal descriptions of areas corresponding to rooms. To have a well defined interface to access the knowledge level, a query language exists. The symbol level contains interface procedures and a planning module. The coordination unit is not responsible for coordinating the interaction of the different modules of the robot but to supervise the inference processes within the symbolic level. For example, messages sent to the symbolic processor have to be added to the knowledge base where procedural attachments trigger their evaluation.

3 Interaction of the modules

To initiate an action of the robot, a description of the state to be reached is sent to the symbolic processor where it is entered into the symbolic representation of the robot status. Then within the symbolic processor the difference between the current state and the goal state is detected which initiates the symbolic planning modul to derive a sequence of operators to eliminate those differences. If this was successful further messages will activate the plan execution. For example, the next room to be reached as described in the symbolic plan translates into the polygon which defines the room. Now the subsymbolic path planning module is able to use the cells inside the polygon as the goal area. The actual laser range data and knowledge available so far, for example from the longterm map, are the basis for the path planner which determines the intermediate goal points for the driver. On the basis of the messages containing the intermediate goal points, the driver generates the final path. If the robot reaches the goal area a message will be generated that triggers the symbolic processor to execute the next operator of the plan. When the room detection module generates a message that a room was entered, the current plan execution can be verified. If the entered room is not part of the symbolic plan, e.g. the symbolic planning module will be activated to cope with this new situation.

Due to a precise tuning of the competences, e.g. of the modules mapper, planner and driver, the robot is able to react online to a changing environment. For instance, it is able to move around unknown obstacles. Because of the distribution of competences over different modules, which can influence each other, it is possible to use very simple and fast algorithms. Within this approach, single modules are allowed to decide for themselves but are supervised by each other. This allows weaknesses of single algorithms to be considerably compensated by other modules. For example, the fuzzy controller is responsible for generating a path which is suitable to reach the given intermediate goal point. The requirements on the fuzzy controller are comparably easy to fulfil because it does not care about the current environmental situation. This is possible because it is the mapper's responsibility to provide a current map to the planner. The planner has to determine intermediate goal points in such a way that there are no collisions. If e.g. the driver's path leads into an obstacle, the permanent processing of the sensor data within the mapper in combination with the planner will result in a new intermediate goal point for the driver. The driver now generates another path which from outside looks like a bypass maneuver. Because not every cell on the path found by the planner has to be transmitted to the driver, a lot of problems with the updating of via-points do not occur. Especially the shortening heuristic, which results in seldom changes of the current goal point, allows the fuzzy controller to apply its path generating strategy without permanent disturbance by goal point updates. This yields to a robust control behavior of the fuzzy controller. Because of the permanent supervision of the current path generated by the fuzzy controller, it is possible to do path planning without considering the exact dynamics of the vehicle. This leads to a very clear structure of the distribution of responsibilities, drastically reducing the complexity of every involved module.

Due to the loose coupling of the modules and the simple and fast algorithms used, there are situations where the given goal point can not be reached by a path as it is typically generated by the fuzzy controller. Therefore, it is necessary to detect such problems. Another difficult situation can arise when the path planner is not able to find a path from the current position to the goal area as determined by the symbolic plan. This happens when a passage is unexpectedly blocked. In those situations, a plan break down arises because expectation differs substantially from real world experience. Modules which detect a plan break down signal this by messages to other modules. In the case of a closed door, the symbolic planning module has to modify the current plan whereas in the case of an unreachable intermediate goal point caused by the final path generated by the driver, it is eventually possible to solve the problem by a local maneuver and then to go ahead as planned before (local maneuvers such as turning round are implemented at the moment).

There are a lot of other situations where a plan break down arises. In general it will occur if expectation differs substantially from what is perceived. The differences will be relevant (and therefore perceivable) if they have effects with regard to the actions of the robot. Therefore, actually a plan break down will arise at the subsymbolic level if the path planner can not find a path, the current position of the robot is too close to an obstacle or all the cells of the goal area are marked as obstacles. At the symbolic level a plan break down will arise if the postconditions of an operator are not valid after its execution.

In the case of a plan break down, the region of interest is detected. A typical situation is shown in figure 5, where the path planner can not find a path on the current map although such a path exists within the environmental model. Especially in this case, the region of interest is described by x/y-coordinates pointing to those differences between both maps, which could have caused the plan break down. As shown in figure 5, the region of interest is detected by first searching a path on the map representing the incomplete environmental model. Then this path is followed on the map representing the current environment until the path hits an occupied grid cell. The position of the grid cell now serves as region of interest. This gives a hint to where changes within the world took place which are of relevance to the robot and allows to ignore insignificant changes. Therefore the region of interest is particularly interesting for active vision.

Due to the ability to detect plan break downs it is possible to use simple and fast algorithms as long as they are able to cope with the situations which occur when the robot is acting in its environment. Only in those situations where problems arise which ask for more sophisticated algorithms those can be activated as local strategies. Afterwards, control is given back to the default algorithms. Depending on the arising problems, sometimes symbolic mechanisms are more suited than subsymbolic approaches and vice versa.

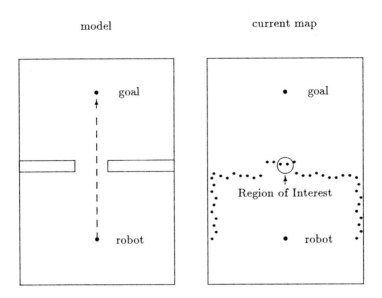

Fig. 5. Detection of the Region of Interest

4 Real world experiments

In figure 6 the ground floor of the FAW is shown. One of the typical tasks of the robot is to reach the marked goal within the cafeteria from the current position in the corridor.

The knowledge of the robot is restricted to what until now has been accumulated in the longterm map. The symbolic level only has knowledge about polygons bordering different areas of the ground floor and about operators describing the possible actions. The polygons roughly correspond to the rooms like corridor, cafeteria, lecture room and so on, and the operators describe the possible transitions from one room to another. Especially, there is no knowledge available about the chairs and tables above the chosen path which make it impossible for the robot to move around in this area. In addition, nothing is known about the pillars and the staircase.

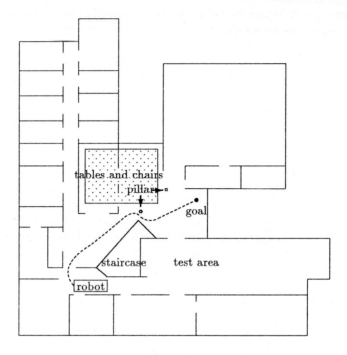

Fig. 6. The FAW ground floor

As a first experiment the robot has to reach the goal area. At the symbolic level the task is solved by first changing from the corridor to the first part of the cafeteria and then to change to the second part of it wherein the goal area is. During plan execution the current environment is recognized resulting in the path as shown in figure 6.

The robot is able to pass the bottleneck between the pillar and the staircase by a continuous motion with a speed of approximately 200 mm/s. With this experiment it is of no relevance if there are obstacles within the cafeteria as long as there is enough space to bypass them. Then the only difference is a slightly different path. With these results it has to be mentioned that the current implementation is not optimized with regard to execution speed.

In a second experiment, the link between corridor and cafeteria is blocked by an obstacle or the door is closed. Then the robot is not able to execute the symbolic plan and the path planner generates a plan break down message. Because there are no other direct links between the corridor and the cafeteria, it is the symbolic planner which has to find another plan. For example, it is possible to take the path through the test area. At this situation the reaction to the difference between expectation and real world is performed by the symbolic level. While executing the new plan another plan break down, generated at the subsymbolic level, occurs. Because the path planner is able to find a path from the current robot positon in the corridor to the test area, the driver gets a waypoint somewhere in the direction of the test area. Now the fuzzy controller generates a path to turn round the robot to reach the waypoint. Because there is not enough space for a turn round maneuver as proposed by the fuzzy controller, the robot gets too close to the wall and therefore is stopped. Because the path planner is able to find a path from the corridor to the test area, there is no need to replan the abstract symbolic plan but just to activate a local turn round maneuver, which can be much more complex than the simple version running into troubles. Afterwards, normal plan execution continues.

In the third experiment, it is demonstrated how the nonsymbolic modules can support the symbolic modules if at the symbolic level no plan can be derived that allows to reach the goal configuration. For example, this will be the case if the robot knows nothing about a link between the corridor and other rooms. Instead of not being able to derive a plan and then just to do nothing, the symbolic level can ask the nonsymbolic path planner to reactively move the robot until a new situation occurs and henceforward a plan can be derived. In this case, the nonsymbolic path planner operates on a map which is big enough to hold both the current robot position as well as the goal area. This enables the robot to move around even if there is no plan derivable in the symbolic level, but with lower speed as if the path planner would operate on a smaller map, as it is the usual case. When the robot reaches the first part of the cafeteria the message generated by the room detection module triggers the symbolic planner, which now succesfully can derive a plan. Then speed is increased because the path planner has not any longer to operate on the huge map but can benefit from the symbolically derived sequence of operators. In particular, the path planner does not at once have to cope with the complexity of the whole task but only with a part of it.

5   Summary

As figured out with the previously described experiments the proposed architecture allows to integrate both symbolic and nonsymbolic algorithms in such a way that the system can benefit from both. With this integration it is possible to use powerful symbolic inference mechanisms without renouncing the robustness of subsymbolic approaches. The interaction of both provides a basic autonomy which in most cases ensures the "survival" of the robot within its environment. Therefore it serves as a basic platform for the future work of the AMOS project focusing on learning aspects of autonomous systems. In particular, plan break downs, as they are detectable within this architecture, are considered as situations wherein the robot gets hints on important structural aspects of its environment. The region of interest is seen as a main key to evaluate plan break downs resulting from inadequate environmental models.

Acknowledgements

The members of the AMOS team are especially grateful to BMFT, the German Ministry for Research and Technology, which started funding the AMOS project as a confederate project with industrial partners (grant 01 IW 302A|1) since March 1993. We would also like to acknowledge our partner "NOELL Autonome Roboter", Hamburg, who supplied valuable support.

References

1. R. E. Fikes and N. J. Nilsson. STRIPS: A new approach to the application of theorem proving to problem solving. Artificial Intelligence, 2(3-4):189–208, 1971.
2. Jörg Illmann. Klassifikation von Abstandsbildern. Diploma thesis, Universität Karlsruhe, 1992.
3. M. Knick and F. J. Radermacher. Integration of subsymbolic and symbolic information processing in robot control. In Third annual Conference on AI, Simulation and Planning in High Autonomy Systems, pages 238–243. IEEE Computer Society Press, 1992.
4. Seon-Gon Kong and Bart Kosko. Comparison of fuzzy and neural truck backer-upper control systems. In IJCNN, volume 3, pages 349–358, 1990.
5. C. Y. Lee. An algorithm for path connections and its applications. IRE Transactions on Electronic Computers, pages 346–365, 1961.
6. N. J. Nilsson. A mobile automaton: An application of artificial intelligence techniques. In First International Joint Conference on Artificial Intelligence, pages 509–520, Washington, D.C., 1969.
7. Franco P. Preparata and Michael Ian Shamos. Computational Geometry. Text and Monographs in Computer Science. Springer Verlag, New York, 1985.
8. Christian Schlegel. Fuzzy Control. Technical Report FAW-TR-92005, FAW Ulm, März 1992.
9. Christian Schlegel. Verteiltes Betriebssystem. Technical Report FAW-TR-92019, FAW Ulm, September 1992.
10. Christian Schlegel. Entwurf und Realisierung eines autonomen mobilen Systems. Diploma thesis, Universität Ulm, 1993.
11. Christian Schlegel and Manfred Knick. Die Roboterarchitektur im Projekt AMOS. In Günther Schmidt, editor, Autonome Mobile Systeme, pages 95–100, October 1993. 9. Fachgespräch an der Technischen Universität München.

List of Contributors

Lecturers

Brooks, Rodney Allen (U.S.A.)
Massachusetts Institute of Technology

Artificial Intelligence Laboratory
545 Technology Square
Cambridge MA 02139

Babloyantz, Angessa (Belgium)
Université Libre de Bruxelles

Department Chemie Physique
Boulevard du Triomphe
1050 Bruxelles

Douglas, Rodney (U.K.)
MRC

Anatomical Neuropharmacology Unit
Mansfield Rd.
Oxford OX1 3TH

McFarland, David (U.K.)*
Balliol College & Department of
Zoology

Broad Street
Oxford OX1 3BJ

Kaelbling, Leslie Pack (U.S.A.)*
Brown University

Computer Science Department,
Box 1910
115 Waterman Street
Providence, RI 02912-1910
lpk@cs.brown.edu

Mitchell, Tom M. (U.S.A.)*
Carnegie Mellon University

School of Computer Science and
Robotics Institute
5000 Forbes Ave
Pittsburgh, PA 15213

Pfeifer, Rolf (Switzerland)*
University of Zürich

Artificial Intelligence Laboratory
Department of Computer Science
Winterthurerstrasse 190
8057 Zürich
pfeifer@ifi.unizh.ch

Schuster, Peter (Germany)
Institut für Molekulare Biotechnologie

Beutenbergstraße 11
69900 Jena
pks@imb-jena.de

Smithers, Tim (Spain)*
Universidad del Pais Vasco

Facultad de Informática
Apartado 649
20080 San Sebastia'n
ccpsmsmt@si.ehu.es

Steels, Luc (Belgium)*
Vrije Universiteit Brussel

Artificial Intelligence Laboratory
Department of Computer Science
Pleinlaan 2
1050 Brussel
steels@arti.vub.ac.be

Torras, Carme (Spain)*
Institut de Cibernetica (CSIC-UPC)

Diagonal 647
08028 - Barcelona
torras@ic.upc.es

Van de Velde, Walter (Belgium)*
Vrije Universiteit Brussel

Artificial Intelligence Laboratory
Department of Computer Science
Pleinlaan 2
1050 Brussel
walter@arti.vub.ac.be

* paper included

Contributing Participants

Alpaydin, Ethem (Turkey)
Boğaziçi University

Department of Computer Engineering
TR-80815 Istanbul
alpaydin@boun.edu.tr

Antoniol, Giulio (Italy)
IRST

Istituto per la Ricerca Scientifica e
Tecnologica
38050 Povo, Trento
antoniol@irst.it

Bryson, Joanna (U.S.A.)
MIT

Artificial Intelligence Lab
545 Technology Square
Cambridge, MA 02139
joanna@ai.mit.edu

Caprile, Bruno (Italy)
IRST

Istituto per la Ricerca Scientifica e
Tecnologica
38050 Povo, Trento
caprile@irst.it

Chialvo, Dante R. (U.S.A.)
The Santa Fe Institute

Fluctuations and Biophysics Group
1399 Hyde Park Road
NM 87501 Santa Fe

Cimatti, Alessandro (Italy)
IRST

Istituto per la Ricerca Scientifica e
Tecnologica
38050 Povo, Trento
cimatti@irst.it

Cliff, Dave (U.K.)
University of Sussex

School of Cognitive and Computing
Sciences (COGS)
Brighton BN1 9QH
davec@cogs.susx.ac.uk

Correia, Luís (Portugal)
Universidade Nova de Lisboa

Computer Science Department
Quinta da Torre
2825 Mte de Caparica
lc@fct.unl.pt

Fiutem, Roberto (Italy)
IRST

Istituto per la Ricerca Scientifica e
Tecnologica
38050 Povo, Trento
fiutem@irst.it

Harvey, Inman (U.K.)
University of Sussex

School of Cognitive and Computing
Sciences (COGS)
Brighton BN1 9QH
inmanh@cogs.susx.ac.uk

Hexmoor, Henry H. (U.S.A.)
State University of
New York at Buffalo

Computer Science Department
NY 14260
hexmoor@cs.buffalo.edu

Husbands, Philip (U.K.)
University of Sussex

School of Cognitive and Computing
Sciences (COGS)
Brighton BN1 9QH
philh@cogs.susx.ac.uk

Knick, Manfred (Germany)
FAW Ulm

Postfach 2060
89010 Ulm
knick@faw.uni-ulm.de

Lammens, Johan M. (Italy)
Advanced Robotics Technology and
Systems (ARTS) Laboratory

Scuola Superiore S. Anna
via Carducci, 40
57127 Pisa
lammens@arts.sssup.it

Littman, Michael L. (U.S.A.)
Brown University

Computer Science Department, Box 1910
115 Waterman Street
Providence, RI 02912-1910

Matarić, Maja J. (U.S.A.)
Brandeis University

Volen Center for Complex Systems
Computer Science Department
Waltham, MA 02254
maja@cs.brandeis.edu

Millonas, Mark M. (U.S.A.)
Los Alamos National Laboratory

Theoretical Division and CNLS
NM 87545 Los Alamos
marko@goshawk.lanl.gov

Moore, Andrew W. (U.S.A.)
Carnegie Mellon University

School of Computer Science and
Robotics Institute
5000 Forbes Ave
Pittsburgh PA 15213

Nehmzow, Ulrich (U.K.)
Manchester University

Department of Computer Science
Manchester M13 9PL
u.nehmzow@cs.man.ac.uk

Nilsson, Martin (Sweden)
Swedish Institute of Computer Science

Box 1263
S-164 28 Kista
mn@sics.se

Pebody, Miles (U.K.)
University College London

Department of Computer Science
Gower St
London WC1E 6BT
M.Pebody@cs.ucl.ac.uk

Pinhanez, Claudio S. (Brazil)
University of São Paulo

Institute of Mathematics and Statistics
(MIT Media Laboratory
Cambridge MA 02139
pinhanez@media.mit.edu)

Prem, Erich (Austria)
Austrian Research Institute for
Artificial Intelligence

Schottengasse 3
A-1010 Vienna
erich@ai.univie.ac.at

Schlegel, Christian (Germany)
FAW Ulm

Postfach 2060
89010 Ulm
schlegel@faw.uni-ulm.de

Shapiro, Stuart C. (U.S.A.)
State University of New York at
Buffalo

Computer Science Department
NY 14260
shapiro@cs.buffalo.edu

Simsarian, Kristian T.(Sweden)
Swedish Institute of Computer Science

Box 1263
S-164 28 Kista
kristian@sics.se

Steiger-Garção, A. (Portugal)
Universidade Nova de Lisboa

Electrotechnical Engineering Dept.
Quinta da Torre
2825 Mte de Caparica

Thrun, Sebastian (Germany)
University of Bonn

Dept. for Computer Science III
Römerstr. 164
53117 Bonn
thrun@uran.informatik.uni-bonn.de

Vertommen, Filip (Belgium)
Vrije Universiteit Brussel

Artificial Intelligence Laboratory
Department of Computer Science
Pleinlaan 2
1050 Brussel
filip@arti.vub.ac.be

Weiß, Gerhard (Germany)
Technische Universität München

Institut für Informatik
D-80290 München
weissg@informatik.tu-muenchen.de

NATO ASI Series F

Including Special Programmes on Sensory Systems for Robotic Control (ROB) and on Advanced Educational Technology (AET)

Vol. 98: Medical Images: Formation, Handling and Evaluation. Edited by A. E. Todd-Pokropek and M. A. Viergever. IX, 700 pages. 1992.

Vol. 99: Multisensor Fusion for Computer Vision. Edited by J. K. Aggarwal. XI, 456 pages. 1993. *(ROB)*

Vol. 100: Communication from an Artificial Intelligence Perspective. Theoretical and Applied Issues. Edited by A. Ortony, J. Slack and O. Stock. XII, 260 pages. 1992.

Vol. 101: Recent Developments in Decision Support Systems. Edited by C. W. Holsapple and A. B. Whinston. XI, 618 pages. 1993.

Vol. 102: Robots and Biological Systems: Towards a New Bionics? Edited by P. Dario, G. Sandini and P. Aebischer. XII, 786 pages. 1993.

Vol. 103: Parallel Computing on Distributed Memory Multiprocessors. Edited by F. Özgüner and F. Erçal. VIII, 332 pages. 1993.

Vol. 104: Instructional Models in Computer-Based Learning Environments. Edited by S. Dijkstra, H. P. M. Krammer and J. J. G. van Merriënboer. X, 510 pages. 1993. *(AET)*

Vol. 105: Designing Environments for Constructive Learning. Edited by T. M. Duffy, J. Lowyck and D. H. Jonassen. VIII, 374 pages. 1993. *(AET)*

Vol. 106: Software for Parallel Computation. Edited by J. S. Kowalik and L. Grandinetti. IX, 363 pages. 1993.

Vol. 107: Advanced Educational Technologies for Mathematics and Science. Edited by D. L. Ferguson. XII, 749 pages. 1993. *(AET)*

Vol. 108: Concurrent Engineering: Tools and Technologies for Mechanical System Design. Edited by E. J. Haug. XIII, 998 pages. 1993.

Vol. 109: Advanced Educational Technology in Technology Education. Edited by A. Gordon, M. Hacker and M. de Vries. VIII, 253 pages. 1993. *(AET)*

Vol. 110: Verification and Validation of Complex Systems: Human Factors Issues. Edited by J. A. Wise, V. D. Hopkin and P. Stager. XIII, 704 pages. 1993.

Vol. 111: Cognitive Models and Intelligent Environments for Learning Programming. Edited by E. Lemut, B. du Boulay and G. Dettori. VIII, 305 pages. 1993. *(AET)*

Vol. 112: Item Banking: Interactive Testing and Self-Assessment. Edited by D. A. Leclercq and J. E. Bruno. VIII, 261 pages. 1993. *(AET)*

Vol. 113: Interactive Learning Technology for the Deaf. Edited by B. A. G. Elsendoorn and F. Coninx. XIII, 285 pages. 1993. *(AET)*

Vol. 114: Intelligent Systems: Safety, Reliability and Maintainability Issues. Edited by O. Kaynak, G. Honderd and E. Grant. XI, 340 pages. 1993.

Vol. 115: Learning Electricity and Electronics with Advanced Educational Technology. Edited by M. Caillot. VII, 329 pages. 1993. *(AET)*

Vol. 116: Control Technology in Elementary Education. Edited by B. Denis. IX, 311 pages. 1993 *(AET)*

Vol. 118: Program Design Calculi. Edited by M. Broy. VIII, 409 pages. 1993.

Vol. 119: Automating Instructional Design, Development, and Delivery. Edited by. R. D. Tennyson. VIII, 266 pages. 1994 *(AET)*

Vol. 120: Reliability and Safety Assessment of Dynamic Process Systems. Edited by T. Aldemir, N. O. Siu, A. Mosleh, P. C. Cacciabue and B. G. Göktepe. X, 242 pages. 1994.

Vol. 121: Learning from Computers: Mathematics Education and Technology. Edited by C. Keitel and K. Ruthven. XIII, 332 pages. 1993. *(AET)*

NATO ASI Series F

Including Special Programmes on Sensory Systems for Robotic Control (ROB) and on Advanced Educational Technology (AET)

Vol. 122: Simulation-Based Experiential Learning. Edited by D. M. Towne, T. de Jong and H. Spada. XIV, 274 pages. 1993. *(AET)*

Vol. 123: User-Centred Requirements for Software Engineering Environments. Edited by D. J. Gilmore, R. L. Winder and F. Détienne. VII, 377 pages. 1994.

Vol. 124: Fundamentals in Handwriting Recognition. Edited by S. Impedovo. IX, 496 pages. 1994.

Vol. 125: Student Modelling: The Key to Individualized Knowledge-Based Instruction. Edited by J. E. Greer and G. I. McCalla. X, 383 pages. 1994. *(AET)*

Vol. 126: Shape in Picture. Mathematical Description of Shape in Grey-level Images. Edited by Y.-L. O, A. Toet, D. Foster, H. J. A. M. Heijmans and P. Meer. XI, 676 pages. 1994.

Vol. 127: Real Time Computing. Edited by W. A. Halang and A. D. Stoyenko. XXII, 762 pages. 1994.

Vol. 128: Computer Supported Collaborative Learning. Edited by C. O'Malley. X, 303 pages. 1994. *(AET)*

Vol. 129: Human-Machine Communication for Educational Systems Design. Edited by M. D. Brouwer-Janse and T. L. Harrington. X, 342 pages. 1994. *(AET)*

Vol. 130: Advances in Object-Oriented Database Systems. Edited by A. Dogac, M. T. Özsu, A. Biliris and T. Sellis. XI, 515 pages. 1994.

Vol. 131: Constraint Programming. Edited by B. Mayoh, E. Tyugu and J. Penjam. VII, 452 pages. 1994.

Vol. 132: Mathematical Modelling Courses for Engineering Education. Edited by Y. Ersoy and A. O. Moscardini. X, 246 pages. 1994. *(AET)*

Vol. 133: Collaborative Dialogue Technologies in Distance Learning. Edited by M. F. Verdejo and S. A. Cerri. XIV, 296 pages. 1994. *(AET)*

Vol. 134: Computer Integrated Production Systems and Organizations. The Human-Centred Approach. Edited by F. Schmid, S. Evans, A. W. S. Ainger and R. J. Grieve. X, 347 pages. 1994.

Vol. 135: Technology Education in School and Industry. Emerging Didactics for Human Resource Development. Edited by D. Blandow and M. J. Dyrenfurth. XI, 367 pages. 1994. *(AET)*

Vol. 136: From Statistics to Neural Networks. Theory and Pattern Recognition Applications. Edited by V. Cherkassky, J. H. Friedman and H. Wechsler. XII, 394 pages. 1994.

Vol. 137: Technology-Based Learning Environments. Psychological and Educational Foundations. Edited by S. Vosniadou, E. De Corte and H. Mandl. X, 302 pages. 1994. *(AET)*

Vol. 138: Exploiting Mental Imagery with Computers in Mathematics Education. Edited by R. Sutherland and J. Mason. VIII, 326 pages. 1995. *(AET)*

Vol. 139: Proof and Computation. Edited by H. Schwichtenberg. VII, 470 pages. 1995.

Vol. 140: Automating Instructional Design: Computer-Based Development and Delivery Tools. Edited by R. D. Tennyson and A. E. Barron. IX, 618 pages. 1995. *(AET)*

Vol. 141: Organizational Learning and Technological Change. Edited by C. Zucchermaglio, S. Bagnara and S. U. Stucky. X, 368 pages. 1995. *(AET)*

Vol. 142: Dialogue and Instruction. Modeling Interaction in Intelligent Tutoring Systems. Edited by R.-J. Beun, M. Baker and M. Reiner. IX, 368 pages. 1995. *(AET)*

Vol. 144: The Biology and Technology of Intelligent Autonomous Agents. Edited by Luc Steels. VIII, 517 pages. 1995.

Vol. 146: Computers and Exploratory Learning. Edited by A. A. diSessa, C. Hoyles and R. Noss. VIII, 482 pages. 1995. *(AET)*